ENGINEERING ECONOMIC ANALYSIS

An Introduction

Michael R. Lindeburg, PE

The Power to Pass
www.ppi2pass.com

Professional Publications, Inc. • Belmont, California

ENGINEERING ECONOMIC ANALYSIS

Current printing of this edition: 9

Printing History

edition number	printing number	update
1	7	Minor corrections. Copyright update.
1	8	Minor corrections.
1	9	Minor corrections.

Printed in the United States of America.

PPI
1250 Fifth Avenue, Belmont, CA 94002
(650) 593-9119
www.ppi2pass.com

ISBN: 978-0-912045-60-3

TABLE OF CONTENTS

PREFACE

The subject of economic analysis (also known as *engineering economy* and *engineering economic analysis*) is frustrating for most engineers. There are a lot of misleading terms, some assumptions are arbitrary and illogical, the data go on forever, projections of future data are imprecise and illusive, there are multiple competing techniques to choose from, and the solutions tend to be long, iterative, and tedious.

As an engineer, I can assure you that there is neither any engineering nor any economics in the subject of engineering economy, yet economic analysis is one of my favorite subjects. Perhaps because of this, I've always enjoyed teaching this topic to engineers and presenting the concepts in an intuitive manner. This intuitive approach to economic analysis is the focus of this book.

In developing this book, I paid particular attention to two functional aspects: (1) the explanatory text and practice problems, and (2) the interest tables. The explanatory text comes from my lectures and other writings, and from research into engineering economy, revised for current applications. After teaching engineers economic analysis for more than 15 years and writing hundreds of pages on the topic in my reference books, I've learned some approaches to presenting the information that are consistent with an engineer's problem-solving experience. I've attempted to include the latest information and standards, along with practice problems and solutions that really reinforce the material covered.

The interest tables in this book originated in 1975 when I first generated them from a FORTRAN program fed card-by-card into an IBM-360 mainframe. Fewer than 10 interest tables were all I needed then. Over the years, though, the need has become more refined. For example, as inflation decreased, tables of lower interest rates were needed. Odd interest rates (e.g., 5% and 7%) became commonplace. And it became clear that solving rate-of-return problems would be simplified if fractional interest rate tables were available. The end result was *Expanded Interest Tables*, published in 1988. All of the tables from *Expanded Interest Tables* have been included in the book you are now holding.

No other book contains as many interest tables as this book; it includes tables for both odd and even interest rates over a wide range of values. Furthermore, factors are calculated for up to 100 compounding periods for use with long-range projections and nonannual compounding. The factors in this book were calculated using double-precision variables throughout. All numbers have been rounded at the last decimal place. These interest tables will simplify solutions to your economic analysis problems by reducing (if not eliminating) the need for interpolation and use of factor formulas.

I hope that this book is a useful addition to your library.

Michael R. Lindeburg, P.E.
Belmont, CA

ACKNOWLEDGMENTS

As a derivative work, this book was by far the easiest one I ever wrote. However, that placed a greater fraction of the total workload on others.

Thank you, Joanne Bergeson, for resurrecting and perfecting my ancient FORTRAN program that generated the factors in this book. I know the difficulties seemed, at least initially, insurmountable. You had to first learn a new programming language, and then you struggled with a problem with elusive accuracy.

Thank you, Jennifer Pasqual, for the cover design. You put as much creativity and effort into this little book as you would have for a *New York Times* best-seller. Thank you, Jessica Whitney, for the page formatting. Watching you manipulate TEX on the Macintosh always amazes me. Many decisions had to be made on the fly using personal judgment. Your decisions determined how beautiful the final product would appear to the reader. And I owe a special thank you to Kurt Stephan for proofing this book and holding me to his highest editorial standards. He has shouldered much of the burden that I used to carry for getting things right.

Finally, a big thank you to Lisa Rominger, manager of PPI's Production Department. You managed to fit this impromptu project into your work schedule. As I wrote in the earlier *Expanded Interest Tables* book, "The project was in good hands."

ENGINEERING ECONOMIC ANALYSIS

Nomenclature

A	annual amount	$
B	present worth of all benefits	$
BV_j	book value at end of the jth year	$
C	cost, or present worth of all costs	$
d	declining balance depreciation rate	decimal
D	demand	various
D	depreciation	$
DR	present worth of after-tax depreciation recovery	$
e	constant inflation rate	decimal
E_0	initial amount of an exponentially growing cash flow	$
EAA	equivalent annual amount	$
EUAC	equivalent uniform annual cost	$
f	federal income tax rate	decimal
F	forecasted quantity	various
F	future worth	$
g	exponential growth rate	decimal
G	uniform gradient amount	$
i	effective rate per period (usually per year)	decimal per unit time
i'	effective interest rate corrected for inflation	decimal
k	number of compounding periods per year	–
m	an integer	–
n	number of compounding periods, or life of asset	–
P	present worth	$
r	nominal rate per year (rate per annum)	decimal per unit time
ROI	return on investment	$
ROR	rate of return	decimal per unit time
s	state income tax rate	decimal
S_n	expected salvage value in year n	$
t	composite tax rate	decimal
t	time	years (typical)
T	a quantity equal to $\frac{1}{2}n(n+1)$	–
TC	tax credit	$
z	a quantity equal to $\dfrac{1+i}{1-d}$	decimal

Symbols

α	smoothing coefficient for forecasts	–
ϕ	effective rate per period (r/k)	decimal
$\mathcal{E}$	expected value	various

Subscripts

0	initial
j	at time j
n	at time n
t	at time t

1 IRRELEVANT CHARACTERISTICS

In its simplest form, an *engineering economic analysis* is a study of the desirability of making an investment.[1] The decision-making principles in this book can be applied by individuals as well as companies. The nature of the spending opportunity or industry is not important. Farming equipment, personal investments, and multi-million dollar factory improvements can all be evaluated using the same principles.

Similarly, the applicable principles are largely insensitive to the monetary units. Although *dollars* are used in this book, it is equally convenient to use British pounds, Japanese yen, or German marks.

Finally, this book may give the impression that investment alternatives must be evaluated on a year-by-year basis. Actually, the *effective period* can be defined as a day, month, century, or any other convenient period of time.

2 MULTIPLICITY OF SOLUTION METHODS

Most economic conclusions can be reached in more than one manner. There are usually several different analyses that will eventually result in identical answers.[2] Other than the pursuit of elegant solutions in a timely manner, there is no reason to favor one procedural method over another.[3]

---

[1]This subject is also known as *engineering economics* and *engineering economy*. There is very little, if any, true economics in this subject.

[2]Because of round-off errors, particularly when factors are taken from tables, these different calculations will produce slightly different numerical results (e.g., $49.49 versus $49.50). However, this type of divergence is well known and accepted in engineering economic analysis.

[3]This is not meant to imply that approximate methods, simplifications, and rules of thumb are acceptable.

3 PRECISION AND SIGNIFICANT DIGITS

The full potential of electronic calculators will never be realized in engineering economic analyses. Considering that calculations are based on estimates of far-future cash flows, and that unrealistic assumptions (e.g., no inflation, identical cost structures of replacement assets, etc.) are routinely made, it makes little sense to carry cents along in calculations.

The calculations in this book have been designed to illustrate and review the principles presented. Because of this, greater precision than is normally necessary in everyday problems may be used. Though used, such precision is not warranted.

Unless there is some compelling reason to strive for greater precision, the following rules are presented for use in reporting final answers to engineering economic analysis problems.

- Omit fractional parts of the dollar (i.e., cents).

- Report and record a number to a maximum of four significant digits unless the first digit of that number is 1, in which case, a maximum of five significant digits should be written. For example,

$49	not	$49.43
$93,450	not	$93,453
$1,289,700	not	$1,289,673

4 YEAR-END CONVENTION

Except in short-term transactions, it is simpler to assume that all receipts and disbursements (cash flows) take place at the end of the year in which they occur.[4] This is known as the *year-end convention*. The exceptions to the year-end convention are initial project cost (purchase cost), trade-in allowance, and other cash flows that are associated with the inception of the project at $t = 0$.

On the surface, such a convention appears grossly inappropriate, since repair expenses, interest payments, corporate taxes, etc., seldom coincide with the end of a year. However, the convention greatly simplifies engineering economic analysis problems, and it is justifiable on the basis that the increased precision associated with a more rigorous analysis is not warranted (due to the numerous other assumptions and estimates made in the problem to start with).

There are various established procedures, known as *rules* or *conventions*, imposed by the Internal Revenue Service on U.S. taxpayers. An example is the *half-year rule*,

---

[4] A *short-term transaction* typically has a lifetime of five years or less and has payments or compounding that are more frequent than once per year.

which permits only half of the first-year depreciation to be taken in the first year of an asset's life when certain methods of depreciation are used. These rules are subject to constantly changing legislation and are not covered in this book. The implementation of such rules is outside the scope of engineering practice and is best left to accounting professionals.

5 NONQUANTIFIABLE FACTORS

An engineering economic analysis is a quantitative analysis. Some factors cannot be introduced as numbers into the calculations. Such factors are known as *nonquantitative factors, judgment factors*, and *irreducible factors*. Typical nonquantifiable factors are

- preferences
- political ramifications
- urgency
- goodwill
- prestige
- utility
- corporate strategy
- environmental effects
- health and safety rules
- reliability
- political risks

Since these factors are not included in the calculations, the policy is to disregard the issues entirely. Of course, the factors should be discussed in a final report. The factors are particularly useful in breaking ties between competing alternatives that are economically equivalent.

6 CASH FLOW DIAGRAMS

Although they are not always necessary in simple problems (and they are often unwieldy in very complex problems), *cash flow diagrams* can be drawn to help visualize and simplify problems having diverse receipts and disbursements.

The following conventions are used to standardize cash flow diagrams.

- The horizontal (time) axis is marked off in equal increments, one per period, up to the *duration* (*life* or *horizon*) of the project.

- Two or more transfers in the same period are placed end-to-end, and these may be combined.

- Expenses incurred before $t = 0$ are called *sunk costs*. Sunk costs are not relevant to the problem unless they have tax consequences in an after-tax analysis.

- *Receipts* are represented by arrows directed upward. *Disbursements* are represented by arrows directed downward. The arrow length is proportional to the magnitude of the cash flow.

Example 1

A mechanical device will cost $20,000 when purchased. Maintenance will cost $1000 each year. The device will generate revenues of $5000 each year for five years, after which the salvage value is expected to be $7000. Draw and simplify the cash flow diagram.

(solution)

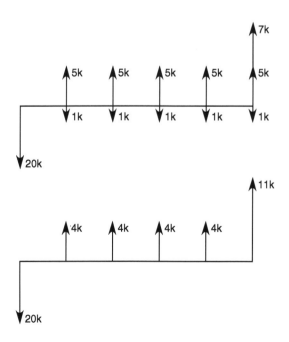

7 TYPES OF CASH FLOWS

In order to evaluate a real-world project, it is necessary to present the project's cash flows in terms of standard cash flows that can be handled by engineering economic analysis techniques. The standard cash flows are single payment cash flow, uniform series cash flow, gradient series cash flow, and the rare exponential series cash flow.

A *single payment cash flow* can occur at the beginning of the time line (designated as $t = 0$), at the end of the time line (designated as $t = n$), or any time in between.

The *uniform series cash flow* consists of a series of equal transactions starting at $t = 1$ and ending at $t = n$. The symbol A is typically given to the magnitude of each individual cash flow.[5]

The *gradient series cash flow* starts with a cash flow (typically given the symbol G) at $t = 2$ and increases by G each year until $t = n$, at which time the final cash flow is $(n - 1)G$.

An *exponential gradient cash flow* is based on a phantom value (typically given the symbol E_0) at $t = 0$ and grows or decays exponentially according to the following relationship.[6]

$$\text{amount at time } t = E_t = E_0(1 + g)^t \quad [t = 1, 2, 3, \ldots, n] \tag{1}$$

In Eq. 1, g is the *exponential growth rate*, which can be either positive or negative. Exponential gradient cash flows are rarely seen in economic justification projects assigned to engineers.[7]

---

[5]Notice that the cash flows do not begin at $t = 0$. This is an important concept with all of the series cash flows. This convention has been established to accommodate the timing of annual maintenance (and similar) cash flows for which the year-end convention is applicable.

[6]Notice the convention for an exponential cash flow series: the first cash flow E_0 is at $t = 1$, as in the uniform annual series. However, the first cash flow is $E_0(1 + g)$. The cash flow of E_0 at $t = 0$ is absent (i.e., is a *phantom cash flow*).

[7]For one of the few discussions on exponential cash flow, see *Capital Budgeting*, Robert V. Oakford, The Ronald Press Company, New York, 1970.

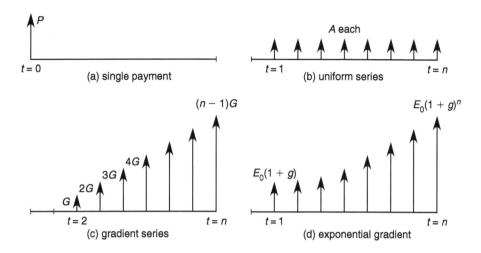

Figure 1 Standard Cash Flows

8 TYPICAL PROBLEM TYPES

There is a wide variety of problem types that, collectively, are considered to be engineering economic analysis problems.

By far, the majority of engineering economic analyis problems are *alternative comparisons*. In these problems, two or more mutually exclusive investments compete for limited funds. A variation of this is a *replacement/retirement analysis*, which is repeated each year to determine if an existing asset should be replaced. Finding the percentage return on an investment is a *rate of return problem*, one of the alternative comparison solution methods.

Investigating interest and principal amounts in loan payments is a *loan repayment problem*. An *economic life analysis* will determine when an asset should be retired. In addition, there are miscellaneous problems involving economic order quantity, learning curves, break-even points, product costs, etc.

9 IMPLICIT ASSUMPTIONS

Several assumptions are implicitly made when solving engineering economic analysis problems. Some of these assumptions are made with the knowledge that they are or will be poor approximations of what really will happen. The assumptions are made, regardless, for the benefit of obtaining a solution.

The most common assumptions are

- The year-end convention is applicable.

- There is no inflation now, nor will there be any during the lifetime of the project.

- Unless specifically called for otherwise, a before-tax analysis is needed.

- The effective interest rate used in the problem will be constant during the lifetime of the project.

- Nonquantifiable factors can be disregarded.

- Funds invested in a project are available and are not urgently needed elsewhere.

- Excess funds continue to earn interest at the effective rate used in the analysis.

This last assumption, like most of the assumptions listed, is almost never specifically mentioned in the body of a solution. However, it is a key assumption when comparing two alternatives that have different initial costs.

For example, suppose two investments, one costing $10,000 and the other costing $8000, are to be compared at 10%. It is obvious that $10,000 in funds is available, otherwise the costlier investment would not be under consideration. If the smaller investment is chosen, what is done with the remaining $2000? The last assumption yields the answer: the $2000 is put to work in some investment earning (in this case) 10%.

10 EQUIVALENCE

Industrial decision makers using engineering economic analysis are concerned with the magnitude and timing of a project's cash flow as well as with the total profitability of that project. In this situation, a method is required to compare projects involving receipts and disbursements occurring at different times.

By way of illustration, consider $100 placed in a bank account that pays 5% effective annual interest at the end of each year. After the first year, the account will have grown to $105. After the second year, the account will have grown to $110.25.

Assume that you will have no need for money during the next two years, and any money received will immediately go into your 5% bank account. Then, which of the following options would be more desirable?

Option a: $100 now.

Option b: $105 to be delivered in one year.

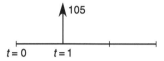

Option c: $110.25 to be delivered in two years.

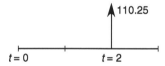

In light of the previous illustration, none of the options is superior under the assumptions given. If the first option is chosen, you will immediately place $100 into a 5% account, and in two years the account will have grown to $110.25. In fact, the account will contain $110.25 at the end of two years regardless of the option chosen. Therefore, these alternatives are said to be *equivalent*.

It is obvious that equivalence may or may not be the case, depending on the interest rate. Thus, an alternative that is acceptable to one decision maker may be unacceptable to another decision maker. The interest rate that is used in actual calculations is known as the *effective interest rate*.[8] If compounding is once a year, it is known as the *effective annual interest rate*. However, effective quarterly, monthly, daily, etc., interest rates are also used.

The fact that $100 today grows to $105 in one year (at 5% annual interest) is an example of what is known as the *time value of money* principle. This principle simply articulates what is obvious: funds placed in a secure investment will increase to an equivalent future amount. The procedure for determining the present investment from the equivalent future amount is known as *discounting*.

---

[8]The adjective *effective* distinguishes this interest rate from other interest rates (e.g., nominal interest rates) that are not meant to be used in calculations.

11 SINGLE-PAYMENT EQUIVALENCE

The equivalence of any present amount, P, at $t = 0$, to any future amount, F, at $t = n$, is called the *future worth* and can be calculated from Eq. 2.

$$F = P(1 + i)^n \qquad [2]$$

The factor $(1 + i)^n$ is known as the single payment *compound amount factor* and has been tabulated at the end of this book for various combinations of i and n.

Similarly, the equivalence of any future amount to any present amount is called the *present worth* and can be calculated from Eq. 3.

$$P = F(1 + i)^{-n} = \frac{F}{(1 + i)^n} \qquad [3]$$

The factor $(1 + i)^{-n}$ is known as the *single payment present worth factor*.[9]

The interest rate used in Eqs. 2 and 3 must be the effective rate per period. Also, the basis of the rate (e.g., annually, monthly, etc.) must agree with the type of period used to count n. Thus, it would be incorrect to use an effective annual interest rate if n was the number of compounding periods in months.

Example 2

How much should you put into a 10% (effective annual rate) savings account in order to have $10,000 in five years?

(solution)

This problem could also be stated: What is the equivalent present worth of $10,000 five years from now if money is worth 10% per year?

$$P = F(1 + i)^{-n} = (\$10{,}000)(1 + 0.10)^{-5}$$
$$= \$6209$$

The factor 0.6209 would usually be obtained from the tables.

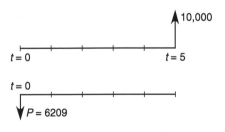

---

[9]The *present worth* is also called the *present value* and *net present value*. These terms are used interchangeably and no significance should be attached to the terms *value, worth,* and *net.*

12 STANDARD CASH FLOW FACTORS AND SYMBOLS

Equations 2 and 3 may give the impression that solving engineering economic analysis problems involves a lot of calculator use, and, in particular, a lot of exponentiation. Such calculations may be necessary from time to time, but most problems are simplified by the use of tabulated values of the factors.

Rather than actually writing the formula for the compound amount factor (which converts a present amount to a future amount), it is common convention to substitute the standard functional notation of $(F/P, i\%, n)$. Thus, the future value in n periods of a present amount would be symbolically written as

$$F = P(F/P, i\%, n) \qquad [4]$$

Similarly, the present worth factor has a functional notation of $(P/F, i\%, n)$. The present worth of a future amount n periods hence would be symbolically written as

$$P = F(P/F, i\%, n) \qquad [5]$$

Values of these *cash flow (discounting) factors* are tabulated at the end of this book. There is often initial confusion about whether the (F/P) or (P/F) column should be used in a particular problem. There are several ways of remembering what the functional notations mean.

One method of remembering which factor should be used is to think of the factors as conditional probabilities. The conditional probability of event **A** given that event **B** has occurred is written as $p\{A|B\}$, where the given event comes after the vertical bar. In the standard notational form of discounting factors, the given amount is similarly placed after the slash. What you want comes before the slash. (F/P) would be a factor to find F given P.

Another method of remembering the notation is to interpret the factors algebraically. Thus, the (F/P) factor could be thought of as the fraction F/P. Algebraically, Eq. 4 would be

$$F = P \times \frac{F}{P} \qquad [6]$$

This algebraic approach is actually more than an interpretation. The numerical values of the discounting factors are consistent with this algebraic manipulation. Thus, the (F/A) factor could be calculated as $(F/P) \times (P/A)$. This consistent relationship can be used to calculate other factors that might be occasionally needed, such as (F/G) or (G/P). For instance, the annual cash flow that would be equivalent to a uniform gradient may be found from

$$A = G(P/G, i\%, n)(A/P, i\%, n) \qquad [7]$$

Formulas for the compounding and discounting factors are contained in Table 1. Normally, it will not be necessary to calculate factors from the formulas. The tables at the end of this book are adequate for solving most problems.

Table 1
Discount Factors for Discrete Compounding

factor name	converts	symbol	formula
single payment compound amount	P to F	$(F/P, i\%, n)$	$(1+i)^n$
single payment present worth	F to P	$(P/F, i\%, n)$	$(1+i)^{-n}$
uniform series sinking fund	F to A	$(A/F, i\%, n)$	$\dfrac{i}{(1+i)^n - 1}$
capital recovery	P to A	$(A/P, i\%, n)$	$\dfrac{i(1+i)^n}{(1+i)^n - 1}$
uniform series compound amount	A to F	$(F/A, i\%, n)$	$\dfrac{(1+i)^n - 1}{i}$
uniform series present worth	A to P	$(P/A, i\%, n)$	$\dfrac{(1+i)^n - 1}{i(1+i)^n}$
uniform gradient present worth	G to P	$(P/G, i\%, n)$	$\dfrac{(1+i)^n - 1}{i^2(1+i)^n} - \dfrac{n}{i(1+i)^n}$
uniform gradient future worth	G to F	$(F/G, i\%, n)$	$\dfrac{(1+i)^n - 1}{i^2} - \dfrac{n}{i}$
uniform gradient uniform series	G to A	$(A/G, i\%, n)$	$\dfrac{1}{i} - \dfrac{n}{(1+i)^n - 1}$

Example 3

What factor will convert a gradient cash flow ending at $t = 8$ to a future value at $t = 8$? (That is, what is the $(F/G, i\%, 8)$ factor?) The effective annual interest rate is 10%.

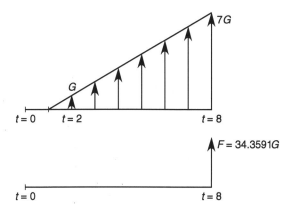

(solution)

method 1:   From Table 1, the $(F/G, 10\%, 8)$ factor is

$$(F/G, 10\%, 8) = \frac{(1+i)^n - 1}{i^2} - \frac{n}{i}$$

$$= \frac{(1+0.10)^8 - 1}{(0.10)^2} - \frac{8}{0.10} = 34.3589$$

method 2:   The tabulated values of (P/G) and (F/P) in the tables at the end of this book can be used to calculate the factor.

$$(F/G, 10\%, 8) = (P/G, 10\%, 8)(F/P, 10\%, 8)$$
$$= (16.0287)(2.1436) = 34.3591$$

The (F/G) factor could also have been calculated as the product of the (A/G) and (F/A) factors.

13 CALCULATING UNIFORM SERIES EQUIVALENCE

A cash flow that repeats each year for n years without change in amount is known as an *annual amount* and is given the symbol A. As an example, a piece of equipment may require annual maintenance, and the maintenance cost will be an annual amount. Although the equivalent value for each of the n annual amounts could be calculated and then summed, it is more expedient to use one of the uniform series factors. For example, it is possible to convert from an annual amount to a future amount by use of the (F/A) factor.

$$F = A(F/A, i\%, n) \qquad [8]$$

A *sinking fund* is a fund or account into which annual deposits of A are made in order to accumulate F at $t = n$ in the future. Since the annual deposit is calculated as $A = F(A/F, i\%, n)$, the (A/F) factor is known as the *sinking fund factor*.

An *annuity* is a series of equal payments (A) made over a period of time.[10] Usually, it is necessary to buy into an investment (e.g., a bond, an insurance policy, etc.) in order to ensure the annuity. In the simplest case of an annuity that starts at the end of the first year and continues for n years, the purchase price (P) is

$$P = A(P/A, i\%, n) \qquad [9]$$

The present worth of an *infinite (perpetual) series* of annual amounts is known as a *capitalized cost*. There is no $(P/A, i\%, \infty)$ factor in the tables, but the capitalized cost can be calculated simply as

$$P = \frac{A}{i} \quad [i \text{ in decimal form}] \qquad [10]$$

Alternatives with different lives will generally be compared by way of *equivalent uniform annual cost* (EUAC), also known as *equivalent annual cost* (EAC) and *equal annual amount* (EAA). An EUAC is the annual amount that is equivalent to all of the cash flows in the alternative. The EUAC differs in sign from all of the other cash flows. Costs and expenses expressed as EUACs, which would normally be considered negative, are actually positive. The term *cost* in the designation EUAC serves to make clear the meaning of a positive number.

Example 4

Maintenance costs for a machine are $250 each year. What is the present worth of these maintenance costs over a 12-year period if the interest rate is 8%?

(solution)

$$P = A(P/A, 8\%, 12) = (-\$250)(7.5361)$$
$$= -\$1884$$

---

[10] An annuity may also consist of a lump sum payment made at some future time. However, this rare interpretation is not considered in this book.

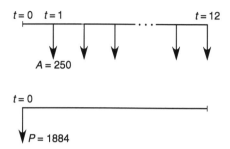

14 FINDING PAST VALUES

From time to time, it will be necessary to determine an amount in the past equivalent to some current (or future) amount. For example, you might have to calculate the original investment made 15 years ago given a current annuity payment.

Such problems are solved by placing the $t = 0$ point at the time of the original investment, and then calculating the past amount as a P value. For example, the original investment, P, can be extracted from the annuity, A, by using the standard cash flow factors.

$$P = A(P/A, i\%, n) \tag{11}$$

The choice of $t = 0$ is flexible. As a general rule, the $t = 0$ point should be selected for convenience in solving a problem.

Example 5

You are currently paying $250 per month to lease your office phone equipment. You have three years (36 months) left on the five-year (60-month) lease. What would have been an equivalent purchase price two years ago? The effective interest rate per month is 1%.

(solution)

The solution of this example is not affected by the fact that investigation is being performed in the middle of the horizon. This is a simple calculation of present worth.

$$P = A(P/A, 1\%, 60)$$
$$= (-\$250)(44.9550) = -\$11,239$$

15 TIMES TO DOUBLE AND TRIPLE AN INVESTMENT

If an investment doubles in value (in n compounding periods and with $i\%$ effective interest), the ratio of current value to past investment will be 2.

$$\frac{F}{P} = (1 + i)^n = 2 \tag{12}$$

Similarly, the ratio of current value to past investment will be 3 if an investment triples in value. This can be written as

$$\frac{F}{P} = (1 + i)^n = 3 \qquad [13]$$

It is a simple matter to extract the number of periods, n, from Eqs. 12 and 13 to determine the *doubling time* and *tripling time*, respectively. For example, the doubling time is

$$n = \frac{\log 2}{\log (1 + i)} \qquad [14]$$

When a quick estimate of the doubling time is needed, the *rule of 72* can be used. The doubling time is approximately $72/i$.

The tripling time is

$$n = \frac{\log 3}{\log (1 + i)} \qquad [15]$$

Equations 14 and 15 form the basis of Table 2.

Table 2
Doubling and Tripling Times for Various Interest Rates

interest rate ($i\%$)	doubling time (periods)	tripling time (periods)
1	69.7	110.4
2	35.0	55.5
3	23.4	37.2
4	17.7	28.0
5	14.2	22.5
6	11.9	18.9
7	10.2	16.2
8	9.01	14.3
9	8.04	12.7
10	7.27	11.5
11	6.64	10.5
12	6.12	9.69
13	5.67	8.99
14	5.29	8.38
15	4.96	7.86
16	4.67	7.40
17	4.41	7.00
18	4.19	6.64
19	3.98	6.32
20	3.80	6.03

16 VARIED AND NONSTANDARD CASH FLOWS

A. The Gradient Cash Flow

A common situation involves a uniformly increasing cash flow. If the cash flow has the proper form, its present worth can be determined by using the *uniform gradient factor*, $(P/G, i\%, n)$. The uniform gradient factor finds the present worth of a uniformly increasing cash flow that starts in year two (not year one).

There are three common difficulties associated with the form of the uniform gradient. The first is the fact that the first cash flow starts at $t = 2$. This convention recognizes that annual costs, if they increase uniformly, begin with some value at $t = 1$ (due to the year-end convention) but do not begin to increase until $t = 2$. The tabulated values of (P/G) have been calculated to find the present worth of only the increasing part of the annual expense. The present worth of the base expense incurred at $t = 1$ must be found separately with the (P/A) factor.

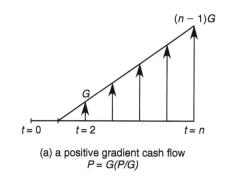

(a) a positive gradient cash flow
$P = G(P/G)$

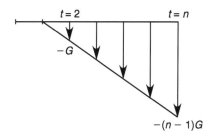

(b) a negative gradient cash flow
$P = -G(P/G)$

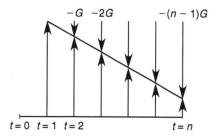

(c) a decreasing revenue incorporating
a negative gradient
$P = A(P/A) - G(P/G)$

Figure 2   Positive and Negative Gradient Cash Flows

PROFESSIONAL PUBLICATIONS, INC.

The second difficulty is that, even though the $(P/G, i\%, n)$ factor is used, there are only $n - 1$ actual cash flows. It is clear that n must be interpreted as the *period number* in which the last gradient cash flow occurs, not the number of gradient cash flows.

Finally, the sign convention used with gradient cash flows may seem confusing. If an expense increases each year (as in Ex. 6), the gradient will be negative, since it is an expense. If a revenue increases each year, the gradient will be positive. In most cases, the sign of the gradient depends on whether the cash flow is an expense or revenue.[11]

Example 6

Maintenance on an old machine is $100 this year but is expected to increase by $25 each year thereafter. What is the present worth of five years of maintenance? Use an interest rate of 10%.

(solution)

In this problem, the cash flow must be broken down into parts. (Notice that the five-year gradient factor is used even though there are only four nonzero gradient cash flows.)

$$P = A(P/A, 10\%, 5) + G(P/G, 10\%, 5)$$
$$= (-\$100)(3.7908) - (\$25)(6.8618)$$
$$= -\$551$$

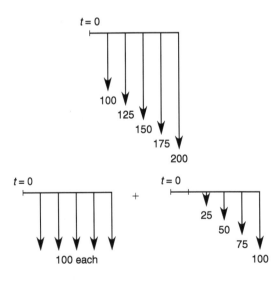

---

[11]This is not a universal rule. It is possible to have a uniformly decreasing revenue. In this case, the gradient would be negative.

B. Stepped Cash Flows

Stepped cash flows are easily handled by the technique of *superposition of cash flows*. This technique is illustrated by Ex. 7.

Example 7

An investment costing $1000 returns $100 for the first five years, and returns $200 for the following five years. How would the present worth of this investment be calculated?

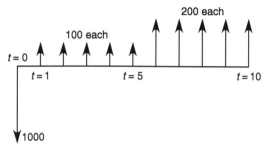

(solution)

Using the principle of superposition, the revenue cash flow can be thought of as $200 each year from $t = 1$ to $t = 10$, with a negative revenue of $100 from $t = 1$ to $t = 5$. Superimposed, these two cash flows make up the actual performance cash flow.

$$P = -\$1000 + \$200(P/A, i\%, 10) - \$100(P/A, i\%, 5)$$

C. Missing and Extra Parts of Standard Cash Flows

A missing or extra part of a standard cash flow can also be handled by superposition. For example, suppose an annual expense is incurred each year for ten years, except in the ninth year. (The cash flow is illustrated in Fig. 3.) The present worth could be calculated as a subtractive process.

$$P = A(P/A, i\%, 10) - A(P/F, i\%, 9) \tag{16}$$

Alternatively, the present worth could be calculated as an additive process.

$$P = A(P/A, i\%, 8) + A(P/F, i\%, 10) \tag{17}$$

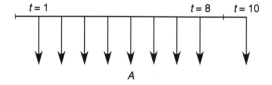

Figure 3   Cash Flow with a Missing Part

PROFESSIONAL PUBLICATIONS, INC.

D. Delayed and Premature Cash Flows

There are cases when a cash flow matches a standard cash flow exactly, except that the cash flow is delayed or starts sooner than it should. Often, such cash flows can be handled with superposition. At other times, it may be more convenient to shift the time axis. This shift is known as the *projection method*. Example 8 illustrates the projection method.

Example 8

An expense of $75 is incurred starting at $t = 3$ and continues until $t = 9$. There are no expenses or receipts until $t = 3$. Use the projection method of determining the present worth of this stream of expenses.

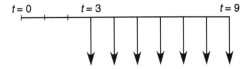

(solution)

First, determine a cash flow at $t = 2$ that is equivalent to the entire expense stream. If $t = 0$ was where $t = 2$ actually is, the present worth of the expense stream would be

$$P' = -\$75(P/A, i\%, 7)$$

P' is a cash flow at $t = 2$. It is now a simple matter to find the present worth (at $t = 0$) of this future amount.

$$P = P'(P/F, i\%, 2) = -\$75(P/A, i\%, 7)(P/F, i\%, 2)$$

E. Cash Flows at Beginnings of Years: The Christmas Club Problem

This type of problem is characterized by a stream of equal payments (or expenses) starting at $t = 0$ and ending at $t = n - 1$. It differs from the standard annual cash flow in the existence of a cash flow at $t = 0$ and the absence of a cash flow at $t = n$. This problem gets its name from the service provided by some savings institutions whereby money is automatically deposited each week or month (starting immediately, when the savings plan is opened) in order to accumulate money to purchase Christmas presents at the end of the year.

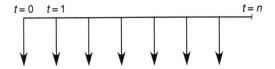

Figure 4   Cash Flows at Beginnings of Years

PROFESSIONAL PUBLICATIONS, INC.

It may seem that the present worth of the savings stream can be determined by directly applying the (P/A) factor. However, this is not the case, since the Christmas Club cash flow and the standard annual cash flow differ. The Christmas Club problem is easily handled by superposition, as is illustrated by Ex. 9.

Example 9

How much can you expect to accumulate by $t = 10$ for a child's college education if you deposit $300 at the beginning of each year for a total of 10 payments?

(solution)

Notice that the first payment is made at $t = 0$ and that there is no payment at $t = 10$. The future worth of the first payment is calculated with the (F/P) factor. The absence of the payment at $t = 10$ is handled by superposition. Notice that this correction is not multiplied by a factor.

$$F = \$300(F/P, i\%, 10) + \$300(F/A, i\%, 10) - \$300$$
$$= \$300(F/A, i\%, 11) - \$300$$

17 THE MEANING OF PRESENT WORTH AND i

By now, it should be obvious that $100 invested in a 5% bank account (using annual compounding) will allow you to remove $105 one year from now. If this investment is made, you will receive a *return on investment* (ROI) of $5. The cash flow diagram and the present worth of the two transactions are

$$P = -\$100 + \$105(P/F, 5\%, 1)$$
$$= -\$100 + (\$105)(0.9524) = 0$$

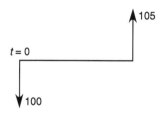

Figure 5 Cash Flow Diagram

Notice that the present worth is zero even though you will receive a $5 return on your investment.

However, if you are offered $120 for the use of $100 over a one-year period, the cash flow diagram and present worth (at 5%) would be

$$P = -\$100 + \$120(P/F, 5\%, 1)$$
$$= -\$100 + (\$120)(0.9524) = \$14.29$$

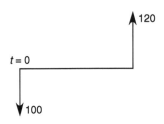

Figure 6   Cash Flow Diagram

Therefore, the present worth of an alternative is seen to be equal to the equivalent value at $t = 0$ of the increase in return above that which you would be able to earn in an investment offering i percent per period. In the above case, $14.29 is the present worth of ($20 − $5), the difference in the two ROIs.

The present worth is also the amount that you would have to be given to dissuade you from making an investment, since placing the initial investment amount along with the present worth into a bank account earning i percent will yield the same eventual return on investment. Relating this to the previous paragraphs, you could be dissuaded from investing $100 in an alternative that would return $120 in one year by a $t = 0$ payment of $14.29. Clearly, ($100 + $14.29) invested at $t = 0$ will also yield $120 in one year at 5%.

It should be obvious that income-producing alternatives with negative present worths are undesirable, and that alternatives with positive present worths are desirable because they increase the average earning power of invested capital. (In some cases, such as municipal and public works projects, the present worths of all alternatives are negative, in which case, the least negative alternative is best.)

The selection of the interest rate is difficult in engineering economics problems. Usually, it is taken as the average rate of return that an individual or business organization has realized in past investments. Alternatively, the interest rate may be associated with a particular level of risk. Usually, i for individuals is the interest rate that can be earned in relatively *risk-free investments*.

18 SIMPLE AND COMPOUND INTEREST

If $100 is invested at 5%, it will grow to $105 in one year. During the second year, 5% interest continues to be accrued, but on $105, not $100. This is the principle of *compound interest*: the interest accrues interest.[12]

If only the original principal accrues interest, the interest is said to be *simple interest*. Simple interest is rarely encountered in engineering economic analyses, but the concept may be incorporated into short-term transactions.

19 EXTRACTING THE INTEREST RATE: RATE OF RETURN

An intuitive definition of the *rate of return* (ROR) is the effective annual interest rate at which an investment accrues income. That is, the rate of return of a project is the interest rate that would yield identical profits if all money was invested at that rate. Although this definition is correct, it does not provide a method of determining the rate of return.

It was previously seen that the present worth of a $100 investment invested at 5% is zero when $i = 5\%$ is used to determine equivalence. Thus, a working definition of rate of return would be the effective annual interest rate that makes the present worth of the investment zero. Alternatively, rate of return could be defined as the effective annual interest rate that will discount all cash flows to a total present worth equal to the required initial investment.

It is tempting, but impractical, to determine a rate of return analytically. It is simply too difficult to extract the interest rate from the equivalence equation. For example, consider a $100 investment that pays back $75 at the end of each of the first two years. The present worth equivalence equation (set equal to zero in order to determine the rate of return) is

$$P = 0 = -\$100 + \$75(1 + i)^{-1} + \$75(1 + i)^{-2} \qquad [18]$$

Solving Eq. 18 requires finding the roots of a quadratic equation. In general, for an investment or project spanning n years, the roots of an nth-order polynomial would have to be found. It should be patently obvious that an analytical solution would be essentially impossible for more complex cash flows. (The rate of return in this example is 31.87%.)

If the rate of return is needed, it can be found from a trial-and-error solution. To find the rate of return of an investment, proceed as follows.

step 1:   Set up the problem as if to calculate the present worth.

step 2:   Arbitrarily select a reasonable value for i. Calculate the present worth.

---

[12]This assumes, of course, that the interest remains in the account. If the interest is removed and spent, only the remaining funds accumulate interest.

step 3: Choose another value of i (not too close to the original value), and again
          solve for the present worth.

step 4: Interpolate or extrapolate the value of i that gives a zero present worth.

step 5: For increased accuracy, repeat steps 2 and 3 with two more values that
          straddle the value found in step 4.

A common, although incorrect, method of calculating the rate of return involves dividing
the annual receipts or returns by the initial investment. (See Sec. 54.) However, this
technique ignores such items as salvage, depreciation, taxes, and the time value of money.
This technique also fails when the annual returns vary.

It is possible that more than one interest rate will satisfy the zero present worth criteria.
This confusing situation occurs whenever there is more than one change in sign in the
investment's cash flow.[13] Table 3 indicates the numbers of possible interest rates as a
function of the number of sign reversals in the investment's cash flow.

Table 3
Multiplicity of Rates of Return

number of sign reversals	number of distinct rates of return
0	0
1	0 or 1
2	0, 1, or 2
3	0, 1, 2, or 3
4	0, 1, 2, 3, or 4
m	$0, 1, 2, 3, \ldots m - 1, m$

Difficulties associated with interpreting the meaning of multiple rates of return can be
handled with the concepts of *external investment* and *external rate of return*. An external
investment is an investment that is distinct from the investment being evaluated (which
becomes known as the *internal investment*). The external rate of return, which is the
rate of return earned by the external investment, does not need to be the same as the rate
earned by the internal investment.

---

[13]There will always be at least one change of sign in the cash flow of a legitimate investment. (This excludes
municipal and other tax-supported functions.) At $t = 0$, an investment is made (a negative cash flow).
Hopefully, the investment will begin to return money (a positive cash flow) at $t = 1$ or shortly thereafter.
Although it is possible to conceive of an investment in which all of the cash flows were negative, such an
investment would probably be classified as a *hobby*.

It can generally be said that the multiple rates of return indicate that the analysis must proceed as though money will be invested outside of the project. The mechanics of how this is done are not covered here.

Example 10

What is the rate of return on invested capital if $1000 is invested now with $500 being returned in year 4 and $1000 being returned in year 8?

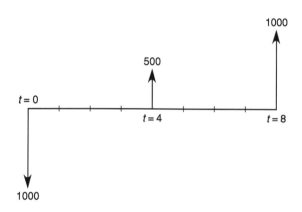

(solution)

First, set up the problem as a present worth calculation.

Try $i = 5\%$.

$$P = -\$1000 + \$500(P/F, 5\%, 4) + \$1000(P/F, 5\%, 8)$$
$$= -\$1000 + (\$500)(0.8227) + (\$1000)(0.6768)$$
$$= \$88$$

Next, try a larger value of i to reduce the present worth. If $i = 10\%$,

$$P = -\$1000 + \$500(P/F, 10\%, 4) + \$1000(P/F, 10\%, 8)$$
$$= -\$1000 + (\$500)(0.6830) + (\$1000)(0.4665)$$
$$= -\$192$$

Using simple interpolation, the rate of return is

$$\text{ROR} = 5\% + \left(\frac{\$88}{\$88 + \$192}\right)(10\% - 5\%)$$
$$= 6.6$$

A second iteration between 6% and 7% yields 6.39%.

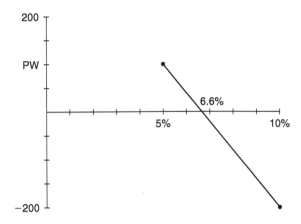

Example 11

An existing biomedical company is developing a new drug. A venture capital firm gives the company $25,000,000 initially and $55,000,000 more at the end of the first year. The drug patent will be sold at the end of year 5 to the highest bidder, and the biomedical company will receive $80,000,000. (The venture capital firm will receive everything in excess of $80,000,000.) The firm invests unused money in short-term commercial paper earning 10% effective interest per year through its bank. In the meantime, the biomedical company incurs development expenses of $50,000,000 for the first three years. The drug is to be evaluated by a government agency and there will be neither expenses nor revenues during the fourth year. What is the biomedical company's rate of return on this investment?

(solution)

Normally, the rate of return is determined by setting up a present worth problem and varying the interest rate until the present worth is zero. Writing the cash flows, though, shows that there are two reversals of sign: one at $t = 2$ (positive to negative) and the other at $t = 5$ (negative to positive). Therefore, there could be two interest rates that produce a zero present worth. (In fact, there actually are two interest rates: 10.7% and 41.4%.)

time	cash flow (millions)
0	+25
1	$-55 - 50 = +5$
2	-50
3	-50
4	0
5	+80

However, this problem can be reduced to one in which there is only one sign reversal in the cash flow series. The initial \$25,000,000 is invested in commercial paper (an *external investment* having nothing to do with the drug development process) during the first year at 10%. The accumulation of interest and principal after one year is

$$(25)(1 + 0.10) = 27.5$$

This 27.5 is combined with the 5 (the money remaining after all expenses are paid at $t = 1$) and invested externally, again at 10%. The accumulation of interest and principal after one year (i.e., at $t = 2$) is

$$(27.5 + 5)(1 + 0.10) = 35.75$$

This 35.75 is combined with the development cost paid at $t = 2$.

The cash flow for the development project (the *internal investment*) is

time	cash flow (millions)
0	0
1	0
2	$35.75 - 50 = -14.25$
3	-50
4	0
5	$+80$

Now, there is only one sign reversal in the cash flow series. The *internal rate of return* on this development project is found by the traditional method to be 10.3%. Notice that this is different from the rate the company can earn from investing externally in commercial paper.

20 ROR VERSUS ROI

Rate of return (ROR) is an effective annual interest rate, typically stated in percent per year. *Return on investment* (ROI) is a dollar amount. Thus, *rate of return* and *return on investment* are not synonymous.

Return on investment can be calculated in two different ways. The accounting method is to subtract the total of all investment costs from the total of all net profits (i.e., revenues less expenses). The time value of money is not considered.

In engineering economic analysis, the return on investment is calculated from equivalent values. Specifically, the present worth (at $t = 0$) of all investment costs is subtracted from the future worth (at $t = n$) of all net profits.

When there are only two cash flows, a single investment amount and a single payback, the two definitions of return on investment yield the same numerical value. When there are more than two cash flows, the returns on investment will be different depending on which definition is used.

21 MINIMUM ATTRACTIVE RATE OF RETURN

A company may not know what effective interest rate, i, to use in engineering economic analysis. In such a case, the company can establish a minimum level of economic performance that it would like to realize on all investments. This criterion is known as the *minimum attractive rate of return* (MARR). Unlike the effective interest rate, i, the minimum attractive rate of return is not used in numerical calculations.[14] It is used only in comparisons with the rate of return.

Once a rate of return for an investment is known, it can be compared to the minimum attractive rate of return. To be a viable alternative, the rate of return must be greater than the minimum attractive rate of return.

The advantage of using comparisons to the minimum attractive rate of return is that an effective interest rate, i, never needs to be known. The minimum attractive rate of return becomes the correct interest rate for use in present worth and equivalent uniform annual cost calculations.

22 TYPICAL ALTERNATIVE-COMPARISON PROBLEM FORMATS

With the exception of some investment and rate of return problems, the typical problem involving engineering economics will have the following characteristics.

- An interest rate will be given.

- Two or more alternatives will be competing for funding.

- Each alternative will have its own cash flows.

- It is necessary to select the best alternative.

Example 12

Investment A costs $10,000 today and pays back $11,500 two years from now. Investment B costs $8000 today and pays back $4500 each year for two years. If an interest rate of 5% is used, which alternative is superior?

(solution)

The solution to this example is not difficult, but it will be postponed until methods of comparing alternatives have been covered.

---

[14]Not everyone adheres to this rule. Some people use minimum attractive rate of return and effective interest rate interchangeably.

23 DURATIONS OF INVESTMENTS

Because they are handled differently, short-term investments and short-lived assets need to be distinguished from investments and assets that constitute an infinitely lived project. Short-term investments are easily identified: a drill press that is needed for three years, or a temporary factory building that is being constructed to last five years.

Investments with perpetual cash flows are also (usually) easily identified: maintenance on a large flood control dam, and revenues from a long-span toll bridge. Furthermore, some items with finite lives can expect renewal on a repeated basis.[15] For example, a major freeway with a pavement life of 20 years is unlikely to be abandoned; it will be resurfaced or replaced every 20 years.

Actually, if an investment's finite lifespan is long enough, it can be considered an infinite investment. The reason for this is that a dollar 50 years from now has little impact on current decisions. The $(P/F, 10\%, 50)$ factor, for example, is 0.0085. Thus, a dollar at $t = 50$ has an equivalent present worth of less than a penny. Since these far-future cash flows are eclipsed by present cash flows, long-term investments can be considered finite or infinite without significant impact on the calculations.

24 CHOICE OF ALTERNATIVES: COMPARING ONE ALTERNATIVE WITH ANOTHER ALTERNATIVE

A variety of methods exist for selecting a superior alternative from among a group of proposals. Each method has its own merits and applications.

A. Present Worth Method

When two or more alternatives are capable of performing the same functions, the superior alternative will have the largest present worth. The *present worth method* is restricted to evaluating alternatives that are mutually exclusive and that have the same lives. This method is suitable for ranking the desirability of alternatives.

Returning to Ex. 12, the present worth of each alternative should be found in order to determine which alternative is superior.

Example 12 continued

(solution)

$$P(A) = -\$10,000 + \$11,500(P/F, 5\%, 2)$$
$$= -\$10,000 + (\$11,500)(0.9070)$$
$$= \$431$$

---

[15]The term *renewal* can be interpreted to mean replacement or repair.

$$P(B) = -\$8000 + \$4500(P/A, 5\%, 2)$$
$$= -\$8000 + (\$4500)(1.8594)$$
$$= \$367$$

Alternative A is superior and should be chosen.

B. Capitalized Cost Method

The present worth of a project with an infinite life is known as the *capitalized cost* or *life cycle cost*. Capitalized cost is the amount of money at $t = 0$ needed to perpetually support the project on the earned interest only. Capitalized cost is a positive number when expenses exceed income.

In comparing two alternatives, each of which is infinitely lived, the superior alternative will have the lowest capitalized cost.

Normally, it would be difficult to work with an infinite stream of cash flows since most economics tables do not list factors for periods in excess of 100 years. However, the (A/P) discounting factor approaches the interest rate as n becomes large. Since the (P/A) and (A/P) factors are reciprocals of each other, it is possible to divide an infinite series of equal cash flows by the interest rate in order to calculate the present worth of the infinite series. This is the basis of Eq. 19.

$$\text{capitalized cost} = \text{initial cost} + \frac{\text{annual costs}}{i} \qquad [19]$$

Equation 19 can be used when the annual costs are equal in every year. If the operating and maintenance costs occur irregularly instead of annually, or if the costs vary from year to year, it will be necessary to somehow determine a cash flow of equal annual amounts (EAA) that is equivalent to the stream of original costs.

The equal annual amount may be calculated in the usual manner by first finding the present worth of all the actual costs, and then multiplying the present worth by the interest rate (the (A/P) factor for an infinite series). However, it is not even necessary to convert the present worth to an equal annual amount since Eq. 20 will convert the equal amount back to the present worth.

$$\text{capitalized cost} = \text{initial cost} + \frac{\text{EAA}}{i}$$
$$= \text{initial cost} + \text{present worth of all expenses} \qquad [20]$$

Example 13

What is the capitalized cost of a public works project that will cost $25,000,000 now and will require $2,000,000 in maintenance annually? The effective annual interest rate is 12%.

PROFESSIONAL PUBLICATIONS, INC.

(solution)

Worked in millions of dollars, from Eq. 19, the capitalized cost is

$$\text{capitalized cost} = \$25 + \$2(P/A, 12\%, \infty)$$

$$= \$25 + \frac{\$2}{0.12} = \$41.67$$

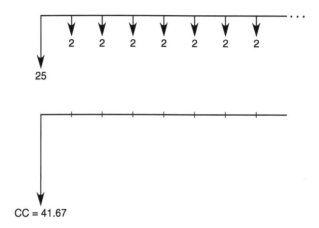

C. Annual Cost Method

Alternatives that accomplish the same purpose but that have unequal lives must be compared by the *annual cost method*.[16] The annual cost method assumes that each alternative will be replaced by an identical twin at the end of its useful life (infinite renewal). This method, which may also be used to rank alternatives according to their desirability, is also called the *annual return method* or *capital recovery method*.

Restrictions are that the alternatives must be mutually exclusive and repeatedly renewed up to the duration of the longest-lived alternative. The calculated annual cost is known as the *equivalent uniform annual cost* (EUAC) or just *equivalent annual cost*. Cost is a positive number when expenses exceed income.

Example 14

Which of the following alternatives is superior over a 30-year period if the interest rate is 7%?

---

[16]Of course, the annual cost method can be used to determine the superiority of assets with identical lives as well.

	alternative A	alternative B
type	brick	wood
life	30 years	10 years
initial cost	$1800	$450
maintenance	$5/year	$20/year

(solution)

$$\text{EUAC(A)} = \$1800(A/P, 7\%, 30) + 5$$
$$= (\$1800)(0.0806) + 5$$
$$= \$150$$
$$\text{EUAC(B)} = \$450(A/P, 7\%, 10) + 20$$
$$= (\$450)(0.1424) + 20$$
$$= \$84$$

Alternative B is superior since its annual cost of operation is the lowest. It is assumed that three wood facilities, each with a life of 10 years and a cost of $450, will be built to span the 30-year period.

25 CHOICE OF ALTERNATIVES: COMPARING AN ALTERNATIVE WITH A STANDARD

With specific economic performance criteria, it is possible to qualify an investment as acceptable or unacceptable without having to compare it with another investment. Two such performance criteria are the *benefit-cost ratio* and the *minimum attractive rate of return*.

A. Benefit-Cost Ratio Method

The benefit-cost ratio method is often used in municipal project evaluations where benefits and costs accrue to different segments of the community. With this method, the present worth of all benefits (irrespective of the beneficiaries) is divided by the present worth of all costs. The project is considered acceptable if the ratio equals or exceeds 1.0; that is, if $B/C \geq 1.0$.

When the benefit-cost ratio method is used, disbursements by the initiators or sponsors are *costs*. Disbursements by the users of the project are known as *disbenefits*. It is often difficult to determine whether a cash flow is a cost or a disbenefit (whether to place it in the numerator or denominator of the benefit-cost ratio calculation).

Regardless of where the cash flow is placed, an acceptable project will always have a benefit-cost ratio greater than or equal to 1.0, although the actual numerical result will depend on the placement. For this reason, the benefit-cost ratio method should not be used to rank competing projects.

The benefit-cost ratio method of comparing alternatives has seen extensive use in transportation engineering where the ratio is often (but not necessarily) written in terms of annual benefits and annual costs instead of present worths. Another characteristic of highway benefit-cost ratios is that the route (road, highway, etc.) is usually already in place and that various alternative upgrades are being considered. There will be existing benefits and costs associated with the current route. Therefore, the *change* (usually an increase) in benefits and costs is used to calculate the benefit-cost ratio.[17]

$$B/C = \frac{\Delta \text{ user benefits}}{\Delta \text{ investment cost} + \Delta \text{ maintenance} - \Delta \text{ residual value}} \quad [21]$$

Notice that the change in *residual value (terminal value)* appears in the denominator as a negative item. An increase in the residual value would decrease the denominator.

Example 15

By building a bridge over a ravine, a state department of transportation can shorten the time it takes to drive through a mountainous area. Estimates of costs and benefits (due to decreased travel time, fewer accidents, reduced gas usage, etc.) have been prepared. Should the bridge be built? Use the benefit-cost ratio method of comparison.

	millions
initial cost	$40
capitalized cost of perpetual annual maintenance	$12
capitalized value of annual user benefits	$49
residual value	0

(solution)

If Eq. 21 is used, the benefit-cost ratio is

$$B/C = \frac{\$49}{\$40 + \$12 + 0} = 0.942$$

Since the benefit-cost ratio is less than 1.00, the bridge should not be built.

---

[17]This discussion of highway benefit-cost ratios is not meant to imply that everyone agrees with Eq. 21. In *Economic Analysis for Highways* (International Textbook Company, Scranton, PA, 1969), author Robley Winfrey takes a strong stand on one aspect of the benefits versus disbenefits issue: highway maintenance. Regular highway maintenance costs (according to Winfrey) should be placed in the numerator as a subtraction from the user benefits. This mandate has been called the *Winfrey method* by some.

If the maintenance costs are placed in the numerator, the benefit-cost ratio value will be different, but the conclusion will not change.

$$B/C_{\text{alternate method}} = \frac{\$49 - \$12}{\$40} = 0.925$$

B. Rate of Return Method

The minimum attractive rate of return (MARR) has already been introduced as a standard of performance against which an investment's actual rate of return (ROR) is compared. If the rate of return is equal to or exceeds the minimum attractive rate of return, the investment is qualified. This is the basis for the *rate of return method* of alternative selection.

Finding the rate of return can be a long, iterative process. Usually, the actual numerical value of rate of return is not needed; it is sufficient to know whether or not the rate of return exceeds the minimum attractive rate of return. This *comparative analysis* can be accomplished without calculating the rate of return simply by finding the present worth of the investment using the minimum attractive rate of return as the effective interest rate (i.e., $i = \text{MARR}$). If the present worth is zero or positive, the investment is qualified. If the present worth is negative, the rate of return is less than the minimum attractive rate of return.

26 RANKING MUTUALLY EXCLUSIVE MULTIPLE PROJECTS

Ranking of multiple investment alternatives is required when there is sufficient funding for more than one investment. Since the best investments should be selected first, it is necessary to be able to place all investments into an ordered list.

Ranking is relatively easy if the present worths, future worths, capitalized costs, or equivalent uniform annual costs have been calculated for all the investments. The highest-ranked investment will be the one with the largest present or future worth, or the smallest capitalized or annual cost. Present worth, future worth, capitalized cost, and equivalent uniform annual cost can all be used to rank multiple investment alternatives.

However, neither rates of return nor benefit-cost ratios should be used to rank multiple investment alternatives. Specifically, if two alternatives both have rates of return exceeding the minimum acceptable rate of return, it is not sufficient to select the alternative with the highest rate of return.

An *incremental analysis*, also known as a *rate of return on added investment study*, should be performed if rate of return is used to select between investments. An incremental analysis starts by ranking the alternatives in order of increasing initial investment. Then, the cash flows for the investment with the lower initial cost are subtracted from the cash flows for the higher-priced alternative on a year-by-year basis. This produces, in effect, a third alternative representing the costs and benefits of the added investment. The added expense of the higher-priced investment is not warranted unless the rate of return of this third alternative exceeds the minimum attractive rate of return as well. The choice criterion is to select the alternative with the higher initial investment if the incremental rate of return exceeds the minimum attractive rate of return.

An incremental analysis is also required if ranking is to be done by the benefit-cost ratio method. The incremental analysis is accomplished by calculating the ratio of differences in benefit to differences in costs for each possible pair of alternatives. If the ratio exceeds 1.0, alternative 2 is superior to alternative 1. Otherwise, alternative 1 is superior.[18]

$$\frac{B_2 - B_1}{C_2 - C_1} \geq 1 \qquad \text{[alternative 2 superior]} \qquad [22]$$

27 ALTERNATIVES WITH DIFFERENT LIVES

Comparison of two alternatives is relatively simple when both alternatives have the same life. For example, a problem might be stated: "Which would you rather have: car A with a life of five years, or car B with a life of five years?"

However, care must be taken to understand what is going on when the two alternatives have different lives. If car A has a life of three years, and car B has a life of five years, what happens at $t = 3$ if the five-year car is chosen? If a car is needed for five years, what happens at $t = 3$ if the three-year car is chosen?

In this type of situation, it is necessary to distinguish between the length of the need (the *analysis horizon*) and the lives of the alternatives or assets intended to meet that need. The lives do not have to be the same as the horizon.

A. Finite Horizon with Incomplete Asset Lives

If an asset with a five-year life is chosen for a three-year need, the disposition of the asset at $t = 3$ must be known in order to evaluate the alternative. If the asset is sold at $t = 3$, the salvage value is entered into the analysis (at $t = 3$), and the alternative is evaluated as a three-year investment. The fact that the asset is sold when it has some useful life remaining does not affect the analysis horizon.

Similarly, if a three-year asset is chosen for a five-year need, something about how the need is satisfied during the last two years must be known. Perhaps a rental asset will be used. Or, perhaps the function will be farmed out to an outside firm. In any

---

[18]It goes without saying that the benefit-cost ratios for all investment alternatives by themselves must also be equal to or greater than 1.0.

case, the costs of satisfying the need during the last two years enter the analysis, and the alternative is evaluated as a five-year investment.

If both alternatives are converted to the same life, any of the alternative selection criteria (e.g., present worth method, annual cost method, etc.) can be used to determine which alternative is superior.

B. Finite Horizon with Integer Multiple Asset Lives

It is common to have a long-term horizon (need) that must be met with short-lived assets. In special instances, the horizon will be an integer number of asset lives. For example, a company may be making a 12-year transportation plan, and may be evaluating two cars: one with a three-year life, and another with a four-year life.

In this example, four of the first car, or three of the second car, are needed to reach the end of the 12-year horizon.

If the horizon is an integer number of asset lives, any of the alternative selection criteria can be used to determine which is superior. If the present worth method is used, all alternatives must be evaluated over the entire horizon. (In this example, the present worth of 12 years of car purchases and use must be determined for both alternatives.)

If the equivalent uniform annual cost method is used, it may be possible to base the calculation of annual cost on one lifespan of each alternative only. It may not be necessary to incorporate all of the cash flows into the analysis. (In the running example, the annual cost over three years would be determined for the first car; the annual cost over four years would be determined for the second car.) This simplification is justified if the subsequent asset replacements (renewals) have the same cost and cash flow structure as the original asset. This assumption is typically made implicitly when the annual cost method of comparison is used.

C. Infinite Horizon

If the need horizon is infinite, it is not necessary to impose the restriction that asset lives of alternatives be integer multiples of the horizon. The superior alternative will be replaced (renewed) whenever it is necessary to do so, forever.

Infinite horizon problems are almost always solved with either the annual cost or capitalized cost method. It is common to (implicitly) assume that the cost and cash flow structure of the asset replacements (renewals) are the same as the original asset.

28 OPPORTUNITY COSTS

An *opportunity cost* is an imaginary cost representing what will not be received if a particular strategy is rejected. It is what you will lose if you do or do not do something. As an example, consider a growing company with an existing operational computer system. If the company trades in its existing computer as part of an upgrade plan, it will receive a *trade-in allowance*. (In other problems, a *salvage value* may be involved.)

If one of the alternatives being evaluated is not to upgrade the computer system at all, the trade-in allowance (or, salvage value in other problems) will not be realized. The amount of the trade-in allowance is an opportunity cost that must be included in the problem analysis.

Similarly, if one of the alternatives being evaluated is to wait one year before upgrading the computer, the *difference in trade-in allowances* is an opportunity cost that must be included in the problem analysis.

29 REPLACEMENT STUDIES

An investigation into the retirement of an existing process or piece of equipment is known as a *replacement study*. Replacement studies are similar in most respects to other alternative comparison problems: an interest rate is given, two alternatives exist, and one of the previously mentioned methods of comparing alternatives is used to choose the superior alternative. Usually, the annual cost method is used on a year-by-year basis.

In replacement studies, the existing process or piece of equipment is known as the *defender*. The new process or piece of equipment being considered for purchase is known as the *challenger*.

30 TREATMENT OF SALVAGE VALUE IN REPLACEMENT STUDIES

Since most defenders still have some market value when they are retired, the problem of what to do with the salvage arises. It seems logical to use the salvage value of the defender to reduce the initial purchase cost of the challenger. This is consistent with what would actually happen if the defender were to be retired.

By convention, however, the defender's salvage value is subtracted from the defender's present value. This does not seem logical, but it is done to keep all costs and benefits related to the defender with the defender. In this case, the salvage value is treated as an opportunity cost that would be incurred if the defender is not retired.

If the defender and the challenger have the same lives and a present worth study is used to choose the superior alternative, the placement of the salvage value will have no effect on the net difference between present worths for the challenger and defender. Although the values of the two present worths will be different depending on the placement, the difference in present worths will be the same.

If the defender and the challenger have different lives, an annual cost comparison must be made. Since the salvage value would be spread over a different number of years depending on its placement, it is important to abide by the conventions listed in this section.

There are a number of ways to handle salvage value in retirement studies. The best way is to calculate the cost of keeping the defender one more year. In addition to the usual operating and maintenance costs, that cost includes an opportunity interest cost incurred by not selling the defender, and also a drop in the salvage value if the defender is kept for one additional year. Specifically,

$$\text{EUAC (defender)} = \text{next year's maintenance costs} + i\,(\text{current salvage value})$$
$$+ \text{ current salvage } - \text{ next year's salvage} \qquad [23]$$

It is important in retirement studies not to double count the salvage value. That is, it would be incorrect to add the salvage value to the defender and at the same time subtract it from the challenger.

Equation 23 contains the difference in salvage value between two consecutive years. This calculation shows that the defender/challenger decision must be made on a year-by-year basis. One application of Eq. 23 will not usually answer the question of whether the defender should remain in service. The calculation must be repeatedly made as long as there is a drop in salvage value from one year to the next.

31 ECONOMIC LIFE: RETIREMENT AT MINIMUM COST

As an asset grows older, its operating and maintenance costs typically increase. Eventually, the cost to keep the asset in operation becomes prohibitive, and the asset is retired or replaced. However, it is not always obvious when an asset should be retired or replaced.

As the asset's maintenance cost is increasing each year, the amortized cost of its initial purchase is decreasing. It is the sum of these two costs that should be evaluated to determine the point at which the asset should be retired or replaced. Since an asset's initial purchase price is likely to be high, the amortized cost will be the controlling factor in those years when the maintenance costs are low. Therefore, the EUAC of the asset will decrease in the initial part of its life.

However, as the asset grows older, the change in its amortized cost decreases while maintenance increases. Eventually, the sum of the two costs reaches a minimum and then starts to increase. The age of the asset at the minimum cost point is known as the *economic life* of the asset. The economic life is, generally, less than the length of need and the technological lifetime of the asset.

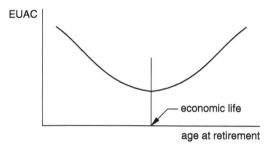

Figure 7   EUAC vs. Age at Retirement

The determination of an asset's economic life is illustrated by Ex. 16.

Example 16

Buses in a municipal transit system have the characteristics listed. When should the city replace its buses if money can be borrowed at 8%?

initial cost of bus: $120,000

year	maintenance cost	salvage value
1	$35,000	$60,000
2	$38,000	$55,000
3	$43,000	$45,000
4	$50,000	$25,000
5	$65,000	$15,000

(solution)

The annual maintenance is different each year. Each maintenance cost must be spread over the life of the bus. This is done by first finding the present worth and then amortizing the maintenance costs. If a bus is kept for one year and then sold, the annual cost will be

$$\begin{aligned}
\text{EUAC}(1) &= \$120,000(A/P, 8\%, 1) + \$35,000(A/F, 8\%, 1) - \$60,000(A/F, 8\%, 1) \\
&= (\$120,000)(1.0800) + (\$35,000)(1.000) - (\$60,000)(1.000) \\
&= \$104,600
\end{aligned}$$

If a bus is kept for two years and then sold, the annual cost will be

$$\begin{aligned}
\text{EUAC}(2) &= [\$120,000 + \$35,000(P/F, 8\%, 1)](A/P, 8\%, 2) \\
&\quad + (\$38,000 - \$55,000)(A/F, 8\%, 2) \\
&= [\$120,000 + (\$35,000)(0.9259)](0.5608) + (\$38,000 - \$55,000)(0.4808) \\
&= \$77,300
\end{aligned}$$

If a bus is kept for three years and then sold, the annual cost will be

$$
\begin{aligned}
\text{EUAC}(3) &= [\$120,000 + \$35,000(P/F, 8\%, 1) + \$38,000(P/F, 8\%, 2)](A/P, 8\%, 3) \\
&\quad + (\$43,000 - \$45,000)(A/F, 8\%, 3) \\
&= [\$120,000 + (\$35,000)(0.9259) + (\$38,000)(0.8573)](0.3880) \\
&\quad - (\$2000)(0.3080) \\
&= \$71,200
\end{aligned}
$$

This process is continued until the annual cost begins to increase. In this example, EUAC(4) is \$71,700. Therefore, the buses should be retired after three years.

32 LIFE-CYCLE COSTING

The *life-cycle cost* of an alternative is the equivalent value (at $t = 0$) of the alternative's cash flow over the alternative's lifespan. Since the present worth is evaluated using an effective interest rate of i (which would be the interest rate used for all engineering economic analyses), the life-cycle cost is the same as the alternative's present worth. If the alternative has an infinite horizon, the life-cycle cost and capitalized cost will be identical.

33 CAPITALIZED ASSETS VERSUS EXPENSES

High expenses reduce profit, which in turn, reduces income tax. It seems logical to label each and every expenditure, even an asset purchase, as an expense. As an alternative to this *expensing the asset*, it may be decided to capitalize the asset. *Capitalizing the asset* means that the cost of the asset is divided into equal or unequal parts, and only one of these parts is taken as an expense each year. Expensing is clearly the more desirable alternative, since the after-tax profit is increased early in the asset's life.

There are long-standing guidelines as to what can be expensed and what must be capitalized.[19] Some companies capitalize everything, regardless of its cost, with expected lifetimes greater than one year. Most companies, however, expense items whose purchase costs are below a cutoff value. A cutoff value in the range of \$250 to \$500, depending on the size of the company, is chosen as the maximum purchase cost of an expensed asset. Assets costing more than this are capitalized.

It is not necessary for a large corporation to keep track of every typewriter, desk, and chair for which the purchase price is greater than the cutoff value. Such assets, all of which have the same lives and have been purchased in the same year, can be placed into groups or *asset classes*. A group cost, equal to the sum total of the purchase costs of all items in the group, is capitalized as though the group was an identifiable and distinct asset itself.

---

[19]For example, purchased vehicles must be capitalized; payments for leased vehicles can be expensed. Repainting a building with paint that will last five years is an expense, but the replacement cost of a leaking roof must be capitalized.

34 PURPOSE OF DEPRECIATION

Depreciation is an *artificial expense* that spreads the purchase price of an asset or other property over a number of years.[20] Depreciating an asset is an example of capitalization, as previously defined. The inclusion of depreciation in engineering economic analysis problems will increase the after-tax present worth (profitability) of an asset. The larger the depreciation, the greater will be the profitability. Therefore, individuals and companies that are eligible to utilize depreciation desire to maximize and accelerate the depreciation available to them.

Although the entire property purchase price is eventually recognized as an expense, the net recovery from the expense stream never equals the original cost of the asset. That is, depreciation cannot realistically be thought of as a fund (an annuity or sinking fund) that accumulates capital to purchase a replacement at the end of the asset's life. The primary reason for this is that the depreciation expense is reduced significantly by the impact of income taxes, as will be seen in later sections.

35 DEPRECIATION BASIS OF AN ASSET

The *depreciation basis* of an asset is the part of the asset's purchase price that is spread over the *depreciation period*, also known as the *service life*.[21] Usually, the depreciation basis and the purchase price are not the same.

A common depreciation basis is the difference between the purchase price and the expected salvage value at the end of the depreciation period. That is,

$$\text{depreciation basis} = C - S_n \qquad [24]$$

There are several methods of calculating the year-by-year depreciation of an asset. Equation 24 is not universally compatible with all depreciation methods. Some methods do not consider the salvage value. This is known as an *unadjusted basis*. When the depreciation method is known, the depreciation basis can be rigorously defined.[22]

---

[20]The IRS tax regulations allow depreciation on almost all forms of *business property* except land. The following types of property are distinguished: *real* (e.g., buildings used for business), *residential* (e.g., buildings used as rental property), and *personal* (e.g., equipment used for business). Personal property does *not* include items for personal use (such as a personal residence), despite its name. *Tangible personal property* is distinguished from *intangible property* (e.g., goodwill, copyrights, patents, trademarks, franchises, agreements not to compete, etc.).

[21]The *depreciation period* is selected to be as short as possible within recognized limits. This depreciation will not normally coincide with the *economic life* or *useful life* of an asset. For example, a car may be capitalized over a depreciation period of three years. It may become uneconomical to maintain and use at the end of an economic life of nine years. However, the car may be capable of operation over a useful life of 25 years.

[22]For example, with the Accelerated Cost Recovery System (ACRS) the *depreciation basis* is the total purchase cost, regardless of the expected salvage value. With declining balance methods, the depreciation basis is the purchase cost less any previously taken depreciation.

36 DEPRECIATION METHODS

Generally, tax regulations do not allow the cost of an asset to be treated as a deductible expense in the year of purchase. Rather, portions of the depreciation basis must be allocated to each of the n years of the asset's depreciation period. The amount that is allocated each year is called the *depreciation*.

Various methods exist for calculating an asset's depreciation each year.[23] Although the depreciation calculations may be considered independently (for the purpose of determining book value or as an academic exercise), it is important to recognize that depreciation has no effect on engineering economic analyses unless income taxes are also considered.

A. Straight Line Method

With the *straight line method*, depreciation is the same each year. The depreciation basis $(C - S_n)$ is allocated uniformly to all of the n years in the depreciation period. Each year, the depreciation will be

$$D = \frac{C - S_n}{n} \qquad [25]$$

B. Constant Percentage Method

The *constant percentage method*[24] is similar to the straight line method, in that the depreciation is the same each year. If the fraction of the basis used as depreciation is $1/n$, there is no difference between the constant percentage and straight line methods. The two methods differ only in what information is available. (With the straight line method, the life is known. With the constant percentage method, the depreciation fraction is known.)

Each year, the depreciation will be

$$D = \text{(depreciation fraction)} \times \text{(depreciation basis)}$$
$$= \text{(depreciation fraction)} \times (C - S_n) \qquad [26]$$

---

[23]This discussion gives the impression that any form of depreciation may be chosen regardless of the nature and circumstances of the purchase. In reality, the IRS tax regulations place restrictions on the higher-rate (accelerated) methods, such as declining balance and sum-of-the-years' digits methods. Furthermore, the *Economic Recovery Act of 1981* and the Tax Reform Act of 1986 substantially changed the laws relating to personal and corporate income taxes.

[24]The *constant percentage method* should not be confused with the declining balance method, which used to be known as the *fixed percentage on diminishing balance method*.

C. Sum-of-the-Years' Digits Method

In *sum-of-the-years' digits* (SOYD) depreciation, the digits from 1 to n inclusive are summed. The total, T, can also be calculated from

$$T = \tfrac{1}{2}n(n+1) \qquad [27]$$

The depreciation in year j can be found from Eq. 28. Notice that the depreciation in year j, D_j, decreases by a constant amount each year.

$$D_j = \frac{(C - S_n)(n - j + 1)}{T} \qquad [28]$$

D. Double Declining Balance Method[25]

Double declining balance[26] (DDB) depreciation is independent of salvage value. Furthermore, the book value never stops decreasing, although the depreciation decreases in magnitude. Usually, any book value in excess of the salvage value is written off in the last year of the asset's depreciation period. Unlike any of the other depreciation methods, double declining balance depends on accumulated depreciation.

$$D_{\text{first year}} = \frac{2C}{n} \qquad [29]$$

$$D_j = \frac{2\left(C - \sum_{m=1}^{j-1} D_m\right)}{n} \qquad [30]$$

Calculating the depreciation in the middle of an asset's life appears particularly difficult with double declining balance, since all previous years' depreciation amounts seem to be required. It appears that the depreciation in the sixth year (for example) cannot be calculated unless the values of depreciation for the first five years are calculated. However, this is not true.

Depreciation in the middle of an asset's life can be found from the following equations; d is known as the *depreciation rate*.

$$d = \frac{2}{n} \qquad [31]$$

$$D_j = dC(1 - d)^{j-1} \qquad [32]$$

---

[25]The *declining balance method*, has, in the past, also been known as the *fixed percentage of book value* and *fixed percentage on diminishing balance method*.
[26]Double declining balance depreciation is a particular form of *declining balance depreciation,* as defined by the IRS tax regulations. Declining balance depreciation also includes 125% declining balance and 150% declining balance depreciations that can be calculated by substituting 1.25 and 1.50, respectively, for the 2 in Eq. 29.

E. Statutory Depreciation Systems

In the United States, property placed into service in 1981 and thereafter must use the *Accelerated Cost Recovery System* (ACRS), and after 1986, *Modified Accelerated Cost Recovery System* (MACRS) or other statutory method. Other methods (straight line, declining balance, etc.) cannot be used except in special cases.

Property placed into service in 1980 or before must continue to be depreciated according to the method originally chosen (e.g., straight line, declining balance, or sum-of-the-years' digits). ACRS and MACRS cannot be used.

Under ACRS and MACRS, the cost recovery amount in the jth year of an asset's cost recovery period is calculated by multiplying the initial cost by a factor.

$$D_j = C \times \text{factor} \quad\quad [33]$$

The initial cost used is not reduced by the asset's salvage value for ACRS and MACRS calculations. The factor used depends on the asset's cost recovery period. Such factors are subject to continuing legislation changes. Current tax publications should be consulted before using this method.

Table 4
Representative MACRS Depreciation Factors[a]

depreciation rate for recovery period (n)

year(j)	3 years	5 years	7 years	10 years
1	33.33%	20.00%	14.29%	10.00%
2	44.45%	32.00%	24.49%	18.00%
3	14.81%	19.20%	17.49%	14.40%
4	7.41%	11.52%	12.49%	11.52%
5		11.52%	8.93%	9.22%
6		5.76%	8.92%	7.37%
7			8.93%	6.55%
8			4.46%	6.55%
9				6.56%
10				6.55%
11				3.28%

[a]Values are for the half-year convention. This table gives typical values only. Since these factors are subject to continuing revision, they should not be used without consulting an accounting professional.

F. Production or Service Output Method

If an asset has been purchased for a specific task, and that task is associated with a specific lifetime amount of output or production, the depreciation may be calculated by the fraction of total production produced during the year. The depreciation is not expected to be the same each year.

$$D_j = (C - S_n) \times \frac{\text{actual output in year } j}{\text{estimated lifetime output}} \qquad [34]$$

G. Sinking Fund Method

The *sinking fund method* is seldom used in industry because the initial depreciation is low. The formula for sinking fund depreciation (which increases each year) is

$$D_j = (C - S_n)(A/F, i\%, n)(F/P, i\%, j - 1) \qquad [35]$$

H. Disfavored Methods

Three other depreciation methods should be mentioned, not because they are currently accepted or in widespread use, but because they are still occasionally called for by name.[27]

The *sinking fund plus interest on first cost* depreciation method, like the following two methods, is an attempt to include the *opportunity interest cost* on the purchase price with the depreciation. That is, the purchasing company not only incurs an annual expense due to the drop in book value, but it also loses the interest on the purchase price. The formula for this method is

$$D = (C - S_n)(A/F, i\%, n) + Ci \qquad [36]$$

The *straight line plus interest on first cost* method is similar. Its formula is

$$D = \frac{1}{n}(C - S_n) + Ci \qquad [37]$$

The *straight line plus average interest method* assumes that the opportunity interest cost should be based on the book value only, not on the full purchase price. Since the book value changes each year, an average value is used. The depreciation formula is

$$D = \frac{C - S_n}{n}\left[1 + \frac{i(n + 1)}{2}\right] + iS_n \qquad [38]$$

Example 17

An asset is purchased for $9000. Its estimated economic life is 10 years, after which it will be sold for $200. Find the depreciation in the first three years using straight line, double declining balance, and sum-of-the-years' digits depreciation methods.

---

[27]These three depreciation methods should not be used in the usual manner (e.g., in conjunction with the income tax rate). These methods are attempts to calculate a more accurate annual cost of an alternative. Sometimes they give misleading answers. Their use cannot be recommended. They are included in this chapter only for the sake of completeness.

(solution)

$$\text{SL:} \quad D = \frac{\$9000 - \$200}{10} \qquad = \$880 \text{ each year}$$

$$\text{DDB:} \quad D_1 = \frac{(2)(\$9000)}{10} \qquad = \$1800 \text{ in year 1}$$

$$D_2 = \frac{(2)(\$9000 - \$1800)}{10} \qquad = \$1440 \text{ in year 2}$$

$$D_3 = \frac{(2)(\$9000 - \$3240)}{10} \qquad = \$1152 \text{ in year 3}$$

$$\text{SOYD:} \quad T = \left(\frac{1}{2}\right)(10)(11) = 55$$

$$D_1 = \left(\frac{10}{55}\right)(\$9000 - \$200) = \$1600 \text{ in year 1}$$

$$D_2 = \left(\frac{9}{55}\right)(\$8800) \qquad = \$1440 \text{ in year 2}$$

$$D_3 = \left(\frac{8}{55}\right)(\$8800) \qquad = \$1280 \text{ in year 3}$$

37 ACCELERATED DEPRECIATION METHOD

An *accelerated depreciation method* is one that calculates a depreciation amount greater than a straight line amount. Double declining balance and sum-of-the-years' digits methods are accelerated methods. The ACRS and MACRS methods are explicitly accelerated methods. Straight line and sinking fund methods are not accelerated methods.

Use of an accelerated depreciation method may result in unexpected tax consequences when the depreciated asset or property is disposed of. Professional tax advice should be obtained in this area.

38 BOOK VALUE

The difference between original purchase price and accumulated depreciation is known as *book value*.[28] At the end of each year, the book value (which is initially equal to the purchase price) is reduced by the depreciation in that year.

---

[28]The balance sheet of a corporation usually has two asset accounts: the *equipment account* and the *accumulated depreciation account*. There is no book value account on this financial statement, other than the implicit value obtained from subtracting the accumulated depreciation account from the equipment account. The book values of various assets, as well as their original purchase cost, date of purchase, salvage value, etc., and accumulated depreciation appear on detail sheets or other peripheral records for each asset.

It is important to properly synchronize depreciation calculations. It is difficult to answer the question, "What is the book value in the fifth year?" unless the timing of the book value change is mutually agreed upon. It is better to be specific about an inquiry by identifying when the book value change occurs. For example, the following question is unambiguous: "What is the book value at the end of year 5, after subtracting depreciation in the fifth year?" or "What is the book value after five years?"

Unfortunately, this type of care is seldom taken in book value inquiries, and it is up to the respondent to exercise reasonable care in distinguishing between beginning-of-year book value and end-of-year book value. To be consistent, the book value equations in this chapter have been written in such a way that the year subscript (j) has the same meaning in book value and depreciation calculations. That is, BV_5 means the book value at the end of the fifth year, after five years of depreciation, including D_5, has been subtracted from the original purchase price.

There can be a great difference between the book value of an asset and the *market value* of that asset. There is no legal requirement for the two values to coincide, and no intent for book value to be a reasonable measure of market value.[29] Therefore, it is apparent that book value is merely an accounting convention with little practical use. Even when a depreciated asset is disposed of, the book value is used to determine the consequences of disposal, not the price the asset should bring at sale.

The calculation of book value is relatively easy, even for the case of the declining balance depreciation method.

For the straight line depreciation method, the book value at the end of the jth year, after the jth depreciation deduction has been made, is

$$BV_j = C - \frac{j(C - S_n)}{n} = C - jD \qquad [39]$$

For the sum-of-the-years' digits method, the book value is

$$BV_j = (C - S_n)\left(1 - \frac{j(2n + 1 - j)}{n(n + 1)}\right) + S_n \qquad [40]$$

For the declining balance method, the book value is

$$BV_j = C(1 - d)^j \qquad [41]$$

---

[29]Common examples of assets with great divergences of book and market values are buildings (rental houses, apartment complexes, factories, etc.) and company luxury automobiles (Porsches, Mercedes, etc.) during periods of inflation. Book values decrease, but actual values increase.

It is not necessary to calculate depreciation for interim years when finding the book value of an asset depreciated by the sinking fund method. The book value can be calculated directly as

$$\text{BV}_j = C - (C - S_n)(A/F, i\%, n)(F/A, i\%, j) \qquad [42]$$

Of course, the book value at the end of year j can always be calculated for any method by successive subtractions (i.e., subtraction of the accumulated depreciation), as Eq. 43 illustrates.

$$\text{BV}_j = C - \sum_{m=1}^{j} D_m \qquad [43]$$

Figure 8 illustrates the book value of a hypothetical asset depreciated using several depreciation methods. Notice that the double declining balance method initially produces the fastest write-off, while the sinking fund method produces the slowest write-off. Note also that the book value does not automatically equal the salvage value at the end of an asset's depreciation period with the double declining balance method.[30]

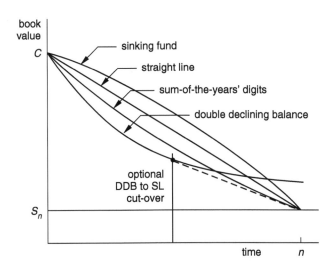

Figure 8   Book Value with Different Depreciation Methods

Example 18

For the asset described in Ex. 17, calculate the book value at the end of the first three years if sum-of-the-years' digits depreciation is used. The book value at the beginning of year 1 is $9000.

---

[30]This means that the straight line method of depreciation may result in a lower book value at some point in the depreciation period than if double declining balance is used. Such a *cut-over* for double-declining balance to straight-line may be permitted in certain cases. Finding the *cut-over point*, however, is usually done by comparing book values determined by both methods. The analytical method is complicated.

(solution)

From Eq. 43,

$$BV_1 = \$9000 - \$1600 = \$7400$$
$$BV_2 = \$7400 - \$1440 = \$5960$$
$$BV_3 = \$5960 - \$1280 = \$4680$$

39 AMORTIZATION

Amortization and depreciation are similar in that they both divide up the cost basis or value of an asset. In fact, in certain cases, the term *amortization* may be used in place of the term *depreciation*. However, depreciation is a specific form of amortization.

Amortization spreads the cost basis or value of an asset over some base. The base can be time, units of production, number of customers, etc. The asset can be tangible (e.g., a delivery truck or building) or intangible (e.g., goodwill or a patent).

If the asset is tangible, if the base is time, and if the length of time is consistent with accounting standards and taxation guidelines, the term depreciation is appropriate. However, if the asset is intangible, if the base is some other variable, or if some length of time other than the customary period is used, then the term amortization is more appropriate.[31]

Example 19

A company purchases complete and exclusive patent rights to an invention for $1,200,000. It is estimated that, once commercially produced, the invention will have a specific but limited market of 1200 units. For the purpose of allocating the patent right cost to production cost, what is the amortization rate in dollars per unit?

(solution)

The patent should be amortized at the rate of

$$\frac{\$1,200,000}{1200} = \$1000 \text{ per unit}$$

40 DEPLETION

Depletion is another artificial deductible operating expense designed to compensate mining organizations for decreasing mineral reserves. Since original and remaining

---

[31]From time to time, the U.S. Congress has allowed certain types of facilities (e.g., emergency, grain storage, and pollution control) to be written off in 60 months in order to encourage investment in such facilities. The term amortization has been used with such write-off periods.

quantities of minerals are seldom known accurately, the *depletion allowance* is calculated as a fixed percentage of the organization's gross income. These percentages are usually in the 10–20% range and apply to such mineral deposits as oil, natural gas, coal, uranium, and most metal ores.

41 BASIC INCOME TAX CONSIDERATIONS

The issue of income taxes is often overlooked in academic engineering economic analysis exercises. Such a position is justifiable when an organization (e.g., a nonprofit school, a church, or the government) pays no income taxes. However, if an individual or organization is subject to income taxes, the income taxes must be included in an economic analysis of investment alternatives.

Assume that an organization pays a fraction f of its profits to the federal government as income taxes. If the organization also pays a fraction s of its profits as state income tax and if state taxes paid are recognized by the federal government as tax-deductible expenses, then the composite tax rate is

$$t = s + f - sf \qquad [44]$$

For simplicity, most engineering economics practice problems involving income taxes specify a single income tax rate. In practice, however, the federal and most state tax rates depend on the income level. Each range of incomes and its associated tax rate are known as an *income bracket* and *tax bracket*, respectively. For example, the state income tax rate may be 4% for incomes up to and including \$30,000 and 5% for incomes above \$30,000. The income tax for a taxpaying entity with an income of \$50,000 would have to be calculated in two parts.

$$\text{tax} = (0.04)(\$30{,}000) + (0.05)(\$50{,}000 - \$30{,}000)$$
$$= \$2200$$

The basic principles used to incorporate taxation into engineering economic analyses are listed here.[32]

- Initial purchase expenditures are unaffected by income taxes.

- Salvage revenues are unaffected by income taxes.

- Deductible expenses, such as operating costs, maintenance costs, and interest payments, are reduced by the fraction t (e.g., multiplied by the quantity $1 - t$).

---

[32]There are various established procedures, known as *rules* or *conventions*, imposed by the Internal Revenue Service on U.S. taxpayers. An example is the *half-year rule* or *half-year convention*, which permits only half of the first-year depreciation to be taken in the first year of an asset's life when certain methods of depreciation are used. The rules are subject to constantly changing legislation and are not covered in this book. The implementation of such rules is outside the scope of engineering practice and is best left to accounting professionals.

- Revenues are reduced by the fraction t (e.g., multiplied by the quantity $1 - t$).

- Since tax regulations allow the depreciation in any year to be handled as if it were an actual operating expense, and since operating expenses are deductible from the income base prior to taxation, the after-tax profits will be increased. If D is the depreciation, the net result to the after-tax cash flow will be the addition of tD. Depreciation is multiplied by t and added to the appropriate year's cash flow, increasing that year's present worth.

Income taxes and depreciation have no bearing on municipal or governmental projects since municipalities, states, and the U.S. government pay no taxes.

Example 20

A corporation that pays 53% of its profit in income taxes invests \$10,000 in an asset that will produce \$3000 annual revenue for eight years. If the annual expenses are \$700, salvage after eight years is \$500, and 9% interest is used, what is the after-tax present worth? Disregard depreciation.

(solution)

$$P = -\$10,000 + \$3000(P/A, 9\%, 8)(1 - 0.53) - \$700(P/A, 9\%, 8)(1 - 0.53)$$
$$+ \$500(P/F, 9\%, 8)$$
$$= -\$10,000 + (\$3000)(5.5348)(0.47) - (\$700)(5.5348)(0.47) + (\$500)(0.5019)$$
$$= -\$3766$$

42 TAXATION AT THE TIMES OF ASSET PURCHASE AND SALE

There are numerous rules and conventions that governmental tax codes and the accounting profession impose on organizations. Engineers are not normally expected to be aware of most of the rules and conventions, but occasionally it may be necessary to incorporate their effects into an engineering economic analysis.

A. Tax Credit

A *tax credit* (also known as an *investment tax credit* or *investment credit*) is a one-time credit against income taxes.[33] Therefore, it is added to the after-tax present worth as a last step in an engineering economic analysis. Such tax credits may be allowed by the government from time to time for equipment purchases, employment of various classes of workers, rehabilitation of historic landmarks, etc.

---

[33] Strictly, *tax credit* is the more general term, and applies to a credit for doing anything creditable. An *investment tax credit* requires an investment in something (usually real property or equipment).

A tax credit (TC) is usually calculated as a fraction of the initial purchase price or cost of an asset or activity. That is,

$$TC = \text{fraction} \times \text{initial cost} \qquad [45]$$

When the tax credit is applicable, the fraction used is subject to legislation. A professional tax expert or accountant should be consulted prior to applying the tax credit concept to engineering economic analysis problems.

Since the investment tax credit reduces the buyer's tax liability, a tax credit should only be included in after-tax engineering economic analyses. The credit is assumed to be received at the end of the year.

B. Gain or Loss on the Sale of a Depreciated Asset

If an asset that has been depreciated over a number of prior years is sold for more than its current book value, the difference between the book value and selling price is taxable income in the year of the sale. Alternatively, if the asset is sold for less than its current book value, the difference between the selling price and book value is an expense in the year of the sale.

Example 21

One year, a company makes a $5000 investment in a historic building. The investment is not depreciable, but it does qualify for a one-time 20% tax credit. In that same year, revenue is $45,000 and expenses (exclusive of the $5000 investment) are $25,000. The company pays a total of 40% in income taxes. What is the after-tax present worth of this year's activities if the company's interest rate for investment is 10%?

(solution)

The tax credit is

$$TC = (0.20)(\$5000) = \$1000$$

This tax credit is assumed to be received at the end of the year. The after-tax present worth is

$$P = -\$5000 + (\$45,000 - \$25,000)(1 - 0.40)(P/F, 10\%, 1) + \$1000(P/F, 10\%, 1)$$
$$= -\$5000 + (\$20,000)(0.60)(0.9091) + (\$1000)(0.9091)$$
$$= \$6818$$

43 DEPRECIATION RECOVERY

The economic effect of depreciation is to reduce the income tax in year j by tD_j. The present worth of the asset is also affected: the present worth is increased by $tD_j(P/F, i\%, j)$. The after-tax present worth of all depreciation effects over the depreciation period of the asset is called the *depreciation recovery* (DR).[34]

$$\text{DR} = t\sum_{j=1}^{n} D_j(P/F, i\%, j) \qquad [46]$$

Straight line depreciation recovery from an asset is easily calculated, since the depreciation is the same each year. Assuming the asset has a constant depreciation of D and depreciation period of n years, the depreciation recovery is

$$\text{DR} = tD(P/A, i\%, n) \qquad [47]$$

$$D = \frac{C - S_n}{n} \qquad [48]$$

Sum-of-the-years' digits depreciation recovery is also relatively easily calculated, since the depreciation decreases uniformly each year.

$$\text{DR} = \frac{t(C - S_n)}{T}\left[n(P/A, i\%, n) - (P/G, i\%, n)\right] \qquad [49]$$

Finding *declining balance depreciation recovery* is more involved. There are three difficulties. The first (that of an apparent need to calculate all previous depreciations in order to determine the subsequent depreciation) has already been addressed by Eq. 32.

The second difficulty with calculating the declining balance depreciation recovery is that there is no way to force the book value to be S_n at $t = n$. Therefore, it is common to write off the remaining book value (down to S_n) at $t = n$ in one lump sum. This assumes $\text{BV}_n \geq S_n$.

The third difficulty is that of finding the present worth of an *exponentially decreasing cash flow*. Although the proof is omitted here, such exponential cash flows can be handled with the *exponential gradient factor*, (P/EG).[35]

$$(P/EG, z - 1, n) = \frac{z^n - 1}{z^n(z - 1)} \qquad [50]$$

$$z = \frac{1 + i}{1 - d} \qquad [51]$$

---

[34]Since the depreciation benefit is reduced by taxation, depreciation cannot be thought of as an annuity to fund a replacement asset.

[35]The (*P/A*) columns in the tables at the end of this book can be used for (*P/EG*) as long as the interest rate is assumed to be $z - 1$.

Then, as long as $\mathrm{BV}_n > S_n$, the declining balance depreciation recovery is

$$\mathrm{DR} = tC\left(\frac{d}{1-d}\right)(P/EG, z-1, n) \qquad [52]$$

Example 22

For the asset described in Ex. 17, calculate the after-tax depreciation recovery with straight line, sum-of-the-years' digits, and double declining balance depreciation methods. Use 6% interest with 40% income taxes.

(solution)

SL:
$$\begin{aligned}
\mathrm{DR} &= (0.40)(\$880)(P/A, 6\%, 10) \\
&= (0.40)(\$880)(7.3601) \\
&= \$2591
\end{aligned}$$

SOYD: The depreciation series can be thought of as a constant 1600 term with a negative 160 gradient.

$$\begin{aligned}
\mathrm{DR} &= (0.48)(\$1600)(P/A, 6\%, 10) - (0.48)(\$160)(P/G, 6\%, 10) \\
&= (0.48)(\$1600)(7.3601) - (0.48)(\$160)(29.6023) \\
&= \$3379
\end{aligned}$$

Notice that the ten-year (P/G) factor is used even though there are only nine years in which the gradient reduces the initial $1600 amount.

Example 23

What is the after-tax present worth of the asset described in Ex. 20 if straight line, sum-of-the-years' digits, and double declining balance depreciation methods are used?

(solution)

Using SL, the depreciation recovery is

$$\begin{aligned}
\mathrm{DR} &= (0.53)\left(\frac{\$10,000 - \$500}{8}\right)(P/A, 9\%, 8) \\
&= (0.53)\left(\frac{\$9500}{8}\right)(5.5348) \\
&= \$3483
\end{aligned}$$

Table 5
Depreciation Calculation Summary

method	depreciation basis	depreciation in year j (D_j)	book value after jth depreciation (BV_j)	after-tax depreciation recovery (DR)	supplementary formulas
straight line (SL)	$C - S_n$	$\dfrac{C - S_n}{n}$ [constant]	$C - jD$	$tD(P/A, i\%, n)$	
constant percentage	$C - S_n$	fraction $\times (C - S_n)$ [constant]	$C - jD$	$tD(P/A, i\%, n)$	
sum-of-the-years' digits (SOYD)	$C - S_n$	$\dfrac{(C - S_n)(n - j + 1)}{T}$	$(C - S_n) \times \left(1 - \dfrac{j(2n + 1 - j)}{n(n+1)}\right) + S_n$	$\dfrac{t(C - S_n)}{T} \times [n(P/A, i\%, n) - (P/G, i\%, n)]$	$T = \dfrac{1}{2}n(n+1)$
double declining balance (DDB)	C	$dC(1 - d)^{j-1}$	$C(1 - d)^j$	$tC\left(\dfrac{d}{1 - d}\right)(P/EG, z - 1, n)$	$d = \dfrac{2}{n} \quad (P/EG, z - 1, n)$ $= \dfrac{z^n - 1}{z^n(z - 1)}$ $z = \dfrac{1 + i}{1 - d}$
sinking fund (SF)	$C - S_n$	$(C - S_n)(A/F, i\%, n) \times (F/P, i\%, j - 1)$	$C - (C - S_n)(A/F, i\%, n) \times (F/A, i\%, j)$	$\dfrac{t(C - S_n)(A/F, i\%, n)}{1 + i}$	
accelerated cost recovery system (ACRS/MACRS)	C	$C \times$ factor	$C - \displaystyle\sum_{m=1}^{j} D_m$	$t\displaystyle\sum_{j=1}^{n} D_j(P/F, i\%, j)$	
production or service output	$C - S_n$	$(C - S_n)$ $\times \dfrac{\text{actual output in year } j}{\text{lifetime output}}$	$C - \displaystyle\sum_{m=1}^{j} D_m$	$t\displaystyle\sum_{j=1}^{n} D_j(P/F, i\%, j)$	

Using SOYD, the depreciation recovery is calculated as follows.

$$T = \left(\tfrac{1}{2}\right)(8)(9) = 36$$

$$\text{depreciation base} = \$10{,}000 - \$500 = 9500$$

$$D_1 = \left(\frac{8}{36}\right)(\$9500) = \$2111$$

$$G = \text{gradient} = \left(\frac{1}{36}\right)(\$9500)$$

$$= 264$$

$$\text{DR} = (0.40)[\$2111(P/A, 9\%, 8) - \$264(P/G, 9\%, 8)]$$

$$= (0.40)[(\$2111)(5.5348) - (\$264)(16.8877)]$$

$$= \$2890$$

Using DDB, the depreciation recovery is calculated as follows.[36]

$$d = \frac{2}{8} = 0.25$$

$$z = \frac{1 + 0.09}{1 - 0.25} = 1.4533$$

$$(P/EG, z - 1, n) = \frac{(1.453)^8 - 1}{(1.453)^8(0.453)} = 2.095$$

From Eq. 52,

$$\text{DR} = (0.40)\left[\frac{(0.25)(\$10{,}000)}{0.75}\right](2.095)$$

$$= \$3701$$

The after-tax present worth, neglecting depreciation, was previously found to be $-\$2111$.

The after-tax present worths, including depreciation recovery, are

$$
\begin{aligned}
\text{SL:} &\quad P = -\$3766 + \$3483 = -\$283 \\
\text{SOYD:} &\quad P = -\$3766 + \$3830 = \quad\ \$64 \\
\text{DDB:} &\quad P = -\$3766 + \$3701 = \ \ -\$65
\end{aligned}
$$

---

[36]This method should start by checking that the book value at the end of the depreciation period is greater than the salvage value. In this example, such is the case. However, the step is not shown.

44 OTHER INTEREST RATES

The *effective interest rate per period*, i, (also called *yield* by banks) is the only interest rate that should be used in equivalence equations. The interest rates at the top of the factor tables in the Expanded Interest Tables are implicitly all effective interest rates. Usually, the period will be one year, hence the name *effective annual interest rate*. However, there are other interest rates in use as well.

The term *nominal interest rate*, r, (*rate per annum*) is encountered when compounding is more than once per year. The nominal rate does not include the effect of compounding and is not the same as the effective rate. And, since the effective interest rate can be calculated from the nominal rate only if the number of compounding periods per year is known, nominal rates cannot be compared unless the method of compounding is specified. The only practical use for a nominal rate is for calculating the effective rate.

45 RATE AND PERIOD CHANGES

If there are k compounding periods during the year (e.g., two for semiannual compounding, four for quarterly compounding, twelve for monthly compounding, etc.), the *effective rate per compounding period* is

$$\phi = \frac{r}{k} \qquad [53]$$

The effective annual rate, i, can be calculated from the effective rate per period, ϕ, by using Eq. 54.

$$i = (1 + \phi)^k - 1$$
$$= \left(1 + \frac{r}{k}\right)^k - 1 \qquad [54]$$

Sometimes, only the effective rate per period (e.g., per month) is known. However, that will be a simple problem since compounding for n periods at an effective rate per period is not affected by the definition or length of the period.

The following rules may be used to determine which interest rate is given in a problem.

- Unless specifically qualified in the problem, the interest rate given is an annual rate.

- If the compounding is annual, the rate given is the effective rate. If compounding is other than annual, the rate given is the nominal rate.

The effective annual interest rate determined on a *daily compounding basis* will not be significantly different than if *continuous compounding* is assumed.[37] In the case of continuous (or daily) compounding, the discounting factors can be calculated directly from the nominal interest rate and number of years, without having to find the effective interest rate per period. Table 6 can be used to determine the discount factors for continuous compounding.

Table 6

Discount Factors for Continuous Compounding

(n is the number of years)

symbol	formula
$(F/P, r\%, n)$	e^{rn}
$(P/F, r\%, n)$	e^{-rn}
$(A/F, r\%, n)$	$\dfrac{e^r - 1}{e^{rn} - 1}$
$(F/A, r\%, n)$	$\dfrac{e^{rn} - 1}{e^r - 1}$
$(A/P, r\%, n)$	$\dfrac{e^r - 1}{1 - e^{-rn}}$
$(P/A, r\%, n)$	$\dfrac{1 - e^{-rn}}{e^r - 1}$

Example 24

A savings and loan offers $5\frac{1}{4}\%$ compounded daily over 365 days in a year. What is the effective annual rate?

(solution)

    method 1:   Use Eq. 54.

$$r = 0.0525$$
$$k = 365$$
$$i = \left(1 + \frac{0.0525}{365}\right)^{365} - 1 = 0.0539$$

---

[37] The number of *banking days in a year* (e.g., 250, 360, etc.) must be specifically known.

method 2: Assume daily compounding is the same as continuous compounding.

$$i = (F/P, r\%, 1) - 1$$
$$= e^{0.0525} - 1 = 0.0539$$

Example 25

A real estate investment trust pays $7,000,000 for a 100-unit apartment complex. The trust expects to sell the complex in 10 years for $15,000,000. In the meantime, it expects to receive an average rent of $900 per month from each apartment. Operating expenses are expected to be $200 per month per occupied apartment. A 95% occupancy rate is predicted. In similar investments, the trust has realized a 15% effective annual return on its investment. Compare the expected present worth of this investment when calculated assuming (a) annual compounding (i.e., the year-end convention) and (b) monthly compounding. Disregard taxes, depreciation, and other factors.

(solution)

(a) The net annual income will be

$$(0.95)(100\,\text{units}) \left(\$900 \, \frac{\$}{\text{unit-month}} - \$200 \, \frac{\$}{\text{unit-month}} \right)$$
$$\times \left(12 \, \frac{\text{months}}{\text{year}} \right) = \$798{,}000/\text{year}$$

The present worth of 10 years of operation is

$$P = -\$7{,}000{,}000 + (\$798{,}000)(P/A, 15\%, 10) + (\$15{,}000{,}000)(P/F, 15\%, 10)$$
$$= -\$7{,}000{,}000 + (\$798{,}000)(5.0188) + (\$15{,}000{,}000)(0.2472)$$
$$= \$713{,}000$$

(b) The net monthly income is

$$(0.95)(100\,\text{units}) \left(\$900 \, \frac{\$}{\text{unit-month}} - \$200 \, \frac{\$}{\text{unit-month}} \right) = \$66{,}500/\text{month}$$

Equation 54 is used to calculate the effective monthly rate, ϕ, from the effective annual rate, $i = 15\%$, and the number of compounding periods per year, $k = 12$.

$$\phi = (1+i)^{\frac{1}{k}} - 1$$
$$= (1 + 0.15)^{\frac{1}{12}} - 1$$
$$= 0.011715 \quad (1.1715\%)$$

The number of compounding periods in 10 years is

$$n = (10 \text{ years}) \left(12 \ \frac{\text{months}}{\text{year}} \right) = 120 \text{ months}$$

The present worth of 120 months of operation is

$$P = -\$7,000,000 + (\$66,500)(P/A, 1.1715\%, 120)$$
$$+ (\$15,000,000)(P/F, 1.1715\%, 120)$$

Since table values for 1.1715% discounting factors are not available, the factors are calculated from Table 1.

$$(P/A, 1.1715\%, 120) = \frac{(1+i)^n - 1}{i(1+i)^n} = \frac{(1.011715)^{120} - 1}{(0.011715)(1.011715)^{120}} = 64.261$$
$$(P/F, 1.1715\%, 120) = (1+i)^{-n} = (1.011715)^{-120} = 0.2472$$

The present worth over 120 monthly compounding periods is

$$P = -\$7,000,000 + (\$66,500)(64.261) + (\$15,000,000)(0.2472)$$
$$= \$981,400$$

46 BONDS

A *bond* is a method of long-term financing commonly used by governments, states, municipalities, and very large corporations.[38] The bond represents a contract to pay the bondholder specific amounts of money at specific times. The holder purchases the bond in exchange for specific payments of interest and principal. Typical municipal bonds call for quarterly or semiannual interest payments, and a payment of the *face value of the bond* on the *date of maturity* (end of the bond period).[39] Due to the practice of discounting in the bond market, a bond's face value and its purchase price will generally not coincide.

In the past, a bondholder had to submit a coupon or ticket in order to receive an interim interest payment. This has given rise to the term *coupon rate*, which is the nominal annual interest rate on which the interest payments are made. Coupon books are seldom used with modern bonds, but the term survives in any case. The coupon rate determines the magnitude of the semiannual (or otherwise) interest payments during the life of the bond. The bondholder's own effective interest rate should be used for economic decisions about the bond.

---

[38]In the past, 30-year bonds were typical. Shorter-term 10-year, 15-year, 20-year, and 25-year bonds are also commonly issued.

[39]A *fully amortized bond* pays back interest and principal throughout the life of the bond. There is no balloon payment.

Actual *bond yield* is the bondholder's actual rate of return of the bond, considering the purchase price, interest payments, and face value payment (or, value realized if the bond is sold before it matures). By convention, bond yield is calculated as a nominal rate (rate per annum), not an effective rate per year. The bond yield should be determined by finding the effective rate of return per payment period (e.g., per semiannual interest payment) as a conventional rate of return problem. Then, the nominal rate can be found by multiplying the effective rate per period by the number of payments per year, as in Eq. 54.

Example 26

What is the maximum amount an investor should pay for a 25-year bond with a $20,000 face value and 8% coupon rate (interest only paid semiannually)? The bond will be kept to maturity. The investor's effective annual interest rate for economic decisions is 10%.

(solution)

For this problem, take the compounding period to be six months. Then, there are 50 compounding periods. Since 8% is a nominal rate, the effective bond rate per period is calculated from Eq. 53 as $\phi_{bond} = r/k = 8\%/2 = 4\%$.

The bond payment received semiannually is

$$(0.04)(\$20{,}000) = \$800$$

10% is the investor's effective rate per year, so Eq. 54 is again used to calculate the effective analysis rate per period.

$$0.10 = (1 + \phi)^2 - 1$$
$$\phi = 0.04881 \quad (4.88\%)$$

The maximum amount that the investor should be willing to pay is the present worth of the investment.

$$P = \$800(P/A, 4.88\%, 50) + \$20{,}000(P/F, 4.88\%, 50)$$

Table 1 can be used to calculate the factors.

$$(P/A, 4.88\%, 50) = \frac{(1 + 0.0488)^{50} - 1}{(0.0488)(1.0488)^{50}} = 18.600$$

$$(P/F, 4.88\%, 50) = \frac{1}{(1 + 0.0488)^{50}} = 0.09233$$

Then, the present worth is

$$P = (\$800)(18.600) + (\$20{,}000)(0.09233)$$
$$= \$16{,}727$$

47 PROBABILISTIC PROBLEMS

If an alternative's cash flows are specified by an implicit or explicit probability distribution rather than being known exactly, the problem is *probabilistic*.

Probabilistic problems typically possess the following characteristics.

- There is a chance of loss that must be minimized (or, rarely, a chance of gain that must be maximized) by selection of one of the alternatives.

- There are multiple alternatives. Each alternative offers a different degree of protection from the loss. Usually, the alternatives with the greatest protection will be the most expensive.

- The magnitude of loss or gain is independent of the alternative selected.

Probabilistic problems are typically solved using annual costs and expected values. An *expected value* is similar to an *average value* since it is calculated as the mean of the given probability distribution. If cost 1 has a probability of occurrence, p_1, cost 2 has a probability of occurrence, p_2, and so on, the expected value is

$$\mathcal{E}\{\text{cost}\} = p_1 \times (\text{cost } 1) + p_2 \times (\text{cost } 2) + \cdots \qquad [55]$$

Example 27

Flood damage in any year is given according to the following table. What is the present worth of flood damage for a 10-year period? Use 6% as the effective annual interest rate.

damage	probability
0	0.75
$10,000	0.20
$20,000	0.04
$30,000	0.01

(solution)

The expected value of flood damage in any given year is

$$\mathcal{E}\{\text{damage}\} = (0)(0.75) + (\$10,000)(0.20) + (\$20,000)(0.04) + (\$30,000)(0.01)$$
$$= \$3100$$

The present worth of 10 years of expected flood damage is

$$\text{present worth} = \$3100(P/A, 6\%, 10)$$
$$= (\$3100)(7.3601)$$
$$= \$22,816$$

Example 28

A dam is being considered on a river that periodically overflows and causes $600,000 damage. (The damage is essentially the same each time the river causes flooding.) The project horizon is 40 years. A 10% interest rate is being used.

Three different designs are available, each with different costs and storage capacities.

design alternative	cost	maximum capacity
A	$500,000	1 unit
B	$625,000	1.5 units
C	$900,000	2.0 units

The national weather service has provided a statistical analysis of annual rainfall in the area draining into the river.

units annual rainfall	probability
0	0.10
0.1 to 0.5	0.60
0.6 to 1.0	0.15
1.1 to 1.5	0.10
1.6 to 2.0	0.04
2.1 or more	0.01

Which design alternative would you choose assuming the dam is essentially empty at the start of each rainfall season?

(solution)

The sum of the construction cost and the expected damage should be minimized. If alternative A is chosen, it will have a capacity of 1 unit. Its capacity will be exceeded (causing $600,000 damage) when the annual rainfall exceeds 1 unit. Therefore, the expected value of the annual cost of alternative A is

$$\mathcal{E}\{EUAC(A)\} = \$500,000(A/P, 10\%, 40) + (\$600,000)(0.10 + 0.04 + 0.01)$$
$$= (\$500,000)(0.1023) + (\$600,000)(0.15)$$
$$= \$141,150$$

Similarly,

$$\mathcal{E}\{EUAC(B)\} = \$625,000(A/P, 10\%, 40) + \$(600,000)(0.04 + 0.01)$$
$$= (\$625,000)(0.1023) + (\$600,000)(0.05)$$
$$= \$93,940$$

$$\mathcal{E}\{EUAC(C)\} = \$900{,}000(A/P, 10\%, 40) + (\$600{,}000)(0.01)$$
$$= (\$900{,}000)(0.1023) + (\$600{,}000)(0.01)$$
$$= \$98{,}070$$

Alternative B should be chosen.

48 FIXED AND VARIABLE COSTS

The distinction between fixed and variable costs depends on how these costs vary when an independent variable changes. For example, factory or machine production is frequently the independent variable. However, it could just as easily be vehicle miles driven, hours of operation, or quantity (mass, volume, etc.).

If a cost is a function of the independent variable, the cost is said to be a *variable cost*. The change in cost per unit variable change (i.e., what is usually called the *slope*) is known as the *incremental cost*. Material and labor costs are examples of variable costs. They increase in proportion to the number of product units manufactured.

If a cost is not a function of the independent variable, the cost is said to be a *fixed cost*. Rent and lease payments are typical fixed costs. These costs will be incurred regardless of production levels.

Some costs have both fixed and variable components, as Fig. 9 illustrates. The fixed portion can be determined by calculating the cost at zero production.

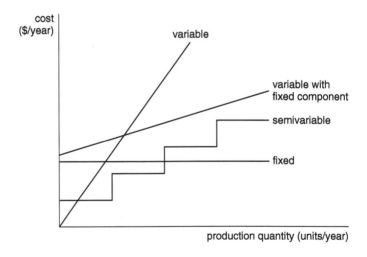

Figure 9   Fixed and Variable Costs

An additional category of cost is the *semivariable cost*. This type of cost increases stepwise. Semivariable cost structures are typical of situations where *excess capacity* exists. For example, supervisory cost is a stepwise function of the number of production shifts. Also, labor cost for truck drivers is a stepwise function of weight (volume) transported. As long as a truck has room left (i.e., excess capacity), no additional driver is needed. As soon as the truck is filled, labor cost will increase.

Table 7
Summary of Fixed and Variable Costs

fixed costs
> rent
> property taxes
> interest on loans
> insurance
> janitorial service expense
> tooling expense
> set-up, clean-up, and tear-down expenses
> depreciation expense
> marketing and selling costs
> cost of utilities
> general burden and overhead expense

variable costs
> direct material costs
> direct labor costs
> cost of miscellaneous supplies
> payroll benefit costs
> income taxes
> supervision costs

49 ACCOUNTING COST AND EXPENSE TERMS

The accounting profession has developed special terms for certain groups of costs. When annual costs are incurred due to the functioning of a piece of equipment, they are known as *operating and maintenance* (O&M) *costs*. The annual costs associated with operating a business (other than the costs directly attributable to production) are known as *general, selling, and administrative* (GS&A) *expenses*.

Direct labor costs are costs incurred in the factory, such as assembly, machining, and painting labor costs. *Direct material costs* are the costs of all materials that go into production.[40] Typically, both direct labor and direct material costs are given on a per-unit or per-item basis. The sum of the direct labor and direct material costs is known as the *prime cost*.

---

[40]There may be problems with pricing the material when it is purchased from an outside vendor and the stock on hand derives from several shipments purchased at different prices.

There are certain additional expenses incurred in the factory, such as the costs of factory supervision, stock-picking, quality control, factory utilities, and miscellaneous supplies (cleaning fluids, assembly lubricants, routing tags, etc.) that are not incorporated into the final product. Such costs are known as *indirect manufacturing expenses* (IME) or *indirect material and labor costs*.[41] The sum of the per-unit indirect manufacturing expense and prime cost is known as the *factory cost*.

Research and development (R&D) *costs* and *administrative expenses* are added to the factory cost to give the *manufacturing cost* of the product.

Additional costs are incurred in marketing the product. Such costs are known as *selling expenses* or *marketing expenses*. The sum of the selling expenses and manufacturing cost is the *total cost* of the product.

Mark-up is the difference between the product's selling price and its cost. Companies are not in agreement as to which cost to use, although using total cost is the most logical.

Figure 10 illustrates these terms.[42]

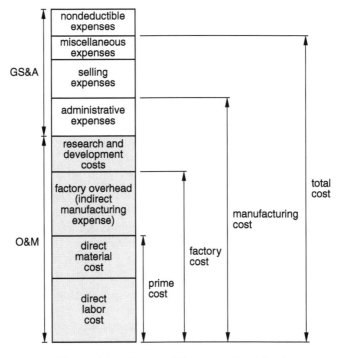

Figure 10   Costs and Expenses Combined

---

[41]The *indirect material* and *labor costs* usually exclude costs incurred in the office area.
[42]Notice that *total cost* does not include income taxes.

The distinctions among the various forms of cost (particularly with overhead costs) are not standardized. Each company must develop a classification system to deal with the various cost factors in a consistent manner. There are also other terms in use (e.g., *raw materials, operating supplies, general plant overhead*), but these terms must be interpreted within the framework of each company's classification system. Table 8 is typical of such classification systems.

50 ACCOUNTING PRINCIPLES

A. Basic Bookkeeping

An accounting or *bookkeeping system* is used to record historical financial transactions. The resultant records are used for product costing, satisfaction of statutory requirements, reporting of profit for income tax purposes, and general company management.

Bookkeeping consists of two main steps: recording the transactions, followed by categorization of the transactions.[43] The transactions (receipts and disbursements) are recorded in a *journal (book of original entry)* to complete the first step. Such a journal is organized in a simple chronological and sequential manner.[44] The transactions are then categorized (e.g., into interest income, advertising expense, etc.) and posted (i.e., entered or written) into the appropriate *ledger account*.

The ledger accounts together constitute the *general ledger* or *ledger*. All ledger accounts can be classified into one of three types: *asset accounts, liability accounts*, and *owners' equity accounts*. Strictly speaking, income and expense accounts, kept in a separate journal, are included within the classification of liability accounts.

Together, the journal and ledger are known simply as *the books* of the company.

B. Balancing the Books

In a business environment, *balancing the books* means more than reconciling the checkbook and bank statements. All accounting entries must be posted in such a way as to maintain the equality of the *basic accounting equation*.

$$\text{assets} = \text{liability} + \text{owners' equity} \qquad [56]$$

In a *double-entry bookkeeping system*, the equality is maintained within the ledger system by entering each transaction into two balancing ledger accounts. For example, paying a utility bill would decrease the cash account (an asset account) and decrease the utility expense account (a liability account) by the same amount.

---

[43] These two steps are not to be confused with the *double-entry bookkeeping method*.

[44] The two-step process is more typical of a *manual bookkeeping system* than a computerized *general ledger system*. However, even most computerized systems produce reports in journal entry order, as well as account summaries.

Table 8
Typical Classification of Expenses

direct material expenses
    items purchased from other vendors
    manufactured assemblies
direct labor expenses
    machining and forming
    assembly
    finishing
    inspection
    testing
factory overhead expenses
    supervision
    benefits
        pension
        medical insurance
        vacations
    wages overhead
        unemployment compensation taxes
        social security taxes
        disability taxes
    stock-picking
    quality control and inspection
    expediting
    rework
    maintenance
    miscellaneous supplies
        routing tags
        assembly lubricants
        cleaning fluids
        wiping cloths
        janitorial supplies
    packaging (materials and labor)
    factory utilities
    laboratory
    depreciation on factory equipment
research and development expenses
    engineering (labor)
    patents
    testing
    prototypes (material and labor)
    drafting
    O&M of R&D facility

administrative expenses
    corporate officers
    accounting
    secretarial/clerical/reception
    security (protection)
    medical (nurse)
    employment (personnel)
    reproduction
    data processing
    production control
    depreciation on nonfactory
     equipment
    office supplies
    office utilities
    O&M of offices
selling expenses
    marketing (labor)
    advertising
    transportation (if not
     paid by customer)
    outside sales force
     (labor and expenses)
    demonstration units
    commissions
    technical service and support
    order processing
    branch office expenses
miscellaneous expenses
    insurance
    property taxes
    interest on loans
nondeductible expenses
    federal income taxes
    fines and penalties

Transactions are either *debits* or *credits*, depending on their sign. Increases in asset accounts are debits; decreases are credits. For liability and equity accounts, the opposite is true: increases are credits, and decreases are debits.[45]

C. Cash and Accrual Systems[46]

The simplest form of bookkeeping is based on the *cash system*. The only transactions that are entered into the journal are those that represent cash receipts and disbursements. In effect, a checkbook register or bank deposit book could serve as the journal.

During a given period (e.g., month or quarter), expense liabilities may be incurred even though the payments for those expenses have not been made. For example, an invoice (bill) may have been received but not paid. Under the *accrual system*, the obligation is posted into the appropriate expense account before it is paid.[47] Analogous to expenses, under the accrual system, income will be claimed before payment is received. Specifically, a sales transaction can be recorded as income when the customer's order is received, the outgoing invoice is generated, or the merchandise is shipped.

D. Financial Statements

Each period, two types of financial statements are typically generated: the *balance sheet* and *profit and loss* (P&L) *statements*.[48] The profit and loss statement, also known as a *statement of income and retained earnings*, is a summary of sources of *income* or *revenue* (e.g., interest, sales, fees charged, etc.) and *expenses* (e.g., utilities, advertising, repairs, etc.) for the period. The expenses are subtracted from the revenues to give a *net income* (generally, before taxes).[49] Figure 11 illustrates a simple profit and loss statement.

---

[45]There is a difference in sign between asset and liability accounts. Thus, an increase in an expense account is actually a decrease. The accounting profession, apparently, is comfortable with the common confusion that exists between debits and credits.

[46]There is also a distinction made between cash flows that are known and those that are expected. It is a *standard accounting principle* to record losses in full, at the time they are recognized, even before their occurrence. In the construction industry, for example, losses are recognized in full and projected to the end of a project as soon as they are foreseeable. Profits, on the other hand, are recognized only as they are realized (typically, as a percentage of project completion). The difference between cash and accrual systems is a matter of *bookkeeping*. The difference between loss and profit recognition is a matter of *accounting convention*. Engineers seldom need to be concerned with the accounting tradition.

[47]The expense for an item or service might be accrued even *before* the invoice is received. It might be recorded when the purchase order for the item or service is generated, or when the item or service is received.

[48]Other types of financial statements (*statements of changes in financial position, cost of sales statements*, inventory and asset reports, etc.) will also be generated, depending on the needs of the company.

[49]Financial statements can also be prepared with percentages (of total assets and net revenue) instead of dollars, in which case they are known as *common size financial statements*.

revenue

interest	2000	
sales	237,000	
returns	⟨23,000⟩	
net revenue		216,000

expenses

salaries	149,000	
utilities	6000	
advertising	28,000	
insurance	4000	
supplies	1000	
net expenses		188,000

period net income   28,000

beginning retained earnings   63,000

net year-to-date earnings   91,000

Figure 11   Simplified Profit and Loss Statement

The *balance sheet* presents the *basic accounting equation* in tabular form. The balance sheet lists the major categories of assets and outstanding liabilities. The difference between asset values and liabilities is the *equity*, as defined in Eq. 56. This equity represents what would be left over after satisfying all debts by liquidating the company.

There are several terms that appear regularly on balance sheets.

- *current assets:* cash and other assets that can be converted quickly into cash, such as accounts receivable, notes receivable, and merchandise (inventory). Also known as *liquid assets*.

- *fixed assets:* relatively permanent assets used in the operation of the business and relatively difficult to convert into cash. Examples are land, buildings, and equipment. Also known as *nonliquid assets*.

- *current liabilities:* liabilities due within a short period of time (e.g., within one year) and typically paid out of current assets. Examples are accounts payable, notes payable, and other accrued liabilities.

- *long-term liabilities:* obligations that are not totally payable within a short period of time (e.g., within one year).

Figure 12 is a simplified balance sheet.

ASSETS

current assets

cash	14,000	
accounts receivable	36,000	
notes receivable	20,000	
inventory	89,000	
prepaid expenses	3000	
total current assets		162,000

plant, property, and equipment

land and buildings	217,000	
motor vehicles	31,000	
equipment	94,000	
accumulated depreciation	⟨52,000⟩	
total fixed assets		290,000
total assets		452,000

LIABILITIES AND OWNERS' EQUITY

current liabilities

accounts payable	66,000	
accrued income taxes	17,000	
accrued expenses	8000	
total current liabilities		91,000

long-term debt

notes payable	117,000	
mortgage	23,000	
total long-term debt		140,000

owners' and stockholders' equity

stock	130,000	
retained earnings	91,000	
total owners' equity		221,000

total liabilities and owners' equity   452,000

Figure 12   Simplified Balance Sheet

E. Analysis of Financial Statements

Financial statements are evaluated by management, lenders, stockholders, potential in-vestors, and many other groups for the purpose of determining the *health of the com-pany*. The health can be measured in terms of *liquidity* (ability to convert assets to cash quickly), *solvency* (ability to meet debts as they become due), and *relative risk* (of which one measure is *leverage*—the portion of total capital contributed by owners).

The analysis of financial statements involves several common ratios, usually expressed as percentages. The following are some frequently encountered ratios.

- *current ratio:* an index of short-term paying ability.

$$\text{current ratio} = \frac{\text{current assets}}{\text{current liabilities}}$$

- *quick (*or *acid-test) ratio:* a more stringent measure of short-term debt-paying ability. The *quick assets* are defined to be current assets minus inventories and prepaid expenses.

$$\text{acid-test ratio} = \frac{\text{quick assets}}{\text{current liabilities}}$$

- *receivable turnover:* a measure of the average speed with which accounts re-ceivable are collected.

$$\text{receivable turnover} = \frac{\text{net credit sales}}{\text{average net receivables}}$$

- *average age of receivables:* number of days, on the average, in which receivables are collected.

$$\text{average age of receivables} = \frac{365}{\text{receivable turnover}}$$

- *inventory turnover:* a measure of the speed with which inventory is sold, on the average.

$$\text{inventory turnover} = \frac{\text{cost of goods sold}}{\text{average cost of inventory on hand}}$$

- *days supply of inventory on hand:* number of days, on the average, that the current inventory would last.

$$\text{days supply of inventory on hand} = \frac{365}{\text{inventory turnover}}$$

- *book value per share of common stock:* number of dollars represented by the balance sheet owners' equity for each share of common stock outstanding.

$$\text{book value per share of common stock} = \frac{\text{common shareholders' equity}}{\text{number of outstanding shares}}$$

- *gross margin:* gross profit as a percentage of sales. (Gross profit is sales less cost of goods sold.)

$$\text{gross margin} = \frac{\text{gross profit}}{\text{net sales}}$$

- *profit margin ratio:* percentage of each dollar of sales that is net income.

$$\text{profit margin} = \frac{\text{net income before taxes}}{\text{net sales}}$$

- *return on investment ratio:* shows the percent return on owners' investment.

$$\text{return on investment} = \frac{\text{net income}}{\text{owners' equity}}$$

- *price-earnings ratio:* indication of relationship between earnings and market price per share of common stock, useful in comparisons between alternative investments.

$$\text{price-earnings} = \frac{\text{market price per share}}{\text{earnings per share}}$$

51 COST ACCOUNTING

Cost accounting is the system that determines the cost of manufactured products. Cost accounting is called *job cost accounting* if costs are accumulated by part number or contract. It is called *process cost accounting* if costs are accumulated by departments or manufacturing processes.

Cost accounting is dependent on historical and recorded data. The unit product cost is determined from actual expenses and numbers of units produced. Allowances (i.e., budgets) for future costs are based on these historical figures. Any deviation from historical figures is called a *variance*. Where adequate records are available, variances can be divided into *labor variance* and *material variance*.

When determining a unit product cost, the direct material and direct labor costs are generally clear-cut and easily determined. Furthermore, these costs are 100% variable costs. However, the indirect cost per unit of product is not as easily determined. Indirect costs (burden, overhead, etc.) can be fixed or semivariable costs. The amount of indirect cost allocated to a unit will depend on the unknown future overhead expense as well as the unknown future production (*vehicle size*).

A typical method of allocating indirect costs to a product is as follows.

step 1:   Estimate the total expected indirect (and overhead) costs for the upcoming year.

step 2:   Determine the most appropriate vehicle (basis) for allocating the overhead to production. Usually, this vehicle is either the number of units expected to be produced or the number of direct hours expected to be worked in the upcoming year.

step 3:   Estimate the quantity or size of the overhead vehicle.

step 4:   Divide expected overhead costs by the expected overhead vehicle to obtain the unit overhead.

step 5:   Regardless of the true size of the overhead vehicle during the upcoming year, one unit of overhead cost is allocated per unit of overhead vehicle.

Once the prime cost has been determined and the indirect cost calculated based on projections, the two are combined into a *standard factory cost* or *standard cost*, which remains in effect until the next budgeting period (usually a year).

During the subsequent manufacturing year, the standard cost of a product is not generally changed merely because it is found that an error in projected indirect costs or production quantity (vehicle size) has been made. The allocation of indirect costs to a product is assumed to be independent of errors in forecasts. Rather, the difference between the expected and actual expenses, known as the *burden (overhead) variance*, experienced during the year is posted to one or more *variance accounts*.

Burden (overhead) variance is caused by errors in forecasting both the actual indirect expense for the upcoming year and the overhead vehicle size. In the former case, the variance is called *burden budget variance*; in the latter, it is called *burden capacity variance*.

Example 29

A company expects to produce 8000 items in the coming year. The current material cost is $4.54 each. Sixteen minutes of direct labor are required per unit. Workers are paid $7.50 per hour. 2133 direct labor hours are forecasted for the product. Miscellaneous overhead costs are estimated at $45,000.

Find the per unit (a) expected direct material cost, (b) direct labor cost, (c) prime cost, (d) burden as a function of production and direct labor, and (e) total cost if the burden vehicle is direct labor hours.

(solution)

(a) The direct material cost was given as $4.54.

(b) The direct labor cost is

$$\left(\frac{16 \text{ min}}{60 \frac{\text{min}}{\text{hr}}}\right)\left(\$7.50 \frac{1}{\text{hr}}\right) = \$2.00$$

(c) The prime cost is

$$\$4.54 + \$2.00 = \$6.54$$

(d) If the burden vehicle is production, the burden rate is $45,000/8000 = \$5.63$ per item, making the total cost

$$\$4.54 + \$2.00 + \$5.63 = \$12.17$$

(e) If the burden vehicle is direct labor hours, the burden rate is $45,000/2133 = \$21.10$ per hour, making the total cost

$$\$4.54 + \$2.00 + \left(\frac{16 \text{ min}}{60 \frac{\text{min}}{\text{hr}}}\right)\left(\$21.10 \frac{1}{\text{hr}}\right) = \$12.17$$

Example 30

The actual performance of the company in Ex. 29 is given by the following figures.

actual production: 7560
actual overhead costs: $47,000

What are the burden budget variance and the burden capacity variance?

(solution)

The burden capacity variance is

$$\$45,000 - (\$7560)(5.63) = \$2437$$

The burden budget variance is

$$\$47,000 - \$45,000 = \$2000$$

The overall burden variance is

$$\$47,000 - (\$7560)(5.63) = \$4437$$

The sum of the burden capacity and burden budget variances should equal the overall burden variance.

$$\$2437 + \$2000 = \$4437$$

52 COST OF GOODS SOLD

Cost of goods sold (COGS) is an accounting term that represents an inventory account adjustment.[50] Cost of goods sold is the difference between the starting and ending inventory valuations. That is,

$$\text{COGS} = \text{starting inventory valuation} - \text{ending inventory valuation} \quad [57]$$

Cost of goods sold is subtracted from *gross profit* to determine the *net profit* of a company. Despite the fact that cost of goods sold can be a significant element in the profit equation, the inventory adjustment may not be made each accounting period (e.g., each month), due to the difficulty in obtaining an accurate inventory valuation.

With a *perpetual inventory system*, a company automatically maintains up-to-date inventory records, either through an efficient stocking and stock-releasing system, or though a *point of sale* (POS) *system* integrated with the inventory records. If a company only counts its inventory (i.e., takes a *physical inventory*) at regular intervals (e.g., once a year), it is said to be operating on a *periodic inventory system*.

Inventory accounting is a source of many difficulties. The inventory value is calculated by multiplying the quantity on hand by the standard cost. In the case of completed items actually assembled or manufactured at the company, this standard cost usually is the manufacturing cost, although factory cost can also be used. In the case of purchased items, the standard cost will be the cost per item charged by the supplying vendor. In some cases, delivery and transportation costs will be included in this standard cost.

It is not unusual for the elements in an item's inventory to come from more than one vendor, or from one vendor in more than one order. Inventory valuation is more difficult if the price paid is different for these different purchases. There are four methods of determining the cost of elements in inventory. Any of these methods can be used (if applicable), but the method must be used consistently from year to year. The four methods are

- *specific identification method:* Each element can be uniquely associated with a cost. Inventory elements with serial numbers fit into this costing scheme. Stock, production, and sales records must include the serial number.

- *average cost method:* The standard cost of an item is the average of (recent or all) purchase costs for that item.

---

[50]The cost of goods sold inventory adjustment is posted to the *COGS expense account.*

- *first-in, first-out* (FIFO) *method:* This method keeps track of how many of each item were purchased each time and the number remaining out of each purchase, as well as the price paid at each purchase. The inventory system assumes that the oldest elements are issued first.[51] Inventory value is a weighted average dependent on the number of elements from each purchase remaining. Items issued no longer contribute to the inventory value.

- *last-in, first-out* (LIFO) *method:* This method keeps track of how many of each item were purchased each time and the number remaining out of each purchase, as well as the price paid at each purchase.[52] The inventory value is a weighted average dependent on the number of elements from each purchase remaining. Items issued no longer contribute to the inventory value.

53 BREAK-EVEN ANALYSIS

Special Nomenclature

f	a fixed cost that does not vary with production
a	the *incremental cost* to produce one additional item. It may also be called the *marginal cost* or *differential cost*.
Q	the quantity sold
p	the *incremental value* (i.e., price)
R	the total revenue
C	the total cost

Break-even analysis is a method of determining when the value of one alternative becomes equal to the value of another. A common application is that of determining when costs exactly equal revenue. If the manufactured quantity is less than the break-even quantity, a loss is incurred. If the manufactured quantity is greater than the break-even quantity, a profit is made.

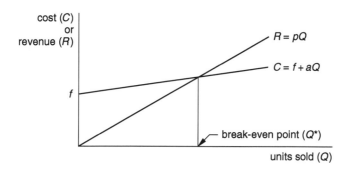

Figure 13 Break-Even Quantity

[51,52]If all elements in an item's inventory are identical, and if all shipments of that item are agglomerated, there will be no way to guarantee that the oldest element in inventory is issued first. But, unless *spoilage* is a problem, it really does not matter.

Assuming no change in the inventory, the *break-even point*, Q^*, can be found from $C = R$, where

$$C = f + aQ \qquad\qquad [58]$$
$$R = pQ \qquad\qquad [59]$$
$$Q^* = \frac{f}{p - a} \qquad\qquad [60]$$

An alternative form of the break-even problem is to find the number of units per period for which two alternatives have the same total costs. Fixed costs are to be spread over a period longer than one year using the equivalent uniform annual cost (EUAC) concept. One of the alternatives will have a lower cost if production is less than the break-even point. The other will have a lower cost for production greater than the break-even point.

Example 31

Two plans are available for a company to obtain automobiles for its sales staff. How many miles must the cars be driven each year for the two plans to have the same costs? Use an interest rate of 10%. (Use the year-end convention for all costs.)

    plan A: Lease the cars and pay $0.15 per mile.

    plan B: Purchase the cars for $5000. Each car has an economic life of three years, after which it can be sold for $1200. Gas and oil cost $0.04 per mile. Insurance is $500 per year.

(solution)

Let x be the number of miles driven per year. Then, the EUAC for both alternatives is

$$\text{EUAC(A)} = 0.15x$$
$$\text{EUAC(B)} = 0.04x + \$500 + \$5000(A/P, 10\%, 3) - \$1200(A/F, 10\%, 3)$$
$$= 0.04x + \$500 + (\$5000)(0.4021) - (\$1200)(0.3021)$$
$$= 0.04x + \$2148$$

Setting EUAC(A) and EUAC(B) equal and solving for x yields 19,527 miles per year as the break-even point.

54 PAY-BACK PERIOD

The *pay-back period* is defined as the length of time, usually in years, for the cumulative net annual profit to equal the initial investment. It is tempting to introduce equivalence into pay-back period calculations, but the convention is not to. If an analysis using equivalence is wanted, the term *pay-back period* should not be used. Other terms, such as *cost recovery period* or *life of an equivalent investment* should be used. Unfortunately, this convention is not always followed in practice.

$$\text{pay-back period} = \frac{\text{initial investment}}{\text{net annual profit}} \qquad [61]$$

Example 32

A ski resort installs two new ski lifts at a cost of $1,800,000. The resort expects annual gross revenue to increase $500,000 while it incurs an annual expense of $50,000 for lift operation and maintenance. What is the pay-back period?

(solution)

From Eq. 61,

$$\text{pay-back period} = \frac{\$1,800,000}{\$500,000 - \$50,000} = 4\,\text{years}$$

55 MANAGEMENT GOALS

Depending on many factors (market position, age of the company, age of the industry, perceived marketing and sales windows, etc.), a company may select one of many production and marketing strategic goals. Three such strategic goals are

- maximization of product demand

- minimization of cost

- maximization of profit

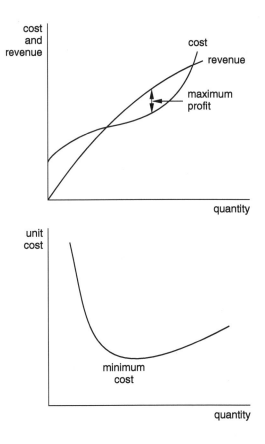

Figure 14   Graphs of Management Goal Functions

Such goals require knowledge of how the dependent variable (e.g., demand quantity or quantity sold) varies as a function of the independent variable (e.g., price). Unfortunately, these three goals are not usually satisfied simultaneously. For example, minimization of product cost may require a large production run to realize economies of scale, while the actual demand is too small to take advantage of such economies of scale.

If sufficient data are available to plot the independent and dependent variables, it may be possible to optimize the dependent variable graphically. Of course, if the relationship between independent and dependent variables is known algebraically, the dependent variable can be optimized by taking derivatives, or by use of other numerical methods.

56 INFLATION

It is important to perform economic studies in terms of *constant value dollars*. One method of converting all cash flows to constant value dollars is to divide the flows by some annual *economic indicator* or price index.[53]

If indicators are not available, cash flows can be adjusted by assuming that inflation is constant at a decimal rate e per year. Then, all cash flows can be converted to $t = 0$ dollars by dividing by $(1 + e)^n$, where n is the year of the cash flow.

An alternative is to replace the effective annual interest rate, i, with a value corrected for inflation. This corrected value, i', is

$$i' = i + e + ie \qquad [62]$$

This method has the advantage of simplifying the calculations. However, precalculated factors may not be available for the noninteger values of i'. Therefore, Table 1 must be used to calculate the factors.

Example 33

What is the uninflated present worth of a $2000 future value in two years if the average inflation rate is 6% and i is 10%?

(solution)

$$P = \frac{\$2000}{(1 + 0.10)^2(1 + 0.06)^2} = \$1471$$

Example 34

Repeat Ex. 33 using i'.

(solution)

$$i' = 0.10 + 0.06 + (0.10)(0.06) = 0.166$$
$$P = \frac{\$2000}{(1 + 0.166)^2} = \$1471$$

---

[53] See Appendix B.

57 CONSUMER LOANS

Special Nomenclature

BAL_j	balance after the jth payment
LV	principal total value loaned (cost minus down payment)
j	payment or period number
N	total number of payments to pay off the loan
PI_j	jth interest payment
PP_j	jth principal payment
PT_j	jth total payment
ϕ	effective rate per period (r/k)

Many different arrangements can be made between a borrower and a lender. With the advent of *creative financing* concepts, it often seems that there are as many variations of loans as there are loans made. Nevertheless, there are several traditional types of transactions. Real estate or investment texts, or a financial consultant, should be consulted for more complex problems.

A. Simple Interest

Interest due does not compound with a *simple interest loan*. The interest due is merely proportional to the length of time that the principal is outstanding. Because of this, simple interest loans are seldom made for long periods (e.g., more than one year). (For loans less than one year, it is commonly assumed that a year consists of 12 months of 30 days each.)

Example 35

A $12,000 simple interest loan is taken out at 16% per annum interest rate. The loan matures in two years with no intermediate payments. How much will be due at the end of the second year?

(solution)

The interest each year is

$$PI = (0.16)(\$12,000) = \$1920$$

The total amount due in two years is

$$PT = \$12,000 + (2)(\$1920) = \$15,840$$

Example 36

$4000 is borrowed for 75 days at 16% per annum simple interest. How much will be due at the end of 75 days?

(solution)

$$\text{amount due} = \$4000 + (0.16) \left(\frac{75 \text{ days}}{360 \frac{\text{days}}{\text{bank year}}} \right) (\$4000) = \$4133$$

B. Loans with Constant Amount Paid Toward Principal

With this loan type, the payment is not the same each period. The amount paid toward the principal is constant, but the interest varies from period to period. The equations that govern this type of loan are

$$\text{BAL}_j = \text{LV} - (j)(\text{PP}) \tag{63}$$

$$\text{PI}_j = \phi(\text{BAL})_{j-1} \tag{64}$$

$$\text{PT}_j = \text{PP} + \text{PI}_j \tag{65}$$

$$\text{PP} = \frac{\text{LV}}{N} \tag{66}$$

$$N = \frac{\text{LV}}{\text{PP}} \tag{67}$$

$$\text{LV} = (\text{PP} + \text{PI}_1)(P/A, \phi, N) - \text{PI}_N(P/G, \phi, N) \tag{68}$$

$$1 = \left(\frac{1}{N} + \phi \right)(P/A, \phi, N) - \left(\frac{\phi}{N} \right)(P/G, \phi, N) \tag{69}$$

Example 37

A \$12,000 six-year loan is taken from a bank that charges 15% effective annual interest. Payments toward the principal are uniform, and repayments are made at the end of each year. Tabulate the interest, total payments, and the balance remaining after each payment is made.

(solution)

The amount of each principal payment is

$$\text{PP} = \frac{\$12,000}{6} = \$2000$$

At the end of the first year (before the first payment is made), the principal balance is 12,000 (i.e., $BAL_0 = 12{,}000$). From Eq. 64, the interest payment is

$$\text{PI}_1 = (0.15)(\$12,000) = \$1800$$

The total first payment is

$$\text{PT}_1 = \text{PP} + \text{PI} = \$2000 + \$1800$$
$$= \$3800$$

The following table is similarly constructed.

j	BAL_j	PP_j	PI_j	PT_j
0	$12,000	–	–	–
1	$10,000	$2000	$1800	$3800
2	$8000	$2000	$1500	$3500
3	$6000	$2000	$1200	$3200
4	$4000	$2000	$900	$2900
5	$2000	$2000	$600	$2600
6	0	$2000	$300	$2300

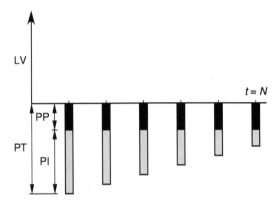

Figure 15   Loan with Constant Amount Paid Toward Principal

C. Direct Reduction Loans

This is the typical interest-paid-on-unpaid-balance loan. The amount of the periodic payment is constant, but the amounts paid toward the principal and interest both vary.

$$BAL_{j-1} = PT \left(\frac{1 - (1 + \phi)^{j-1-N}}{\phi} \right) \qquad [70]$$

$$PI_j = \phi(BAL_{j-1}) \qquad [71]$$

$$PP_j = PT - PI_j \qquad [72]$$

$$BAL_j = BAL_{j-1} - PP_j \qquad [73]$$

$$N = \frac{-\ln\left(1 - \dfrac{\phi(LV)}{PT} \right)}{\ln(1 + \phi)} \qquad [74]$$

Equation 74 calculates the number of payments necessary to pay off a loan. This equation can be solved with effort for the total periodic payment (PT) or the initial value of the loan (LV). It is easier, however, to use the $(A/P, i\%, n)$ factor to find the payment and loan value.

$$PT = LV(A/P, \phi\%, N) \qquad [75]$$

If the loan is repaid in yearly installments, then i is the effective annual rate. If the loan is paid off monthly, then i should be replaced by the effective rate per month (ϕ from Eq. 54). For monthly payments, N is the number of months in the loan period.

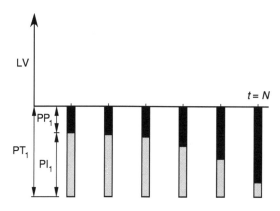

Figure 16   Direct Reduction Loan

Example 38

A \$45,000 loan is financed at 9.25% per annum. The monthly payment is \$385. What are the amounts paid toward interest and principal in the 14th period? What is the remaining principal balance after the 14th payment has been made?

(solution)

The effective rate per month is

$$\phi = \frac{r}{k} = \frac{0.0925}{12}$$
$$= 0.0077083\ldots \text{ (say 0.007708)}$$

$$N = \frac{-\ln\left(1 - \dfrac{(0.007708)(\$45,000)}{\$385}\right)}{\ln(1 + 0.007708)} = 301$$

$$\text{BAL}_{13} = (\$385)\left(\frac{1 - (1 + 0.007708)^{14-1-301}}{0.007708}\right)$$

$$= \$44,476.39$$
$$\text{PI}_{14} = (0.007708)(\$44,476.39) = \$342.82$$
$$\text{PP}_{14} = \$385 - \$342.82 = \$42.18$$
$$\text{BAL}_{14} = \$44,476.39 - \$42.18 = \$44,434.21$$

D. Direct Reduction Loans with Balloon Payments

This type of loan has a constant periodic payment, but the duration of the loan is insufficient to completely pay back the principal (i.e, the loan is not fully amortized). Therefore, all remaining unpaid principal must be paid back in a lump sum when the loan matures. This large payment is known as a *balloon payment*.[54]

Equations 70–74 can also be used with this type of loan. The remaining balance after the last payment is the balloon payment. This balloon payment must be repaid along with the last regular payment calculated.

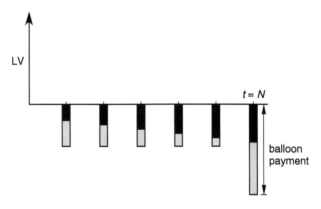

Figure 17   Direct Reduction Loan with Balloon Payment

58 FORECASTING

There are many types of forecasting models, although most are variations of the basic types.[55] All models produce a forecast, F_{t+1}, of some quantity (demand is used in this section) in the next period based on actual measurements, D_j, in current and prior periods. All of the models also try to provide *smoothing* (also known as *damping*) of extreme data points.

A. Forecasts by Moving Averages

The method of *moving average forecasting* weights all previous demand data points equally, and provides some smoothing of extreme data points. The amount of smoothing increases as the number of data points, n, increases.

---

[54]The term *balloon payment* may include the final interest payment as well. Generally, the problem statement will indicate whether the balloon payment is inclusive or exclusive of the regular payment made at the end of the loan period.

[55]For example, forecasting models that take into consideration steady (linear), cyclical annual, and seasonal trends are typically variations of the exponentially weighted model. A truly different forecasting tool, however, is *Monte Carlo simulation*.

$$F_{t+1} = \frac{1}{n} \sum_{m=t+1-n}^{t} D_m \qquad [76]$$

B. Forecasts by Exponentially Weighted Averages

With *exponentially weighted forecasts*, the more current (most recent) data points receive more weight. This method uses a *weighting factor*, α, also known as a *smoothing coefficient*, which typically varies between 0.01 and 0.30. An initial forecast is needed to start the method. Forecasts immediately following are sensitive to the accuracy of this first forecast. It is common to choose $F_0 = D_1$ to get started.

$$F_{t+1} = \alpha D_t + (1 - \alpha)F_t \qquad [77]$$

59 LEARNING CURVES

Special Nomenclature

R	decimal learning curve rate (2^{-b})
T_1	time or cost for the first item
T_n	time or cost for the nth item
n	total number of items produced
b	learning curve constant

The more products that are made, the more efficient the operation becomes due to experience gained. Therefore, direct labor costs decrease.[56] Usually, a *learning curve* is specified by the decrease in cost each time the cumulative quantity produced doubles. If there is a 20% decrease per doubling, the curve is said to be an 80% learning curve (i.e., the *learning curve rate*, R, is 80%).

Then, the time to produce the nth item is

$$T_n = T_1 n^{-b} \qquad [78]$$

The total time to produce units from quantity n_1 to n_2 inclusive is approximately given by Eq. 79. T_1 is a constant, the time for item 1, and does not correspond to n_1, unless $n_1 = 1$.

$$\int_{n_1}^{n_2} T_n \, dn \approx \frac{T_1}{1-b} \left[\left(n_2 + \frac{1}{2}\right)^{1-b} - \left(n_1 - \frac{1}{2}\right)^{1-b} \right] \qquad [79]$$

The *average time per unit* over the production from n_1 to n_2 is the above total time from Eq. 79 divided by the quantity produced, $(n_2 - n_1 + 1)$.

$$T_{\text{ave}} = \frac{\int_{n_1}^{n_2} T_n \, dn}{n_2 - n_1 + 1} \qquad [80]$$

---

[56]It is important to remember that learning curve reductions apply only to direct labor costs. They are not applied to indirect labor or direct material costs.

Table 9 lists representative values of the *learning curve constant*, b. For learning curve rates not listed in the table, Eq. 81 can be used to find b.

$$b = \frac{-\log_{10} R}{\log_{10}(2)} = \frac{-\log_{10} R}{0.301}$$ [81]

Table 9
Learning Curve Constants

learning curve rate, R	b
0.70 (70%)	0.515
0.75 (75%)	0.415
0.80 (80%)	0.322
0.85 (85%)	0.234
0.90 (90%)	0.152
0.95 (95%)	0.074

Example 39

A 70% learning curve is used with an item whose first production time is 1.47 hours. (a) How long will it take to produce the 11th item? (b) How long will it take to produce the 11th through 27th items?

(solution)

(a) From Eq. 78,

$$T_{11} = (1.47)(11)^{-0.515} = 0.428 \text{ hours}$$

(b) The time to produce the 11th item through 27th item is given by Eq. 79.

$$T \approx \left(\frac{1.47}{1 - 0.515}\right)\left[(27.5)^{1-0.515} - (10.5)^{1-0.515}\right]$$
$$= 5.643 \text{ hours}$$

60 ECONOMIC ORDER QUANTITY

Special Nomenclature

a	constant depletion rate (items/unit time)
h	inventory storage cost ($/item-unit time)
H	total inventory storage cost between orders ($)
K	fixed cost of placing an order ($)
Q	order quantity (original quantity on hand)

The *economic order quantity* (EOQ) is the order quantity that minimizes the inventory costs per unit time. Although there are many different EOQ models, the simplest is based on the following assumptions.

- Reordering is instantaneous. The time between order placement and receipt is zero.

- Shortages are not allowed.

- Demand for the inventory item is deterministic (i.e., is not a random variable).

- Demand is constant with respect to time.

- An order is placed when the inventory is zero.

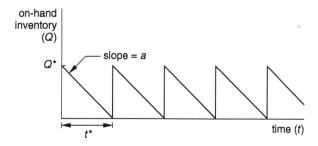

Figure 18   Inventory with Instantaneous Reorder

If the original quantity on hand is Q, the stock will be depleted at

$$t^* = \frac{Q}{a}$$
[82]

The total inventory storage cost between t_0 and t^* is

$$H = \left(\frac{Q}{2}\right)(h)(t^*) = \frac{Q^2 h}{2a}$$
[83]

The total inventory and ordering cost per unit time is

$$C_t = \frac{aK}{Q} + \frac{hQ}{2}$$ [84]

C_t can be minimized with respect to Q. The economic order quantity and time between orders are

$$Q^* = \sqrt{\frac{2aK}{h}}$$ [85]

$$t^* = \frac{Q^*}{a}$$ [86]

61 VALUE ENGINEERING

The *value* of an investment is defined as the ratio of its return (performance or utility) to its cost (effort or investment). The basic object of *value engineering* (VE, also referred to as *value analysis*) is to obtain the maximum per-unit value.[57]

Value engineering concepts are often used to reduce the cost of mass-produced manufactured products. This is done by eliminating unnecessary, redundant, or superfluous features, by redesigning the product for a less expensive manufacturing method, and by including features for easier assembly without sacrificing utility and function.[58] However, the concepts are equally applicable to one-time investments such as buildings, chemical processing plants, and space vehicles. In particular, value engineering has become an important element in all federally funded work.[59]

Typical examples of large-scale value engineering work are using stock-sized bearings and motors instead of custom manufactured units, specifying rectangular concrete columns with round columns (which are easier to form), and substituting custom buildings with prefabricated structures.

Value engineering is usually a team effort. And, while whether to use the original designers on the team is a point of debate, usually outside consultants are utilized. The cost of the consultancy is usually returned many times over by reduced construction and life-cycle costs.

62 SENSITIVITY ANALYSIS

Data analysis and forecasts in economic studies require estimates of costs that will occur in the future. There are always uncertainties about these costs. However, these

---

[57]Value analysis, the methodology that has become today's value engineering, was developed in the early 1950s by Lawrence D. Miles, an analyst at General Electric.

[58]Some people say that value engineering is "the act of going over the plans and taking out everything that is interesting."

[59]*Office of Management and Budget Circular A-131* outlines value engineering for federally funded construction projects.

uncertainties are insufficient reason not to make the best possible estimates of the costs. Nevertheless, a decision between alternatives often can be made more confidently if it is known whether or not the conclusion is sensitive to moderate changes in data forecasts. Sensitivity analysis provides this extra dimension to an economic analysis.

The sensitivity of a decision is determined by inserting a range of estimates for critical cash flows and other parameters. If radical changes can be made to a cash flow without changing the decision, the decision is said to be *insensitive* to uncertainties regarding that cash flow. However, if a small change in the estimate of a cash flow will alter the decision, that decision is said to be very *sensitive* to changes in the estimate. If the decision is sensitive only for a limited range of cash flow values, the term *variable sensitivity* is used. Figure 19 illustrates these terms.

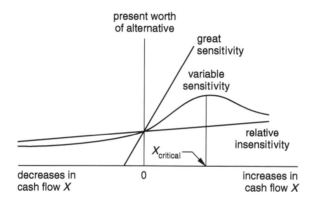

Figure 19   Types of Sensitivity

An established semantic tradition distinguishes between risk analysis and uncertainty analysis. *Risk analysis* addresses variables that have a known or estimated probability distribution. In this regard, statistics and probability theory can be used to determine the probability of a cash flow varying between given limits. On the other hand, *uncertainty analysis* is concerned with situations in which there is not enough information to determine the probability or frequency distribution for the variables involved.

As a first step, sensitivity analysis should be applied one at a time to the dominant factors. Dominant cost factors are those that have the most significant impact on the present value of the alternative.[60] If warranted, additional investigation can be used to determine the sensitivity to several cash flows varying simultaneously. Significant judgment is needed, however, to successfully determine the proper combinations of cash flows to vary. It is common to plot the dependency of the present value on the cash flow being varied in a two-dimensional graph. Simple linear interpolation is used (within reason) to determine the critical value of the cash flow being varied.

---

[60]In particular, engineering economic analysis problems are sensitive to the choice of effective interest rate, i, and to accuracy in cash flows at or near the beginning of the horizon. The problems will be less sensitive to accuracy in far-future cash flows, such as salvage value and subsequent generation replacement costs.

INTEREST TABLES

PROFESSIONAL PUBLICATIONS, INC.

$I = 0.25\%$

n	(P/F)	(P/A)	(P/G)	(F/P)	(F/A)	(A/P)	(A/F)	(A/G)	n
1	0.9975	0.9975	0.0000	1.0025	1.0000	1.0025	1.0000	0.0000	1
2	0.9950	1.9925	0.9950	1.0050	2.0025	0.5019	0.4994	0.4994	2
3	0.9925	2.9851	2.9801	1.0075	3.0075	0.3350	0.3325	0.9983	3
4	0.9901	3.9751	5.9503	1.0100	4.0150	0.2516	0.2491	1.4969	4
5	0.9876	4.9627	9.9007	1.0126	5.0251	0.2015	0.1990	1.9950	5
6	0.9851	5.9478	14.8263	1.0151	6.0376	0.1681	0.1656	2.4927	6
7	0.9827	6.9305	20.7223	1.0176	7.0527	0.1443	0.1418	2.9900	7
8	0.9802	7.9107	27.5839	1.0202	8.0704	0.1264	0.1239	3.4869	8
9	0.9778	8.8885	35.4061	1.0227	9.0905	0.1125	0.1100	3.9834	9
10	0.9753	9.8639	44.1842	1.0253	10.1133	0.1014	0.0989	4.4794	10
11	0.9729	10.8368	53.9133	1.0278	11.1385	0.0923	0.0898	4.9750	11
12	0.9705	11.8073	64.5886	1.0304	12.1664	0.0847	0.0822	5.4702	12
13	0.9681	12.7753	76.2053	1.0330	13.1968	0.0783	0.0758	5.9650	13
14	0.9656	13.7410	88.7587	1.0356	14.2298	0.0728	0.0703	6.4594	14
15	0.9632	14.7042	102.2441	1.0382	15.2654	0.0680	0.0655	6.9534	15
16	0.9608	15.6650	116.6567	1.0408	16.3035	0.0638	0.0613	7.4469	16
17	0.9584	16.6235	131.9917	1.0434	17.3443	0.0602	0.0577	7.9401	17
18	0.9561	17.5795	148.2446	1.0460	18.3876	0.0569	0.0544	8.4328	18
19	0.9537	18.5332	165.4106	1.0486	19.4336	0.0540	0.0515	8.9251	19
20	0.9513	19.4845	183.4851	1.0512	20.4822	0.0513	0.0488	9.4170	20
21	0.9489	20.4334	202.4634	1.0538	21.5334	0.0489	0.0464	9.9085	21
22	0.9466	21.3800	222.3410	1.0565	22.5872	0.0468	0.0443	10.3995	22
23	0.9442	22.3241	243.1131	1.0591	23.6437	0.0448	0.0423	10.8901	23
24	0.9418	23.2660	264.7753	1.0618	24.7028	0.0430	0.0405	11.3804	24
25	0.9395	24.2055	287.3230	1.0644	25.7646	0.0413	0.0388	11.8702	25
26	0.9371	25.1426	310.7516	1.0671	26.8290	0.0398	0.0373	12.3596	26
27	0.9348	26.0774	335.0566	1.0697	27.8961	0.0383	0.0358	12.8485	27
28	0.9325	27.0099	360.2334	1.0724	28.9658	0.0370	0.0345	13.3371	28
29	0.9301	27.9400	386.2776	1.0751	30.0382	0.0358	0.0333	13.8252	29
30	0.9278	28.8679	413.1847	1.0778	31.1133	0.0346	0.0321	14.3130	30
31	0.9255	29.7934	440.9502	1.0805	32.1911	0.0336	0.0311	14.8003	31
32	0.9232	30.7166	469.5696	1.0832	33.2716	0.0326	0.0301	15.2872	32
33	0.9209	31.6375	499.0386	1.0859	34.3547	0.0316	0.0291	15.7736	33
34	0.9186	32.5561	529.3528	1.0886	35.4406	0.0307	0.0282	16.2597	34
35	0.9163	33.4724	560.5076	1.0913	36.5292	0.0299	0.0274	16.7454	35
36	0.9140	34.3865	592.4988	1.0941	37.6206	0.0291	0.0266	17.2306	36
37	0.9118	35.2982	625.3219	1.0968	38.7146	0.0283	0.0258	17.7154	37
38	0.9095	36.2077	658.9727	1.0995	39.8114	0.0276	0.0251	18.1998	38
39	0.9072	37.1149	693.4468	1.1023	40.9109	0.0269	0.0244	18.6838	39
40	0.9050	38.0199	728.7399	1.1050	42.0132	0.0263	0.0238	19.1673	40
41	0.9027	38.9226	764.8476	1.1078	43.1182	0.0257	0.0232	19.6505	41
42	0.9004	39.8230	801.7658	1.1106	44.2260	0.0251	0.0226	20.1332	42
43	0.8982	40.7212	839.4900	1.1133	45.3366	0.0246	0.0221	20.6156	43
44	0.8960	41.6172	878.0162	1.1161	46.4499	0.0240	0.0215	21.0975	44
45	0.8937	42.5109	917.3400	1.1189	47.5661	0.0235	0.0210	21.5789	45
46	0.8915	43.4024	957.4572	1.1217	48.6850	0.0230	0.0205	22.0600	46
47	0.8893	44.2916	998.3637	1.1245	49.8067	0.0226	0.0201	22.5407	47
48	0.8871	45.1787	1040.0552	1.1273	50.9312	0.0221	0.0196	23.0209	48
49	0.8848	46.0635	1082.5276	1.1301	52.0585	0.0217	0.0192	23.5007	49
50	0.8826	46.9462	1125.7767	1.1330	53.1887	0.0213	0.0188	23.9802	50
51	0.8804	47.8266	1169.7983	1.1358	54.3217	0.0209	0.0184	24.4592	51
52	0.8782	48.7048	1214.5885	1.1386	55.4575	0.0205	0.0180	24.9377	52
53	0.8760	49.5809	1260.1430	1.1415	56.5961	0.0202	0.0177	25.4159	53
54	0.8739	50.4548	1306.4577	1.1443	57.7376	0.0198	0.0173	25.8936	54
55	0.8717	51.3264	1353.5286	1.1472	58.8819	0.0195	0.0170	26.3710	55
60	0.8609	55.6524	1600.0845	1.1616	64.6467	0.0180	0.0155	28.7514	60
65	0.8502	59.9246	1864.9427	1.1762	70.4839	0.0167	0.0142	31.1215	65
70	0.8396	64.1439	2147.6111	1.1910	76.3944	0.0156	0.0131	33.4812	70
75	0.8292	68.3108	2447.6069	1.2059	82.3792	0.0146	0.0121	35.8305	75
80	0.8189	72.4260	2764.4568	1.2211	88.4392	0.0138	0.0113	38.1694	80
85	0.8088	76.4901	3097.6963	1.2364	94.5753	0.0131	0.0106	40.4980	85
90	0.7987	80.5038	3446.8700	1.2520	100.7885	0.0124	0.0099	42.8162	90
95	0.7888	84.4677	3811.5311	1.2677	107.0797	0.0118	0.0093	45.1241	95
100	0.7790	88.3825	4191.2417	1.2836	113.4500	0.0113	0.0088	47.4216	100

$$I = 0.50\%$$

n	(P/F)	(P/A)	(P/G)	(F/P)	(F/A)	(A/P)	(A/F)	(A/G)	n
1	0.9950	0.9950	0.0000	1.0050	1.0000	1.0050	1.0000	0.0000	1
2	0.9901	1.9851	0.9901	1.0100	2.0050	0.5038	0.4988	0.4988	2
3	0.9851	2.9702	2.9604	1.0151	3.0150	0.3367	0.3317	0.9967	3
4	0.9802	3.9505	5.9011	1.0202	4.0301	0.2531	0.2481	1.4938	4
5	0.9754	4.9259	9.8026	1.0253	5.0503	0.2030	0.1980	1.9900	5
6	0.9705	5.8964	14.6552	1.0304	6.0755	0.1696	0.1646	2.4855	6
7	0.9657	6.8621	20.4493	1.0355	7.1059	0.1457	0.1407	2.9801	7
8	0.9609	7.8230	27.1755	1.0407	8.1414	0.1278	0.1228	3.4738	8
9	0.9561	8.7791	34.8244	1.0459	9.1821	0.1139	0.1089	3.9668	9
10	0.9513	9.7304	43.3865	1.0511	10.2280	0.1028	0.0978	4.4589	10
11	0.9466	10.6770	52.8526	1.0564	11.2792	0.0937	0.0887	4.9501	11
12	0.9419	11.6189	63.2136	1.0617	12.3356	0.0861	0.0811	5.4406	12
13	0.9372	12.5562	74.4602	1.0670	13.3972	0.0796	0.0746	5.9302	13
14	0.9326	13.4887	86.5835	1.0723	14.4642	0.0741	0.0691	6.4190	14
15	0.9279	14.4166	99.5743	1.0777	15.5365	0.0694	0.0644	6.9069	15
16	0.9233	15.3399	113.4238	1.0831	16.6142	0.0652	0.0602	7.3940	16
17	0.9187	16.2586	128.1231	1.0885	17.6973	0.0615	0.0565	7.8803	17
18	0.9141	17.1728	143.6634	1.0939	18.7858	0.0582	0.0532	8.3658	18
19	0.9096	18.0824	160.0360	1.0994	19.8797	0.0553	0.0503	8.8504	19
20	0.9051	18.9874	177.2322	1.1049	20.9791	0.0527	0.0477	9.3342	20
21	0.9006	19.8880	195.2434	1.1104	22.0840	0.0503	0.0453	9.8172	21
22	0.8961	20.7841	214.0611	1.1160	23.1944	0.0481	0.0431	10.2993	22
23	0.8916	21.6757	233.6768	1.1216	24.3104	0.0461	0.0411	10.7806	23
24	0.8872	22.5629	254.0820	1.1272	25.4320	0.0443	0.0393	11.2611	24
25	0.8828	23.4456	275.2686	1.1328	26.5591	0.0427	0.0377	11.7407	25
26	0.8784	24.3240	297.2281	1.1385	27.6919	0.0411	0.0361	12.2195	26
27	0.8740	25.1980	319.9523	1.1442	28.8304	0.0397	0.0347	12.6975	27
28	0.8697	26.0677	343.4332	1.1499	29.9745	0.0384	0.0334	13.1747	28
29	0.8653	26.9330	367.6625	1.1556	31.1244	0.0371	0.0321	13.6510	29
30	0.8610	27.7941	392.6324	1.1614	32.2800	0.0360	0.0310	14.1265	30
31	0.8567	28.6508	418.3348	1.1672	33.4414	0.0349	0.0299	14.6012	31
32	0.8525	29.5033	444.7618	1.1730	34.6086	0.0339	0.0289	15.0750	32
33	0.8482	30.3515	471.9055	1.1789	35.7817	0.0329	0.0279	15.5480	33
34	0.8440	31.1955	499.7583	1.1848	36.9606	0.0321	0.0271	16.0202	34
35	0.8398	32.0354	528.3123	1.1907	38.1454	0.0312	0.0262	16.4915	35
36	0.8356	32.8710	557.5598	1.1967	39.3361	0.0304	0.0254	16.9621	36
37	0.8315	33.7025	587.4934	1.2027	40.5328	0.0297	0.0247	17.4317	37
38	0.8274	34.5299	618.1054	1.2087	41.7354	0.0290	0.0240	17.9006	38
39	0.8232	35.3531	649.3883	1.2147	42.9441	0.0283	0.0233	18.3686	39
40	0.8191	36.1722	681.3347	1.2208	44.1588	0.0276	0.0226	18.8359	40
41	0.8151	36.9873	713.9372	1.2269	45.3796	0.0270	0.0220	19.3022	41
42	0.8110	37.7983	747.1886	1.2330	46.6065	0.0265	0.0215	19.7678	42
43	0.8070	38.6053	781.0815	1.2392	47.8396	0.0259	0.0209	20.2325	43
44	0.8030	39.4082	815.6087	1.2454	49.0788	0.0254	0.0204	20.6964	44
45	0.7990	40.2072	850.7631	1.2516	50.3242	0.0249	0.0199	21.1595	45
46	0.7950	41.0022	886.5376	1.2579	51.5758	0.0244	0.0194	21.6217	46
47	0.7910	41.7932	922.9252	1.2642	52.8337	0.0239	0.0189	22.0831	47
48	0.7871	42.5803	959.9188	1.2705	54.0978	0.0235	0.0185	22.5437	48
49	0.7832	43.3635	997.5116	1.2768	55.3683	0.0231	0.0181	23.0035	49
50	0.7793	44.1428	1035.6966	1.2832	56.6452	0.0227	0.0177	23.4624	50
51	0.7754	44.9182	1074.4670	1.2896	57.9284	0.0223	0.0173	23.9205	51
52	0.7716	45.6897	1113.8162	1.2961	59.2180	0.0219	0.0169	24.3778	52
53	0.7677	46.4575	1153.7372	1.3026	60.5141	0.0215	0.0165	24.8343	53
54	0.7639	47.2214	1194.2236	1.3091	61.8167	0.0212	0.0162	25.2899	54
55	0.7601	47.9814	1235.2686	1.3156	63.1258	0.0208	0.0158	25.7447	55
60	0.7414	51.7256	1448.6458	1.3489	69.7700	0.0193	0.0143	28.0064	60
65	0.7231	55.3775	1675.0272	1.3829	76.5821	0.0181	0.0131	30.2475	65
70	0.7053	58.9394	1913.6427	1.4178	83.5661	0.0170	0.0120	32.4680	70
75	0.6879	62.4136	2163.7525	1.4536	90.7265	0.0160	0.0110	34.6679	75
80	0.6710	65.8023	2424.6455	1.4903	98.0677	0.0152	0.0102	36.8474	80
85	0.6545	69.1075	2695.6389	1.5280	105.5943	0.0145	0.0095	39.0065	85
90	0.6383	72.3313	2976.0769	1.5666	113.3109	0.0138	0.0088	41.1451	90
95	0.6226	75.4757	3265.3298	1.6061	121.2224	0.0132	0.0082	43.2633	95
100	0.6073	78.5426	3562.7934	1.6467	129.3337	0.0127	0.0077	45.3613	100

$I = 0.75\%$

n	(P/F)	(P/A)	(P/G)	(F/P)	(F/A)	(A/P)	(A/F)	(A/G)	n
1	0.9926	0.9926	0.0000	1.0075	1.0000	1.0075	1.0000	0.0000	1
2	0.9852	1.9777	0.9852	1.0151	2.0075	0.5056	0.4981	0.4981	2
3	0.9778	2.9556	2.9408	1.0227	3.0226	0.3383	0.3308	0.9950	3
4	0.9706	3.9261	5.8525	1.0303	4.0452	0.2547	0.2472	1.4907	4
5	0.9633	4.8894	9.7058	1.0381	5.0756	0.2045	0.1970	1.9851	5
6	0.9562	5.8456	14.4866	1.0459	6.1136	0.1711	0.1636	2.4782	6
7	0.9490	6.7946	20.1808	1.0537	7.1595	0.1472	0.1397	2.9701	7
8	0.9420	7.7366	26.7747	1.0616	8.2132	0.1293	0.1218	3.4608	8
9	0.9350	8.6716	34.2544	1.0696	9.2748	0.1153	0.1078	3.9502	9
10	0.9280	9.5996	42.6064	1.0776	10.3443	0.1042	0.0967	4.4384	10
11	0.9211	10.5207	51.8174	1.0857	11.4219	0.0951	0.0876	4.9253	11
12	0.9142	11.4349	61.8740	1.0938	12.5076	0.0875	0.0800	5.4110	12
13	0.9074	12.3423	72.7632	1.1020	13.6014	0.0810	0.0735	5.8954	13
14	0.9007	13.2430	84.4720	1.1103	14.7034	0.0755	0.0680	6.3786	14
15	0.8940	14.1370	96.9876	1.1186	15.8137	0.0707	0.0632	6.8606	15
16	0.8873	15.0243	110.2973	1.1270	16.9323	0.0666	0.0591	7.3413	16
17	0.8807	15.9050	124.3887	1.1354	18.0593	0.0629	0.0554	7.8207	17
18	0.8742	16.7792	139.2494	1.1440	19.1947	0.0596	0.0521	8.2989	18
19	0.8676	17.6468	154.8671	1.1525	20.3387	0.0567	0.0492	8.7759	19
20	0.8612	18.5080	171.2297	1.1612	21.4912	0.0540	0.0465	9.2516	20
21	0.8548	19.3628	188.3253	1.1699	22.6524	0.0516	0.0441	9.7261	21
22	0.8484	20.2112	206.1420	1.1787	23.8223	0.0495	0.0420	10.1994	22
23	0.8421	21.0533	224.6682	1.1875	25.0010	0.0475	0.0400	10.6714	23
24	0.8358	21.8891	243.8923	1.1964	26.1885	0.0457	0.0382	11.1422	24
25	0.8296	22.7188	263.8029	1.2054	27.3849	0.0440	0.0365	11.6117	25
26	0.8234	23.5422	284.3888	1.2144	28.5903	0.0425	0.0350	12.0800	26
27	0.8173	24.3595	305.6387	1.2235	29.8047	0.0411	0.0336	12.5470	27
28	0.8112	25.1707	327.5416	1.2327	31.0282	0.0397	0.0322	13.0128	28
29	0.8052	25.9759	350.0867	1.2420	32.2609	0.0385	0.0310	13.4774	29
30	0.7992	26.7751	373.2631	1.2513	33.5029	0.0373	0.0298	13.9407	30
31	0.7932	27.5683	397.0602	1.2607	34.7542	0.0363	0.0288	14.4028	31
32	0.7873	28.3557	421.4675	1.2701	36.0148	0.0353	0.0278	14.8636	32
33	0.7815	29.1371	446.4746	1.2796	37.2849	0.0343	0.0268	15.3232	33
34	0.7757	29.9128	472.0712	1.2892	38.5646	0.0334	0.0259	15.7816	34
35	0.7699	30.6827	498.2471	1.2989	39.8538	0.0326	0.0251	16.2387	35
36	0.7641	31.4468	524.9924	1.3086	41.1527	0.0318	0.0243	16.6946	36
37	0.7585	32.2053	552.2969	1.3185	42.4614	0.0311	0.0236	17.1493	37
38	0.7528	32.9581	580.1511	1.3283	43.7798	0.0303	0.0228	17.6027	38
39	0.7472	33.7053	608.5451	1.3383	45.1082	0.0297	0.0222	18.0549	39
40	0.7416	34.4469	637.4693	1.3483	46.4465	0.0290	0.0215	18.5058	40
41	0.7361	35.1831	666.9144	1.3585	47.7948	0.0284	0.0209	18.9556	41
42	0.7306	35.9137	696.8709	1.3686	49.1533	0.0278	0.0203	19.4040	42
43	0.7252	36.6389	727.3297	1.3789	50.5219	0.0273	0.0198	19.8513	43
44	0.7198	37.3587	758.2815	1.3893	51.9009	0.0268	0.0193	20.2973	44
45	0.7145	38.0732	789.7173	1.3997	53.2901	0.0263	0.0188	20.7421	45
46	0.7091	38.7823	821.6283	1.4102	54.6898	0.0258	0.0183	21.1856	46
47	0.7039	39.4862	854.0056	1.4207	56.1000	0.0253	0.0178	21.6280	47
48	0.6986	40.1848	886.8404	1.4314	57.5207	0.0249	0.0174	22.0691	48
49	0.6934	40.8782	920.1243	1.4421	58.9521	0.0245	0.0170	22.5089	49
50	0.6883	41.5664	953.8486	1.4530	60.3943	0.0241	0.0166	22.9476	50
51	0.6831	42.2496	988.0050	1.4639	61.8472	0.0237	0.0162	23.3850	51
52	0.6780	42.9276	1022.5852	1.4748	63.3111	0.0233	0.0158	23.8211	52
53	0.6730	43.6006	1057.5810	1.4859	64.7859	0.0229	0.0154	24.2561	53
54	0.6680	44.2686	1092.9842	1.4970	66.2718	0.0226	0.0151	24.6898	54
55	0.6630	44.9316	1128.7868	1.5083	67.7688	0.0223	0.0148	25.1223	55
60	0.6387	48.1734	1313.5189	1.5657	75.4241	0.0208	0.0133	27.2665	60
65	0.6153	51.2963	1507.0910	1.6253	83.3709	0.0195	0.0120	29.3801	65
70	0.5927	54.3046	1708.6065	1.6872	91.6201	0.0184	0.0109	31.4634	70
75	0.5710	57.2027	1917.2225	1.7514	100.1833	0.0175	0.0100	33.5163	75
80	0.5500	59.9944	2132.1472	1.8180	109.0725	0.0167	0.0092	35.5391	80
85	0.5299	62.6838	2352.6375	1.8873	118.3001	0.0160	0.0085	37.5318	85
90	0.5104	65.2746	2577.9961	1.9591	127.8790	0.0153	0.0078	39.4946	90
95	0.4917	67.7704	2807.5694	2.0337	137.8225	0.0148	0.0073	41.4277	95
100	0.4737	70.1746	3040.7453	2.1111	148.1445	0.0143	0.0068	43.3311	100

$I = 1.00\%$

n	(P/F)	(P/A)	(P/G)	(F/P)	(F/A)	(A/P)	(A/F)	(A/G)	n
1	0.9901	0.9901	0.0000	1.0100	1.0000	1.0100	1.0000	0.0000	1
2	0.9803	1.9704	0.9803	1.0201	2.0100	0.5075	0.4975	0.4975	2
3	0.9706	2.9410	2.9215	1.0303	3.0301	0.3400	0.3300	0.9934	3
4	0.9610	3.9020	5.8044	1.0406	4.0604	0.2563	0.2463	1.4876	4
5	0.9515	4.8534	9.6103	1.0510	5.1010	0.2060	0.1960	1.9801	5
6	0.9420	5.7955	14.3205	1.0615	6.1520	0.1725	0.1625	2.4710	6
7	0.9327	6.7282	19.9168	1.0721	7.2135	0.1486	0.1386	2.9602	7
8	0.9235	7.6517	26.3812	1.0829	8.2857	0.1307	0.1207	3.4478	8
9	0.9143	8.5660	33.6959	1.0937	9.3685	0.1167	0.1067	3.9337	9
10	0.9053	9.4713	41.8435	1.1046	10.4622	0.1056	0.0956	4.4179	10
11	0.8963	10.3676	50.8067	1.1157	11.5668	0.0965	0.0865	4.9005	11
12	0.8874	11.2551	60.5687	1.1268	12.6825	0.0888	0.0788	5.3815	12
13	0.8787	12.1337	71.1126	1.1381	13.8093	0.0824	0.0724	5.8607	13
14	0.8700	13.0037	82.4221	1.1495	14.9474	0.0769	0.0669	6.3384	14
15	0.8613	13.8651	94.4810	1.1610	16.0969	0.0721	0.0621	6.8143	15
16	0.8528	14.7179	107.2734	1.1726	17.2579	0.0679	0.0579	7.2886	16
17	0.8444	15.5623	120.7834	1.1843	18.4304	0.0643	0.0543	7.7613	17
18	0.8360	16.3983	134.9957	1.1961	19.6147	0.0610	0.0510	8.2323	18
19	0.8277	17.2260	149.8950	1.2081	20.8109	0.0581	0.0481	8.7017	19
20	0.8195	18.0456	165.4664	1.2202	22.0190	0.0554	0.0454	9.1694	20
21	0.8114	18.8570	181.6950	1.2324	23.2392	0.0530	0.0430	9.6354	21
22	0.8034	19.6604	198.5663	1.2447	24.4716	0.0509	0.0409	10.0998	22
23	0.7954	20.4558	216.0660	1.2572	25.7163	0.0489	0.0389	10.5626	23
24	0.7876	21.2434	234.1800	1.2697	26.9735	0.0471	0.0371	11.0237	24
25	0.7798	22.0232	252.8945	1.2824	28.2432	0.0454	0.0354	11.4831	25
26	0.7720	22.7952	272.1957	1.2953	29.5256	0.0439	0.0339	11.9409	26
27	0.7644	23.5596	292.0702	1.3082	30.8209	0.0424	0.0324	12.3971	27
28	0.7568	24.3164	312.5047	1.3213	32.1291	0.0411	0.0311	12.8516	28
29	0.7493	25.0658	333.4863	1.3345	33.4504	0.0399	0.0299	13.3044	29
30	0.7419	25.8077	355.0021	1.3478	34.7849	0.0387	0.0287	13.7557	30
31	0.7346	26.5423	377.0394	1.3613	36.1327	0.0377	0.0277	14.2052	31
32	0.7273	27.2696	399.5858	1.3749	37.4941	0.0367	0.0267	14.6532	32
33	0.7201	27.9897	422.6291	1.3887	38.8690	0.0357	0.0257	15.0995	33
34	0.7130	28.7027	446.1572	1.4026	40.2577	0.0348	0.0248	15.5441	34
35	0.7059	29.4086	470.1583	1.4166	41.6603	0.0340	0.0240	15.9871	35
36	0.6989	30.1075	494.6207	1.4308	43.0769	0.0332	0.0232	16.4285	36
37	0.6920	30.7995	519.5329	1.4451	44.5076	0.0325	0.0225	16.8682	37
38	0.6852	31.4847	544.8835	1.4595	45.9527	0.0318	0.0218	17.3063	38
39	0.6784	32.1630	570.6616	1.4741	47.4123	0.0311	0.0211	17.7428	39
40	0.6717	32.8347	596.8561	1.4889	48.8864	0.0305	0.0205	18.1776	40
41	0.6650	33.4997	623.4562	1.5038	50.3752	0.0299	0.0199	18.6108	41
42	0.6584	34.1581	650.4514	1.5188	51.8790	0.0293	0.0193	19.0424	42
43	0.6519	34.8100	677.8312	1.5340	53.3978	0.0287	0.0187	19.4723	43
44	0.6454	35.4555	705.5853	1.5493	54.9318	0.0282	0.0182	19.9006	44
45	0.6391	36.0945	733.7037	1.5648	56.4811	0.0277	0.0177	20.3273	45
46	0.6327	36.7272	762.1765	1.5805	58.0459	0.0272	0.0172	20.7524	46
47	0.6265	37.3537	790.9938	1.5963	59.6263	0.0268	0.0168	21.1758	47
48	0.6203	37.9740	820.1460	1.6122	61.2226	0.0263	0.0163	21.5976	48
49	0.6141	38.5881	849.6237	1.6283	62.8348	0.0259	0.0159	22.0178	49
50	0.6080	39.1961	879.4176	1.6446	64.4632	0.0255	0.0155	22.4363	50
51	0.6020	39.7981	909.5186	1.6611	66.1078	0.0251	0.0151	22.8533	51
52	0.5961	40.3942	939.9175	1.6777	67.7689	0.0248	0.0148	23.2686	52
53	0.5902	40.9844	970.6057	1.6945	69.4466	0.0244	0.0144	23.6823	53
54	0.5843	41.5687	1001.5743	1.7114	71.1410	0.0241	0.0141	24.0945	54
55	0.5785	42.1472	1032.8148	1.7285	72.8525	0.0237	0.0137	24.5049	55
60	0.5504	44.9550	1192.8061	1.8167	81.6697	0.0222	0.0122	26.5333	60
65	0.5237	47.6266	1358.3903	1.9094	90.9366	0.0210	0.0110	28.5217	65
70	0.4983	50.1685	1528.6474	2.0068	100.6763	0.0199	0.0099	30.4703	70
75	0.4741	52.5871	1702.7340	2.1091	110.9128	0.0190	0.0090	32.3793	75
80	0.4511	54.8882	1879.8771	2.2167	121.6715	0.0182	0.0082	34.2492	80
85	0.4292	57.0777	2059.3701	2.3298	132.9790	0.0175	0.0075	36.0801	85
90	0.4084	59.1609	2240.5675	2.4486	144.8633	0.0169	0.0069	37.8724	90
95	0.3886	61.1430	2422.8811	2.5735	157.3538	0.0164	0.0064	39.6265	95
100	0.3697	63.0289	2605.7758	2.7048	170.4814	0.0159	0.0059	41.3426	100

$I = 1.25\%$

n	(P/F)	(P/A)	(P/G)	(F/P)	(F/A)	(A/P)	(A/F)	(A/G)	n
1	0.9877	0.9877	0.0000	1.0125	1.0000	1.0125	1.0000	0.0000	1
2	0.9755	1.9631	0.9755	1.0252	2.0125	0.5094	0.4969	0.4969	2
3	0.9634	2.9265	2.9023	1.0380	3.0377	0.3417	0.3292	0.9917	3
4	0.9515	3.8781	5.7569	1.0509	4.0756	0.2579	0.2454	1.4845	4
5	0.9398	4.8178	9.5160	1.0641	5.1266	0.2076	0.1951	1.9752	5
6	0.9282	5.7460	14.1569	1.0774	6.1907	0.1740	0.1615	2.4638	6
7	0.9167	6.6627	19.6571	1.0909	7.2680	0.1501	0.1376	2.9503	7
8	0.9054	7.5681	25.9949	1.1045	8.3589	0.1321	0.1196	3.4348	8
9	0.8942	8.4623	33.1487	1.1183	9.4634	0.1182	0.1057	3.9172	9
10	0.8832	9.3455	41.0973	1.1323	10.5817	0.1070	0.0945	4.3975	10
11	0.8723	10.2178	49.8201	1.1464	11.7139	0.0979	0.0854	4.8758	11
12	0.8615	11.0793	59.2967	1.1608	12.8604	0.0903	0.0778	5.3520	12
13	0.8509	11.9302	69.5072	1.1753	14.0211	0.0838	0.0713	5.8262	13
14	0.8404	12.7706	80.4320	1.1900	15.1964	0.0783	0.0658	6.2982	14
15	0.8300	13.6005	92.0519	1.2048	16.3863	0.0735	0.0610	6.7682	15
16	0.8197	14.4203	104.3481	1.2199	17.5912	0.0693	0.0568	7.2362	16
17	0.8096	15.2299	117.3021	1.2351	18.8111	0.0657	0.0532	7.7021	17
18	0.7996	16.0295	130.8958	1.2506	20.0462	0.0624	0.0499	8.1659	18
19	0.7898	16.8193	145.1115	1.2662	21.2968	0.0595	0.0470	8.6277	19
20	0.7800	17.5993	159.9316	1.2820	22.5630	0.0568	0.0443	9.0874	20
21	0.7704	18.3697	175.3392	1.2981	23.8450	0.0544	0.0419	9.5450	21
22	0.7609	19.1306	191.3174	1.3143	25.1431	0.0523	0.0398	10.0006	22
23	0.7515	19.8820	207.8499	1.3307	26.4574	0.0503	0.0378	10.4542	23
24	0.7422	20.6242	224.9204	1.3474	27.7881	0.0485	0.0360	10.9056	24
25	0.7330	21.3573	242.5132	1.3642	29.1354	0.0468	0.0343	11.3551	25
26	0.7240	22.0813	260.6128	1.3812	30.4996	0.0453	0.0328	11.8024	26
27	0.7150	22.7963	279.2040	1.3985	31.8809	0.0439	0.0314	12.2478	27
28	0.7062	23.5025	298.2719	1.4160	33.2794	0.0425	0.0300	12.6911	28
29	0.6975	24.2000	317.8019	1.4337	34.6954	0.0413	0.0288	13.1323	29
30	0.6889	24.8889	337.7797	1.4516	36.1291	0.0402	0.0277	13.5715	30
31	0.6804	25.5693	358.1912	1.4698	37.5807	0.0391	0.0266	14.0086	31
32	0.6720	26.2413	379.0227	1.4881	39.0504	0.0381	0.0256	14.4438	32
33	0.6637	26.9050	400.2607	1.5067	40.5386	0.0372	0.0247	14.8768	33
34	0.6555	27.5605	421.8920	1.5256	42.0453	0.0363	0.0238	15.3079	34
35	0.6474	28.2079	443.9037	1.5446	43.5709	0.0355	0.0230	15.7369	35
36	0.6394	28.8473	466.2830	1.5639	45.1155	0.0347	0.0222	16.1639	36
37	0.6315	29.4788	489.0176	1.5835	46.6794	0.0339	0.0214	16.5888	37
38	0.6237	30.1025	512.0952	1.6033	48.2629	0.0332	0.0207	17.0117	38
39	0.6160	30.7185	535.5039	1.6233	49.8662	0.0326	0.0201	17.4326	39
40	0.6084	31.3269	559.2320	1.6436	51.4896	0.0319	0.0194	17.8515	40
41	0.6009	31.9278	583.2681	1.6642	53.1332	0.0313	0.0188	18.2683	41
42	0.5935	32.5213	607.6009	1.6850	54.7973	0.0307	0.0182	18.6832	42
43	0.5862	33.1075	632.2195	1.7060	56.4823	0.0302	0.0177	19.0960	43
44	0.5789	33.6864	657.1130	1.7274	58.1883	0.0297	0.0172	19.5068	44
45	0.5718	34.2582	682.2710	1.7489	59.9157	0.0292	0.0167	19.9156	45
46	0.5647	34.8229	707.6832	1.7708	61.6646	0.0287	0.0162	20.3224	46
47	0.5577	35.3806	733.3393	1.7929	63.4354	0.0283	0.0158	20.7271	47
48	0.5509	35.9315	759.2296	1.8154	65.2284	0.0278	0.0153	21.1299	48
49	0.5441	36.4755	785.3442	1.8380	67.0437	0.0274	0.0149	21.5307	49
50	0.5373	37.0129	811.6738	1.8610	68.8818	0.0270	0.0145	21.9295	50
51	0.5307	37.5436	838.2091	1.8843	70.7428	0.0266	0.0141	22.3263	51
52	0.5242	38.0677	864.9409	1.9078	72.6271	0.0263	0.0138	22.7211	52
53	0.5177	38.5854	891.8604	1.9317	74.5349	0.0259	0.0134	23.1139	53
54	0.5113	39.0967	918.9588	1.9558	76.4666	0.0256	0.0131	23.5048	54
55	0.5050	39.6017	946.2277	1.9803	78.4225	0.0253	0.0128	23.8936	55
60	0.4746	42.0346	1084.8429	2.1072	88.5745	0.0238	0.0113	25.8083	60
65	0.4460	44.3210	1226.5421	2.2422	99.3771	0.0226	0.0101	27.6741	65
70	0.4191	46.4697	1370.4513	2.3859	110.8720	0.0215	0.0090	29.4913	70
75	0.3939	48.4890	1515.7904	2.5388	123.1035	0.0206	0.0081	31.2605	75
80	0.3702	50.3867	1661.8651	2.7015	136.1188	0.0198	0.0073	32.9822	80
85	0.3479	52.1701	1808.0598	2.8746	149.9682	0.0192	0.0067	34.6570	85
90	0.3269	53.8461	1953.8303	3.0588	164.7050	0.0186	0.0061	36.2855	90
95	0.3072	55.4211	2098.6973	3.2548	180.3862	0.0180	0.0055	37.8682	95
100	0.2887	56.9013	2242.2411	3.4634	197.0723	0.0176	0.0051	39.4058	100

ENGINEERING ECONOMIC ANALYSIS

$I = 1.50\%$

n	(P/F)	(P/A)	(P/G)	(F/P)	(F/A)	(A/P)	(A/F)	(A/G)	n
1	0.9852	0.9852	0.0000	1.0150	1.0000	1.0150	1.0000	0.0000	1
2	0.9707	1.9559	0.9707	1.0302	2.0150	0.5113	0.4963	0.4963	2
3	0.9563	2.9122	2.8833	1.0457	3.0452	0.3434	0.3284	0.9901	3
4	0.9422	3.8544	5.7098	1.0614	4.0909	0.2594	0.2444	1.4814	4
5	0.9283	4.7826	9.4229	1.0773	5.1523	0.2091	0.1941	1.9702	5
6	0.9145	5.6972	13.9956	1.0934	6.2296	0.1755	0.1605	2.4566	6
7	0.9010	6.5982	19.4018	1.1098	7.3230	0.1516	0.1366	2.9405	7
8	0.8877	7.4859	25.6157	1.1265	8.4328	0.1336	0.1186	3.4219	8
9	0.8746	8.3605	32.6125	1.1434	9.5593	0.1196	0.1046	3.9008	9
10	0.8617	9.2222	40.3675	1.1605	10.7027	0.1084	0.0934	4.3772	10
11	0.8489	10.0711	48.8568	1.1779	11.8633	0.0993	0.0843	4.8512	11
12	0.8364	10.9075	58.0571	1.1956	13.0412	0.0917	0.0767	5.3227	12
13	0.8240	11.7315	67.9454	1.2136	14.2368	0.0852	0.0702	5.7917	13
14	0.8118	12.5434	78.4994	1.2318	15.4504	0.0797	0.0647	6.2582	14
15	0.7999	13.3432	89.6974	1.2502	16.6821	0.0749	0.0599	6.7223	15
16	0.7880	14.1313	101.5178	1.2690	17.9324	0.0708	0.0558	7.1839	16
17	0.7764	14.9076	113.9400	1.2880	19.2014	0.0671	0.0521	7.6431	17
18	0.7649	15.6726	126.9435	1.3073	20.4894	0.0638	0.0488	8.0997	18
19	0.7536	16.4262	140.5084	1.3270	21.7967	0.0609	0.0459	8.5539	19
20	0.7425	17.1686	154.6154	1.3469	23.1237	0.0582	0.0432	9.0057	20
21	0.7315	17.9001	169.2453	1.3671	24.4705	0.0559	0.0409	9.4550	21
22	0.7207	18.6208	184.3798	1.3876	25.8376	0.0537	0.0387	9.9018	22
23	0.7100	19.3309	200.0006	1.4084	27.2251	0.0517	0.0367	10.3462	23
24	0.6995	20.0304	216.0901	1.4295	28.6335	0.0499	0.0349	10.7881	24
25	0.6892	20.7196	232.6310	1.4509	30.0630	0.0483	0.0333	11.2276	25
26	0.6790	21.3986	249.6065	1.4727	31.5140	0.0467	0.0317	11.6646	26
27	0.6690	22.0676	267.0002	1.4948	32.9867	0.0453	0.0303	12.0992	27
28	0.6591	22.7267	284.7958	1.5172	34.4815	0.0440	0.0290	12.5313	28
29	0.6494	23.3761	302.9779	1.5400	35.9987	0.0428	0.0278	12.9610	29
30	0.6398	24.0158	321.5310	1.5631	37.5387	0.0416	0.0266	13.3883	30
31	0.6303	24.6461	340.4402	1.5865	39.1018	0.0406	0.0256	13.8131	31
32	0.6210	25.2671	359.6910	1.6103	40.6883	0.0396	0.0246	14.2355	32
33	0.6118	25.8790	379.2691	1.6345	42.2986	0.0386	0.0236	14.6555	33
34	0.6028	26.4817	399.1607	1.6590	43.9331	0.0378	0.0228	15.0731	34
35	0.5939	27.0756	419.3521	1.6839	45.5921	0.0369	0.0219	15.4882	35
36	0.5851	27.6607	439.8303	1.7091	47.2760	0.0362	0.0212	15.9009	36
37	0.5764	28.2371	460.5822	1.7348	48.9851	0.0354	0.0204	16.3112	37
38	0.5679	28.8051	481.5954	1.7608	50.7199	0.0347	0.0197	16.7191	38
39	0.5595	29.3646	502.8576	1.7872	52.4807	0.0341	0.0191	17.1246	39
40	0.5513	29.9158	524.3568	1.8140	54.2679	0.0334	0.0184	17.5277	40
41	0.5431	30.4590	546.0814	1.8412	56.0819	0.0328	0.0178	17.9284	41
42	0.5351	30.9941	568.0201	1.8688	57.9231	0.0323	0.0173	18.3267	42
43	0.5272	31.5212	590.1617	1.8969	59.7920	0.0317	0.0167	18.7227	43
44	0.5194	32.0406	612.4955	1.9253	61.6889	0.0312	0.0162	19.1162	44
45	0.5117	32.5523	635.0110	1.9542	63.6142	0.0307	0.0157	19.5074	45
46	0.5042	33.0565	657.6979	1.9835	65.5684	0.0303	0.0153	19.8962	46
47	0.4967	33.5532	680.5462	2.0133	67.5519	0.0298	0.0148	20.2826	47
48	0.4894	34.0426	703.5462	2.0435	69.5652	0.0294	0.0144	20.6667	48
49	0.4821	34.5247	726.6884	2.0741	71.6087	0.0290	0.0140	21.0484	49
50	0.4750	34.9997	749.9636	2.1052	73.6828	0.0286	0.0136	21.4277	50
51	0.4680	35.4677	773.3629	2.1368	75.7881	0.0282	0.0132	21.8047	51
52	0.4611	35.9287	796.8774	2.1689	77.9249	0.0278	0.0128	22.1794	52
53	0.4543	36.3830	820.4986	2.2014	80.0938	0.0275	0.0125	22.5517	53
54	0.4475	36.8305	844.2184	2.2344	82.2952	0.0272	0.0122	22.9217	54
55	0.4409	37.2715	868.0285	2.2679	84.5296	0.0268	0.0118	23.2894	55
60	0.4093	39.3803	988.1674	2.4432	96.2147	0.0254	0.0104	25.0930	60
65	0.3799	41.3378	1109.4752	2.6320	108.8028	0.0242	0.0092	26.8393	65
70	0.3527	43.1549	1231.1658	2.8355	122.3638	0.0232	0.0082	28.5290	70
75	0.3274	44.8416	1352.5600	3.0546	136.9728	0.0223	0.0073	30.1631	75
80	0.3039	46.4073	1473.0741	3.2907	152.7109	0.0215	0.0065	31.7423	80
85	0.2821	47.8607	1592.2095	3.5450	169.6652	0.0209	0.0059	33.2676	85
90	0.2619	49.2099	1709.5439	3.8189	187.9299	0.0203	0.0053	34.7399	90
95	0.2431	50.4622	1824.7224	4.1141	207.6061	0.0198	0.0048	36.1602	95
100	0.2256	51.6247	1937.4506	4.4320	228.8030	0.0194	0.0044	37.5295	100

$I = 1.75\%$

n	(P/F)	(P/A)	(P/G)	(F/P)	(F/A)	(A/P)	(A/F)	(A/G)	n
1	0.9828	0.9828	0.0000	1.0175	1.0000	1.0175	1.0000	0.0000	1
2	0.9659	1.9487	0.9659	1.0353	2.0175	0.5132	0.4957	0.4957	2
3	0.9493	2.8980	2.8645	1.0534	3.0528	0.3451	0.3276	0.9884	3
4	0.9330	3.8309	5.6633	1.0719	4.1062	0.2610	0.2435	1.4783	4
5	0.9169	4.7479	9.3310	1.0906	5.1781	0.2106	0.1931	1.9653	5
6	0.9011	5.6490	13.8367	1.1097	6.2687	0.1770	0.1595	2.4494	6
7	0.8856	6.5346	19.1506	1.1291	7.3784	0.1530	0.1355	2.9306	7
8	0.8704	7.4051	25.2435	1.1489	8.5075	0.1350	0.1175	3.4089	8
9	0.8554	8.2605	32.0870	1.1690	9.6564	0.1211	0.1036	3.8844	9
10	0.8407	9.1012	39.6535	1.1894	10.8254	0.1099	0.0924	4.3569	10
11	0.8263	9.9275	47.9162	1.2103	12.0148	0.1007	0.0832	4.8266	11
12	0.8121	10.7395	56.8489	1.2314	13.2251	0.0931	0.0756	5.2934	12
13	0.7981	11.5376	66.4260	1.2530	14.4565	0.0867	0.0692	5.7573	13
14	0.7844	12.3220	76.6227	1.2749	15.7095	0.0812	0.0637	6.2184	14
15	0.7709	13.0929	87.4149	1.2972	16.9844	0.0764	0.0589	6.6765	15
16	0.7576	13.8505	98.7792	1.3199	18.2817	0.0722	0.0547	7.1318	16
17	0.7446	14.5951	110.6926	1.3430	19.6016	0.0685	0.0510	7.5842	17
18	0.7318	15.3269	123.1328	1.3665	20.9446	0.0652	0.0477	8.0338	18
19	0.7192	16.0461	136.0783	1.3904	22.3112	0.0623	0.0448	8.4805	19
20	0.7068	16.7529	149.5080	1.4148	23.7016	0.0597	0.0422	8.9243	20
21	0.6947	17.4475	163.4013	1.4395	25.1164	0.0573	0.0398	9.3653	21
22	0.6827	18.1303	177.7385	1.4647	26.5559	0.0552	0.0377	9.8034	22
23	0.6710	18.8012	192.5000	1.4904	28.0207	0.0532	0.0357	10.2387	23
24	0.6594	19.4607	207.6671	1.5164	29.5110	0.0514	0.0339	10.6711	24
25	0.6481	20.1088	223.2214	1.5430	31.0275	0.0497	0.0322	11.1007	25
26	0.6369	20.7457	239.1451	1.5700	32.5704	0.0482	0.0307	11.5274	26
27	0.6260	21.3717	255.4210	1.5975	34.1404	0.0468	0.0293	11.9513	27
28	0.6152	21.9870	272.0321	1.6254	35.7379	0.0455	0.0280	12.3724	28
29	0.6046	22.5916	288.9623	1.6539	37.3633	0.0443	0.0268	12.7907	29
30	0.5942	23.1858	306.1954	1.6828	39.0172	0.0431	0.0256	13.2061	30
31	0.5840	23.7699	323.7163	1.7122	40.7000	0.0421	0.0246	13.6188	31
32	0.5740	24.3439	341.5097	1.7422	42.4122	0.0411	0.0236	14.0286	32
33	0.5641	24.9080	359.5613	1.7727	44.1544	0.0401	0.0226	14.4356	33
34	0.5544	25.4624	377.8567	1.8037	45.9271	0.0393	0.0218	14.8398	34
35	0.5449	26.0073	396.3824	1.8353	47.7308	0.0385	0.0210	15.2412	35
36	0.5355	26.5428	415.1250	1.8674	49.5661	0.0377	0.0202	15.6399	36
37	0.5263	27.0690	434.0715	1.9001	51.4335	0.0369	0.0194	16.0357	37
38	0.5172	27.5863	453.2094	1.9333	53.3336	0.0362	0.0187	16.4288	38
39	0.5083	28.0946	472.5264	1.9672	55.2670	0.0356	0.0181	16.8191	39
40	0.4996	28.5942	492.0109	2.0016	57.2341	0.0350	0.0175	17.2066	40
41	0.4910	29.0852	511.6512	2.0366	59.2357	0.0344	0.0169	17.5914	41
42	0.4826	29.5678	531.4363	2.0723	61.2724	0.0338	0.0163	17.9735	42
43	0.4743	30.0421	551.3446	2.1085	63.3446	0.0333	0.0158	18.3528	43
44	0.4661	30.5082	571.3980	2.1454	65.4532	0.0328	0.0153	18.7293	44
45	0.4581	30.9663	591.5540	2.1830	67.5986	0.0323	0.0148	19.1032	45
46	0.4502	31.4165	611.8135	2.2212	69.7816	0.0318	0.0143	19.4743	46
47	0.4425	31.8589	632.1670	2.2600	72.0027	0.0314	0.0139	19.8427	47
48	0.4349	32.2938	652.6054	2.2996	74.2628	0.0310	0.0135	20.2084	48
49	0.4274	32.7212	673.1196	2.3398	76.5624	0.0306	0.0131	20.5714	49
50	0.4200	33.1412	693.7010	2.3808	78.9022	0.0302	0.0127	20.9317	50
51	0.4128	33.5540	714.3413	2.4225	81.2830	0.0298	0.0123	21.2893	51
52	0.4057	33.9597	735.0322	2.4648	83.7055	0.0294	0.0119	21.6442	52
53	0.3987	34.3584	755.7660	2.5080	86.1703	0.0291	0.0116	21.9965	53
54	0.3919	34.7503	776.5351	2.5519	88.6783	0.0288	0.0113	22.3461	54
55	0.3851	35.1354	797.3321	2.5965	91.2302	0.0285	0.0110	22.6931	55
60	0.3531	36.9640	901.4954	2.8318	104.6752	0.0271	0.0096	24.3885	60
65	0.3238	38.6406	1005.3872	3.0884	119.3386	0.0259	0.0084	26.0189	65
70	0.2969	40.1779	1108.3333	3.3683	135.3308	0.0249	0.0074	27.5856	70
75	0.2722	41.5875	1209.7738	3.6735	152.7721	0.0240	0.0065	29.0899	75
80	0.2496	42.8799	1309.2482	4.0064	171.7938	0.0233	0.0058	30.5329	80
85	0.2289	44.0650	1406.3828	4.3694	192.5393	0.0227	0.0052	31.9161	85
90	0.2098	45.1516	1500.8798	4.7654	215.1646	0.0221	0.0046	33.2409	90
95	0.1924	46.1479	1592.5069	5.1972	239.8402	0.0217	0.0042	34.5087	95
100	0.1764	47.0615	1681.0886	5.6682	266.7518	0.0212	0.0037	35.7211	100

I = 2.00%

n	(P/F)	(P/A)	(P/G)	(F/P)	(F/A)	(A/P)	(A/F)	(A/G)	n
1	0.9804	0.9804	0.0000	1.0200	1.0000	1.0200	1.0000	0.0000	1
2	0.9612	1.9416	0.9612	1.0404	2.0200	0.5150	0.4950	0.4950	2
3	0.9423	2.8839	2.8458	1.0612	3.0604	0.3468	0.3268	0.9868	3
4	0.9238	3.8077	5.6173	1.0824	4.1216	0.2626	0.2426	1.4752	4
5	0.9057	4.7135	9.2403	1.1041	5.2040	0.2122	0.1922	1.9604	5
6	0.8880	5.6014	13.6801	1.1262	6.3081	0.1785	0.1585	2.4423	6
7	0.8706	6.4720	18.9035	1.1487	7.4343	0.1545	0.1345	2.9208	7
8	0.8535	7.3255	24.8779	1.1717	8.5830	0.1365	0.1165	3.3961	8
9	0.8368	8.1622	31.5720	1.1951	9.7546	0.1225	0.1025	3.8681	9
10	0.8203	8.9826	38.9551	1.2190	10.9497	0.1113	0.0913	4.3367	10
11	0.8043	9.7868	46.9977	1.2434	12.1687	0.1022	0.0822	4.8021	11
12	0.7885	10.5753	55.6712	1.2682	13.4121	0.0946	0.0746	5.2642	12
13	0.7730	11.3484	64.9475	1.2936	14.6803	0.0881	0.0681	5.7231	13
14	0.7579	12.1062	74.7999	1.3195	15.9739	0.0826	0.0626	6.1786	14
15	0.7430	12.8493	85.2021	1.3459	17.2934	0.0778	0.0578	6.6309	15
16	0.7284	13.5777	96.1288	1.3728	18.6393	0.0737	0.0537	7.0799	16
17	0.7142	14.2919	107.5554	1.4002	20.0121	0.0700	0.0500	7.5256	17
18	0.7002	14.9920	119.4581	1.4282	21.4123	0.0667	0.0467	7.9681	18
19	0.6864	15.6785	131.8139	1.4568	22.8406	0.0638	0.0438	8.4073	19
20	0.6730	16.3514	144.6003	1.4859	24.2974	0.0612	0.0412	8.8433	20
21	0.6598	17.0112	157.7959	1.5157	25.7833	0.0588	0.0388	9.2760	21
22	0.6468	17.6580	171.3795	1.5460	27.2990	0.0566	0.0366	9.7055	22
23	0.6342	18.2922	185.3309	1.5769	28.8450	0.0547	0.0347	10.1317	23
24	0.6217	18.9139	199.6305	1.6084	30.4219	0.0529	0.0329	10.5547	24
25	0.6095	19.5235	214.2592	1.6406	32.0303	0.0512	0.0312	10.9745	25
26	0.5976	20.1210	229.1987	1.6734	33.6709	0.0497	0.0297	11.3910	26
27	0.5859	20.7069	244.4311	1.7069	35.3443	0.0483	0.0283	11.8043	27
28	0.5744	21.2813	259.9392	1.7410	37.0512	0.0470	0.0270	12.2145	28
29	0.5631	21.8444	275.7064	1.7758	38.7922	0.0458	0.0258	12.6214	29
30	0.5521	22.3965	291.7164	1.8114	40.5681	0.0446	0.0246	13.0251	30
31	0.5412	22.9377	307.9538	1.8476	42.3794	0.0436	0.0236	13.4257	31
32	0.5306	23.4683	324.4035	1.8845	44.2270	0.0426	0.0226	13.8230	32
33	0.5202	23.9886	341.0508	1.9222	46.1116	0.0417	0.0217	14.2172	33
34	0.5100	24.4986	357.8817	1.9607	48.0338	0.0408	0.0208	14.6083	34
35	0.5000	24.9986	374.8826	1.9999	49.9945	0.0400	0.0200	14.9961	35
36	0.4902	25.4888	392.0405	2.0399	51.9944	0.0392	0.0192	15.3809	36
37	0.4806	25.9695	409.3424	2.0807	54.0343	0.0385	0.0185	15.7625	37
38	0.4712	26.4406	426.7764	2.1223	56.1149	0.0378	0.0178	16.1409	38
39	0.4619	26.9026	444.3304	2.1647	58.2372	0.0372	0.0172	16.5163	39
40	0.4529	27.3555	461.9931	2.2080	60.4020	0.0366	0.0166	16.8885	40
41	0.4440	27.7995	479.7535	2.2522	62.6100	0.0360	0.0160	17.2576	41
42	0.4353	28.2348	497.6010	2.2972	64.8622	0.0354	0.0154	17.6237	42
43	0.4268	28.6616	515.5253	2.3432	67.1595	0.0349	0.0149	17.9866	43
44	0.4184	29.0800	533.5165	2.3901	69.5027	0.0344	0.0144	18.3465	44
45	0.4102	29.4902	551.5652	2.4379	71.8927	0.0339	0.0139	18.7034	45
46	0.4022	29.8923	569.6621	2.4866	74.3306	0.0335	0.0135	19.0571	46
47	0.3943	30.2866	587.7985	2.5363	76.8172	0.0330	0.0130	19.4079	47
48	0.3865	30.6731	605.9657	2.5871	79.3535	0.0326	0.0126	19.7556	48
49	0.3790	31.0521	624.1557	2.6388	81.9406	0.0322	0.0122	20.1003	49
50	0.3715	31.4236	642.3606	2.6916	84.5794	0.0318	0.0118	20.4420	50
51	0.3642	31.7878	660.5727	2.7454	87.2710	0.0315	0.0115	20.7807	51
52	0.3571	32.1449	678.7849	2.8003	90.0164	0.0311	0.0111	21.1164	52
53	0.3501	32.4950	696.9900	2.8563	92.8167	0.0308	0.0108	21.4491	53
54	0.3432	32.8383	715.1815	2.9135	95.6731	0.0305	0.0105	21.7789	54
55	0.3365	33.1748	733.3527	2.9717	98.5865	0.0301	0.0101	22.1057	55
60	0.3048	34.7609	823.6975	3.2810	114.0515	0.0288	0.0088	23.6961	60
65	0.2761	36.1975	912.7085	3.6225	131.1262	0.0276	0.0076	25.2147	65
70	0.2500	37.4986	999.8343	3.9996	149.9779	0.0267	0.0067	26.6632	70
75	0.2265	38.6771	1084.6393	4.4158	170.7918	0.0259	0.0059	28.0434	75
80	0.2051	39.7445	1166.7868	4.8754	193.7720	0.0252	0.0052	29.3572	80
85	0.1858	40.7113	1246.0241	5.3829	219.1439	0.0246	0.0046	30.6064	85
90	0.1683	41.5869	1322.1701	5.9431	247.1567	0.0240	0.0040	31.7929	90
95	0.1524	42.3800	1395.1033	6.5617	278.0850	0.0236	0.0036	32.9189	95
100	0.1380	43.0984	1464.7527	7.2446	312.2323	0.0232	0.0032	33.9863	100

$I = 2.25\%$

n	(P/F)	(P/A)	(P/G)	(F/P)	(F/A)	(A/P)	(A/F)	(A/G)	n
1	0.9780	0.9780	0.0000	1.0225	1.0000	1.0225	1.0000	0.0000	1
2	0.9565	1.9345	0.9565	1.0455	2.0225	0.5169	0.4944	0.4944	2
3	0.9354	2.8699	2.8273	1.0690	3.0680	0.3484	0.3259	0.9852	3
4	0.9148	3.7847	5.5719	1.0931	4.1370	0.2642	0.2417	1.4722	4
5	0.8947	4.6795	9.1507	1.1177	5.2301	0.2137	0.1912	1.9555	5
6	0.8750	5.5545	13.5258	1.1428	6.3478	0.1800	0.1575	2.4351	6
7	0.8558	6.4102	18.6604	1.1685	7.4906	0.1560	0.1335	2.9110	7
8	0.8369	7.2472	24.5190	1.1948	8.6592	0.1380	0.1155	3.3832	8
9	0.8185	8.0657	31.0672	1.2217	9.8540	0.1240	0.1015	3.8518	9
10	0.8005	8.8662	38.2718	1.2492	11.0757	0.1128	0.0903	4.3166	10
11	0.7829	9.6491	46.1007	1.2773	12.3249	0.1036	0.0811	4.7777	11
12	0.7657	10.4148	54.5231	1.3060	13.6022	0.0960	0.0735	5.2352	12
13	0.7488	11.1636	63.5089	1.3354	14.9083	0.0896	0.0671	5.6889	13
14	0.7323	11.8959	73.0293	1.3655	16.2437	0.0841	0.0616	6.1390	14
15	0.7162	12.6122	83.0565	1.3962	17.6092	0.0793	0.0568	6.5854	15
16	0.7005	13.3126	93.5635	1.4276	19.0054	0.0751	0.0526	7.0282	16
17	0.6851	13.9977	104.5243	1.4597	20.4330	0.0714	0.0489	7.4673	17
18	0.6700	14.6677	115.9139	1.4926	21.8928	0.0682	0.0457	7.9027	18
19	0.6552	15.3229	127.7082	1.5262	23.3853	0.0653	0.0428	8.3345	19
20	0.6408	15.9637	139.8837	1.5605	24.9115	0.0626	0.0401	8.7626	20
21	0.6267	16.5904	152.4180	1.5956	26.4720	0.0603	0.0378	9.1871	21
22	0.6129	17.2034	165.2894	1.6315	28.0676	0.0581	0.0356	9.6080	22
23	0.5994	17.8028	178.4770	1.6682	29.6992	0.0562	0.0337	10.0252	23
24	0.5862	18.3890	191.9607	1.7058	31.3674	0.0544	0.0319	10.4389	24
25	0.5733	18.9624	205.7210	1.7441	33.0732	0.0527	0.0302	10.8489	25
26	0.5607	19.5231	219.7393	1.7834	34.8173	0.0512	0.0287	11.2553	26
27	0.5484	20.0715	233.9974	1.8235	36.6007	0.0498	0.0273	11.6582	27
28	0.5363	20.6078	248.4782	1.8645	38.4242	0.0485	0.0260	12.0575	28
29	0.5245	21.1323	263.1648	1.9065	40.2888	0.0473	0.0248	12.4532	29
30	0.5130	21.6453	278.0412	1.9494	42.1953	0.0462	0.0237	12.8453	30
31	0.5017	22.1470	293.0920	1.9933	44.1447	0.0452	0.0227	13.2339	31
32	0.4907	22.6377	308.3022	2.0381	46.1379	0.0442	0.0217	13.6190	32
33	0.4799	23.1175	323.6576	2.0840	48.1760	0.0433	0.0208	14.0005	33
34	0.4693	23.5868	339.1444	2.1308	50.2600	0.0424	0.0199	14.3786	34
35	0.4590	24.0458	354.7493	2.1788	52.3908	0.0416	0.0191	14.7531	35
36	0.4489	24.4947	370.4598	2.2278	54.5696	0.0408	0.0183	15.1241	36
37	0.4390	24.9337	386.2635	2.2779	56.7974	0.0401	0.0176	15.4917	37
38	0.4293	25.3630	402.1488	2.3292	59.0754	0.0394	0.0169	15.8557	38
39	0.4199	25.7829	418.1045	2.3816	61.4046	0.0388	0.0163	16.2164	39
40	0.4106	26.1935	434.1197	2.4352	63.7862	0.0382	0.0157	16.5736	40
41	0.4016	26.5951	450.1840	2.4900	66.2214	0.0376	0.0151	16.9273	41
42	0.3928	26.9879	466.2877	2.5460	68.7113	0.0371	0.0146	17.2777	42
43	0.3841	27.3720	482.4211	2.6033	71.2574	0.0365	0.0140	17.6246	43
44	0.3757	27.7477	498.5752	2.6619	73.8606	0.0360	0.0135	17.9682	44
45	0.3674	28.1151	514.7412	2.7218	76.5225	0.0356	0.0131	18.3083	45
46	0.3593	28.4744	530.9109	2.7830	79.2443	0.0351	0.0126	18.6452	46
47	0.3514	28.8259	547.0761	2.8456	82.0273	0.0347	0.0122	18.9787	47
48	0.3437	29.1695	563.2293	2.9096	84.8729	0.0343	0.0118	19.3088	48
49	0.3361	29.5057	579.3632	2.9751	87.7825	0.0339	0.0114	19.6357	49
50	0.3287	29.8344	595.4708	3.0420	90.7576	0.0335	0.0110	19.9592	50
51	0.3215	30.1559	611.5454	3.1105	93.7997	0.0332	0.0107	20.2795	51
52	0.3144	30.4703	627.5807	3.1805	96.9102	0.0328	0.0103	20.5965	52
53	0.3075	30.7778	643.5707	3.2520	100.0906	0.0325	0.0100	20.9102	53
54	0.3007	31.0785	659.5095	3.3252	103.3427	0.0322	0.0097	21.2207	54
55	0.2941	31.3727	675.3917	3.4000	106.6679	0.0319	0.0094	21.5280	55
60	0.2631	32.7490	753.7795	3.8001	124.4504	0.0305	0.0080	23.0169	60
65	0.2354	33.9803	830.0710	4.2473	144.3256	0.0294	0.0069	24.4280	65
70	0.2107	35.0821	903.8386	4.7471	166.5396	0.0285	0.0060	25.7635	70
75	0.1885	36.0678	974.7681	5.3058	191.3677	0.0277	0.0052	27.0260	75
80	0.1686	36.9498	1042.6394	5.9301	219.1176	0.0271	0.0046	28.2177	80
85	0.1509	37.7389	1107.3101	6.6280	250.1329	0.0265	0.0040	29.3414	85
90	0.1350	38.4449	1168.7019	7.4080	284.7981	0.0260	0.0035	30.3994	90
95	0.1208	39.0766	1226.7883	8.2797	323.5426	0.0256	0.0031	31.3945	95
100	0.1081	39.6417	1281.5847	9.2540	366.8465	0.0252	0.0027	32.3292	100

$I = 2.50\%$

n	(P/F)	(P/A)	(P/G)	(F/P)	(F/A)	(A/P)	(A/F)	(A/G)	n
1	0.9756	0.9756	0.0000	1.0250	1.0000	1.0250	1.0000	0.0000	1
2	0.9518	1.9274	0.9518	1.0506	2.0250	0.5188	0.4938	0.4938	2
3	0.9286	2.8560	2.8090	1.0769	3.0756	0.3501	0.3251	0.9835	3
4	0.9060	3.7620	5.5269	1.1038	4.1525	0.2658	0.2408	1.4691	4
5	0.8839	4.6458	9.0623	1.1314	5.2563	0.2152	0.1902	1.9506	5
6	0.8623	5.5081	13.3738	1.1597	6.3877	0.1815	0.1565	2.4280	6
7	0.8413	6.3494	18.4214	1.1887	7.5474	0.1575	0.1325	2.9013	7
8	0.8207	7.1701	24.1666	1.2184	8.7361	0.1395	0.1145	3.3704	8
9	0.8007	7.9709	30.5724	1.2489	9.9545	0.1255	0.1005	3.8355	9
10	0.7812	8.7521	37.6032	1.2801	11.2034	0.1143	0.0893	4.2965	10
11	0.7621	9.5142	45.2246	1.3121	12.4835	0.1051	0.0801	4.7534	11
12	0.7436	10.2578	53.4038	1.3449	13.7956	0.0975	0.0725	5.2062	12
13	0.7254	10.9832	62.1088	1.3785	15.1404	0.0910	0.0660	5.6549	13
14	0.7077	11.6909	71.3093	1.4130	16.5190	0.0855	0.0605	6.0995	14
15	0.6905	12.3814	80.9758	1.4483	17.9319	0.0808	0.0558	6.5401	15
16	0.6736	13.0550	91.0801	1.4845	19.3802	0.0766	0.0516	6.9766	16
17	0.6572	13.7122	101.5953	1.5216	20.8647	0.0729	0.0479	7.4091	17
18	0.6412	14.3534	112.4951	1.5597	22.3863	0.0697	0.0447	7.8375	18
19	0.6255	14.9789	123.7546	1.5987	23.9460	0.0668	0.0418	8.2619	19
20	0.6103	15.5892	135.3497	1.6386	25.5447	0.0641	0.0391	8.6823	20
21	0.5954	16.1845	147.2575	1.6796	27.1833	0.0618	0.0368	9.0986	21
22	0.5809	16.7654	159.4556	1.7216	28.8629	0.0596	0.0346	9.5110	22
23	0.5667	17.3321	171.9230	1.7646	30.5844	0.0577	0.0327	9.9193	23
24	0.5529	17.8850	184.6391	1.8087	32.3490	0.0559	0.0309	10.3237	24
25	0.5394	18.4244	197.5845	1.8539	34.1578	0.0543	0.0293	10.7241	25
26	0.5262	18.9506	210.7403	1.9003	36.0117	0.0528	0.0278	11.1205	26
27	0.5134	19.4640	224.0887	1.9478	37.9120	0.0514	0.0264	11.5130	27
28	0.5009	19.9649	237.6124	1.9965	39.8598	0.0501	0.0251	11.9015	28
29	0.4887	20.4535	251.2949	2.0464	41.8563	0.0489	0.0239	12.2861	29
30	0.4767	20.9303	265.1205	2.0976	43.9027	0.0478	0.0228	12.6668	30
31	0.4651	21.3954	279.0739	2.1500	46.0003	0.0467	0.0217	13.0436	31
32	0.4538	21.8492	293.1408	2.2038	48.1503	0.0458	0.0208	13.4166	32
33	0.4427	22.2919	307.3073	2.2589	50.3540	0.0449	0.0199	13.7856	33
34	0.4319	22.7238	321.5602	2.3153	52.6129	0.0440	0.0190	14.1508	34
35	0.4214	23.1452	335.8868	2.3732	54.9282	0.0432	0.0182	14.5122	35
36	0.4111	23.5563	350.2751	2.4325	57.3014	0.0425	0.0175	14.8697	36
37	0.4011	23.9573	364.7135	2.4933	59.7339	0.0417	0.0167	15.2235	37
38	0.3913	24.3486	379.1910	2.5557	62.2273	0.0411	0.0161	15.5734	38
39	0.3817	24.7303	393.6972	2.6196	64.7830	0.0404	0.0154	15.9196	39
40	0.3724	25.1028	408.2220	2.6851	67.4026	0.0398	0.0148	16.2620	40
41	0.3633	25.4661	422.7559	2.7522	70.0876	0.0393	0.0143	16.6007	41
42	0.3545	25.8206	437.2898	2.8210	72.8398	0.0387	0.0137	16.9357	42
43	0.3458	26.1664	451.8150	2.8915	75.6608	0.0382	0.0132	17.2670	43
44	0.3374	26.5038	466.3234	2.9638	78.5523	0.0377	0.0127	17.5946	44
45	0.3292	26.8330	480.8070	3.0379	81.5161	0.0373	0.0123	17.9185	45
46	0.3211	27.1542	495.2586	3.1139	84.5540	0.0368	0.0118	18.2388	46
47	0.3133	27.4675	509.6710	3.1917	87.6679	0.0364	0.0114	18.5554	47
48	0.3057	27.7732	524.0375	3.2715	90.8596	0.0360	0.0110	18.8685	48
49	0.2982	28.0714	538.3519	3.3533	94.1311	0.0356	0.0106	19.1780	49
50	0.2909	28.3623	552.6081	3.4371	97.4843	0.0353	0.0103	19.4839	50
51	0.2838	28.6462	566.8004	3.5230	100.9215	0.0349	0.0099	19.7863	51
52	0.2769	28.9231	580.9234	3.6111	104.4445	0.0346	0.0096	20.0851	52
53	0.2702	29.1932	594.9722	3.7014	108.0556	0.0343	0.0093	20.3805	53
54	0.2636	29.4568	608.9419	3.7939	111.7570	0.0339	0.0089	20.6724	54
55	0.2572	29.7140	622.8280	3.8888	115.5509	0.0337	0.0087	20.9608	55
60	0.2273	30.9087	690.8656	4.3998	135.9916	0.0324	0.0074	22.3518	60
65	0.2009	31.9646	756.2806	4.9780	159.1183	0.0313	0.0063	23.6600	65
70	0.1776	32.8979	818.7643	5.6321	185.2841	0.0304	0.0054	24.8881	70
75	0.1569	33.7227	878.1152	6.3722	214.8883	0.0297	0.0047	26.0393	75
80	0.1387	34.4518	934.2181	7.2096	248.3827	0.0290	0.0040	27.1167	80
85	0.1226	35.0962	987.0269	8.1570	286.2786	0.0285	0.0035	28.1235	85
90	0.1084	35.6658	1036.5499	9.2289	329.1543	0.0280	0.0030	29.0629	90
95	0.0958	36.1692	1082.8381	10.4416	377.6642	0.0276	0.0026	29.9382	95
100	0.0846	36.6141	1125.9747	11.8137	432.5487	0.0273	0.0023	30.7525	100

$I = 2.75\%$

n	(P/F)	(P/A)	(P/G)	(F/P)	(F/A)	(A/P)	(A/F)	(A/G)	n
1	0.9732	0.9732	0.0000	1.0275	1.0000	1.0275	1.0000	0.0000	1
2	0.9472	1.9204	0.9472	1.0558	2.0275	0.5207	0.4932	0.4932	2
3	0.9218	2.8423	2.7909	1.0848	3.0833	0.3518	0.3243	0.9819	3
4	0.8972	3.7394	5.4824	1.1146	4.1680	0.2674	0.2399	1.4661	4
5	0.8732	4.6126	8.9750	1.1453	5.2827	0.2168	0.1893	1.9458	5
6	0.8498	5.4624	13.2239	1.1768	6.4279	0.1831	0.1556	2.4209	6
7	0.8270	6.2894	18.1861	1.2091	7.6047	0.1590	0.1315	2.8916	7
8	0.8049	7.0943	23.8205	1.2424	8.8138	0.1410	0.1135	3.3577	8
9	0.7834	7.8777	30.0874	1.2765	10.0562	0.1269	0.0994	3.8193	9
10	0.7624	8.6401	36.9490	1.3117	11.3328	0.1157	0.0882	4.2765	10
11	0.7420	9.3821	44.3689	1.3477	12.6444	0.1066	0.0791	4.7291	11
12	0.7221	10.1042	52.3124	1.3848	13.9921	0.0990	0.0715	5.1773	12
13	0.7028	10.8070	60.7461	1.4229	15.3769	0.0925	0.0650	5.6210	13
14	0.6840	11.4910	69.6380	1.4620	16.7998	0.0870	0.0595	6.0602	14
15	0.6657	12.1567	78.9577	1.5022	18.2618	0.0823	0.0548	6.4950	15
16	0.6479	12.8046	88.6758	1.5435	19.7640	0.0781	0.0506	6.9253	16
17	0.6305	13.4351	98.7644	1.5860	21.3075	0.0744	0.0469	7.3512	17
18	0.6137	14.0488	109.1966	1.6296	22.8934	0.0712	0.0437	7.7727	18
19	0.5972	14.6460	119.9648	1.6744	24.5230	0.0683	0.0408	8.1897	19
20	0.5813	15.2273	130.9906	1.7204	26.1974	0.0657	0.0382	8.6024	20
21	0.5657	15.7929	142.3045	1.7677	27.9178	0.0633	0.0358	9.0106	21
22	0.5506	16.3435	153.8661	1.8164	29.6856	0.0612	0.0337	9.4145	22
23	0.5358	16.8793	165.6541	1.8663	31.5019	0.0592	0.0317	9.8140	23
24	0.5215	17.4008	177.6481	1.9176	33.3682	0.0575	0.0300	10.2092	24
25	0.5075	17.9083	189.8286	1.9704	35.2858	0.0558	0.0283	10.6000	25
26	0.4939	18.4023	202.1771	2.0245	37.2562	0.0543	0.0268	10.9865	26
27	0.4807	18.8830	214.6757	2.0802	39.2808	0.0530	0.0255	11.3687	27
28	0.4679	19.3508	227.3077	2.1374	41.3610	0.0517	0.0242	11.7467	28
29	0.4553	19.8062	240.0570	2.1962	43.4984	0.0505	0.0230	12.1203	29
30	0.4431	20.2493	252.9082	2.2566	45.6946	0.0494	0.0219	12.4897	30
31	0.4313	20.6806	265.8467	2.3187	47.9512	0.0484	0.0209	12.8549	31
32	0.4197	21.1003	278.8587	2.3824	50.2699	0.0474	0.0199	13.2158	32
33	0.4085	21.5088	291.9309	2.4479	52.6523	0.0465	0.0190	13.5726	33
34	0.3976	21.9064	305.0508	2.5153	55.1002	0.0456	0.0181	13.9252	34
35	0.3869	22.2933	318.2066	2.5844	57.6155	0.0449	0.0174	14.2736	35
36	0.3766	22.6699	331.3868	2.6555	60.1999	0.0441	0.0166	14.6179	36
37	0.3665	23.0364	344.5807	2.7285	62.8554	0.0434	0.0159	14.9581	37
38	0.3567	23.3931	357.7782	2.8036	65.5839	0.0427	0.0152	15.2942	38
39	0.3471	23.7402	370.9697	2.8807	68.3875	0.0421	0.0146	15.6262	39
40	0.3379	24.0781	384.1459	2.9599	71.2681	0.0415	0.0140	15.9542	40
41	0.3288	24.4069	397.2983	3.0413	74.2280	0.0410	0.0135	16.2781	41
42	0.3200	24.7269	410.4187	3.1249	77.2693	0.0404	0.0129	16.5981	42
43	0.3114	25.0384	423.4994	3.2108	80.3942	0.0399	0.0124	16.9140	43
44	0.3031	25.3415	436.5331	3.2991	83.6050	0.0395	0.0120	17.2260	44
45	0.2950	25.6365	449.5130	3.3899	86.9042	0.0390	0.0115	17.5341	45
46	0.2871	25.9236	462.4325	3.4831	90.2940	0.0386	0.0111	17.8383	46
47	0.2794	26.2030	475.2857	3.5789	93.7771	0.0382	0.0107	18.1386	47
48	0.2719	26.4749	488.0669	3.6773	97.3560	0.0378	0.0103	18.4351	48
49	0.2647	26.7396	500.7706	3.7784	101.0333	0.0374	0.0099	18.7277	49
50	0.2576	26.9972	513.3920	3.8823	104.8117	0.0370	0.0095	19.0165	50
51	0.2507	27.2479	525.9262	3.9891	108.6940	0.0367	0.0092	19.3016	51
52	0.2440	27.4918	538.3689	4.0988	112.6831	0.0364	0.0089	19.5829	52
53	0.2374	27.7293	550.7160	4.2115	116.7819	0.0361	0.0086	19.8605	53
54	0.2311	27.9604	562.9638	4.3273	120.9934	0.0358	0.0083	20.1344	54
55	0.2249	28.1853	575.1086	4.4463	125.3207	0.0355	0.0080	20.4046	55
60	0.1964	29.2227	634.1838	5.0923	148.8091	0.0342	0.0067	21.7018	60
65	0.1715	30.1285	690.2945	5.8320	175.7098	0.0332	0.0057	22.9117	65
70	0.1497	30.9194	743.2423	6.6793	206.5184	0.0323	0.0048	24.0381	70
75	0.1307	31.6100	792.9269	7.6496	241.8027	0.0316	0.0041	25.0847	75
80	0.1141	32.2129	839.3240	8.7609	282.2129	0.0310	0.0035	26.0555	80
85	0.0997	32.7394	882.4684	10.0336	328.4935	0.0305	0.0030	26.9543	85
90	0.0870	33.1992	922.4387	11.4912	381.4976	0.0301	0.0026	27.7850	90
95	0.0760	33.6006	959.3459	13.1605	442.2017	0.0298	0.0023	28.5515	95
100	0.0663	33.9510	993.3240	15.0724	511.7244	0.0295	0.0020	29.2575	100

$I = 3.00\%$

n	(P/F)	(P/A)	(P/G)	(F/P)	(F/A)	(A/P)	(A/F)	(A/G)	n
1	0.9709	0.9709	0.0000	1.0300	1.0000	1.0300	1.0000	0.0000	1
2	0.9426	1.9135	0.9426	1.0609	2.0300	0.5226	0.4926	0.4926	2
3	0.9151	2.8286	2.7729	1.0927	3.0909	0.3535	0.3235	0.9803	3
4	0.8885	3.7171	5.4383	1.1255	4.1836	0.2690	0.2390	1.4631	4
5	0.8626	4.5797	8.8888	1.1593	5.3091	0.2184	0.1884	1.9409	5
6	0.8375	5.4172	13.0762	1.1941	6.4684	0.1846	0.1546	2.4138	6
7	0.8131	6.2303	17.9547	1.2299	7.6625	0.1605	0.1305	2.8819	7
8	0.7894	7.0197	23.4806	1.2668	8.8923	0.1425	0.1125	3.3450	8
9	0.7664	7.7861	29.6119	1.3048	10.1591	0.1284	0.0984	3.8032	9
10	0.7441	8.5302	36.3088	1.3439	11.4639	0.1172	0.0872	4.2565	10
11	0.7224	9.2526	43.5330	1.3842	12.8078	0.1081	0.0781	4.7049	11
12	0.7014	9.9540	51.2482	1.4258	14.1920	0.1005	0.0705	5.1485	12
13	0.6810	10.6350	59.4196	1.4685	15.6178	0.0940	0.0640	5.5872	13
14	0.6611	11.2961	68.0141	1.5126	17.0863	0.0885	0.0585	6.0210	14
15	0.6419	11.9379	77.0002	1.5580	18.5989	0.0838	0.0538	6.4500	15
16	0.6232	12.5611	86.3477	1.6047	20.1569	0.0796	0.0496	6.8742	16
17	0.6050	13.1661	96.0280	1.6528	21.7616	0.0760	0.0460	7.2936	17
18	0.5874	13.7535	106.0137	1.7024	23.4144	0.0727	0.0427	7.7081	18
19	0.5703	14.3238	116.2788	1.7535	25.1169	0.0698	0.0398	8.1179	19
20	0.5537	14.8775	126.7987	1.8061	26.8704	0.0672	0.0372	8.5229	20
21	0.5375	15.4150	137.5496	1.8603	28.6765	0.0649	0.0349	8.9231	21
22	0.5219	15.9369	148.5094	1.9161	30.5368	0.0627	0.0327	9.3186	22
23	0.5067	16.4436	159.6566	1.9736	32.4529	0.0608	0.0308	9.7093	23
24	0.4919	16.9355	170.9711	2.0328	34.4265	0.0590	0.0290	10.0954	24
25	0.4776	17.4131	182.4336	2.0938	36.4593	0.0574	0.0274	10.4768	25
26	0.4637	17.8768	194.0260	2.1566	38.5530	0.0559	0.0259	10.8535	26
27	0.4502	18.3270	205.7309	2.2213	40.7096	0.0546	0.0246	11.2255	27
28	0.4371	18.7641	217.5320	2.2879	42.9309	0.0533	0.0233	11.5930	28
29	0.4243	19.1885	229.4137	2.3566	45.2189	0.0521	0.0221	11.9558	29
30	0.4120	19.6004	241.3613	2.4273	47.5754	0.0510	0.0210	12.3141	30
31	0.4000	20.0004	253.3609	2.5001	50.0027	0.0500	0.0200	12.6678	31
32	0.3883	20.3888	265.3993	2.5751	52.5028	0.0490	0.0190	13.0169	32
33	0.3770	20.7658	277.4642	2.6523	55.0778	0.0482	0.0182	13.3616	33
34	0.3660	21.1318	289.5437	2.7319	57.7302	0.0473	0.0173	13.7018	34
35	0.3554	21.4872	301.6267	2.8139	60.4621	0.0465	0.0165	14.0375	35
36	0.3450	21.8323	313.7028	2.8983	63.2759	0.0458	0.0158	14.3688	36
37	0.3350	22.1672	325.7622	2.9852	66.1742	0.0451	0.0151	14.6957	37
38	0.3252	22.4925	337.7956	3.0748	69.1594	0.0445	0.0145	15.0182	38
39	0.3158	22.8082	349.7942	3.1670	72.2342	0.0438	0.0138	15.3363	39
40	0.3066	23.1148	361.7499	3.2620	75.4013	0.0433	0.0133	15.6502	40
41	0.2976	23.4124	373.6551	3.3599	78.6633	0.0427	0.0127	15.9597	41
42	0.2890	23.7014	385.5024	3.4607	82.0232	0.0422	0.0122	16.2650	42
43	0.2805	23.9819	397.2852	3.5645	85.4839	0.0417	0.0117	16.5660	43
44	0.2724	24.2543	408.9972	3.6715	89.0484	0.0412	0.0112	16.8629	44
45	0.2644	24.5187	420.6325	3.7816	92.7199	0.0408	0.0108	17.1556	45
46	0.2567	24.7754	432.1856	3.8950	96.5015	0.0404	0.0104	17.4441	46
47	0.2493	25.0247	443.6515	4.0119	100.3965	0.0400	0.0100	17.7285	47
48	0.2420	25.2667	455.0255	4.1323	104.4084	0.0396	0.0096	18.0089	48
49	0.2350	25.5017	466.3031	4.2562	108.5406	0.0392	0.0092	18.2852	49
50	0.2281	25.7298	477.4803	4.3839	112.7969	0.0389	0.0089	18.5575	50
51	0.2215	25.9512	488.5535	4.5154	117.1808	0.0385	0.0085	18.8258	51
52	0.2150	26.1662	499.5191	4.6509	121.6962	0.0382	0.0082	19.0902	52
53	0.2088	26.3750	510.3742	4.7904	126.3471	0.0379	0.0079	19.3507	53
54	0.2027	26.5777	521.1157	4.9341	131.1375	0.0376	0.0076	19.6073	54
55	0.1968	26.7744	531.7411	5.0821	136.0716	0.0373	0.0073	19.8600	55
60	0.1697	27.6756	583.0526	5.8916	163.0534	0.0361	0.0061	21.0674	60
65	0.1464	28.4529	631.2010	6.8300	194.3328	0.0351	0.0051	22.1841	65
70	0.1263	29.1234	676.0869	7.9178	230.5941	0.0343	0.0043	23.2145	70
75	0.1089	29.7018	717.6978	9.1789	272.6309	0.0337	0.0037	24.1634	75
80	0.0940	30.2008	756.0865	10.6409	321.3630	0.0331	0.0031	25.0353	80
85	0.0811	30.6312	791.3529	12.3357	377.8570	0.0326	0.0026	25.8349	85
90	0.0699	31.0024	823.6302	14.3005	443.3489	0.0323	0.0023	26.5667	90
95	0.0603	31.3227	853.0742	16.5782	519.2720	0.0319	0.0019	27.2351	95
100	0.0520	31.5989	879.8540	19.2186	607.2877	0.0316	0.0016	27.8444	100

$I = 3.25\%$

n	(P/F)	(P/A)	(P/G)	(F/P)	(F/A)	(A/P)	(A/F)	(A/G)	n
1	0.9685	0.9685	0.0000	1.0325	1.0000	1.0325	1.0000	0.0000	1
2	0.9380	1.9066	0.9380	1.0661	2.0325	0.5245	0.4920	0.4920	2
3	0.9085	2.8151	2.7551	1.1007	3.0986	0.3552	0.3227	0.9787	3
4	0.8799	3.6950	5.3948	1.1365	4.1993	0.2706	0.2381	1.4600	4
5	0.8522	4.5472	8.8037	1.1734	5.3357	0.2199	0.1874	1.9361	5
6	0.8254	5.3726	12.9306	1.2115	6.5091	0.1861	0.1536	2.4068	6
7	0.7994	6.1720	17.7271	1.2509	7.7207	0.1620	0.1295	2.8722	7
8	0.7742	6.9462	23.1468	1.2916	8.9716	0.1440	0.1115	3.3323	8
9	0.7499	7.6961	29.1458	1.3336	10.2632	0.1299	0.0974	3.7871	9
10	0.7263	8.4224	35.6823	1.3769	11.5967	0.1187	0.0862	4.2366	10
11	0.7034	9.1258	42.7164	1.4216	12.9736	0.1096	0.0771	4.6808	11
12	0.6813	9.8071	50.2103	1.4678	14.3953	0.1020	0.0695	5.1198	12
13	0.6598	10.4669	58.1283	1.5156	15.8631	0.0955	0.0630	5.5535	13
14	0.6391	11.1060	66.4360	1.5648	17.3787	0.0900	0.0575	5.9820	14
15	0.6189	11.7249	75.1012	1.6157	18.9435	0.0853	0.0528	6.4053	15
16	0.5995	12.3244	84.0930	1.6682	20.5592	0.0811	0.0486	6.8233	16
17	0.5806	12.9049	93.3825	1.7224	22.2273	0.0775	0.0450	7.2362	17
18	0.5623	13.4673	102.9418	1.7784	23.9497	0.0743	0.0418	7.6439	18
19	0.5446	14.0119	112.7449	1.8362	25.7281	0.0714	0.0389	8.0464	19
20	0.5275	14.5393	122.7668	1.8958	27.5642	0.0688	0.0363	8.4438	20
21	0.5109	15.0502	132.9842	1.9575	29.4601	0.0664	0.0339	8.8360	21
22	0.4948	15.5450	143.3747	2.0211	31.4175	0.0643	0.0318	9.2232	22
23	0.4792	16.0242	153.9174	2.0868	33.4386	0.0624	0.0299	9.6053	23
24	0.4641	16.4883	164.5924	2.1546	35.5254	0.0606	0.0281	9.9823	24
25	0.4495	16.9379	175.3808	2.2246	37.6799	0.0590	0.0265	10.3544	25
26	0.4354	17.3732	186.2651	2.2969	39.9045	0.0576	0.0251	10.7214	26
27	0.4217	17.7949	197.2284	2.3715	42.2014	0.0562	0.0237	11.0834	27
28	0.4084	18.2033	208.2550	2.4486	44.5730	0.0549	0.0224	11.4405	28
29	0.3955	18.5988	219.3301	2.5282	47.0216	0.0538	0.0213	11.7927	29
30	0.3831	18.9819	230.4396	2.6104	49.5498	0.0527	0.0202	12.1400	30
31	0.3710	19.3529	241.5705	2.6952	52.1602	0.0517	0.0192	12.4824	31
32	0.3594	19.7123	252.7103	2.7828	54.8554	0.0507	0.0182	12.8199	32
33	0.3480	20.0603	263.8476	2.8732	57.6382	0.0498	0.0173	13.1527	33
34	0.3371	20.3974	274.9714	2.9666	60.5114	0.0490	0.0165	13.4807	34
35	0.3265	20.7239	286.0714	3.0630	63.4780	0.0483	0.0158	13.8039	35
36	0.3162	21.0401	297.1383	3.1626	66.5411	0.0475	0.0150	14.1225	36
37	0.3062	21.3463	308.1631	3.2654	69.7037	0.0468	0.0143	14.4363	37
38	0.2966	21.6429	319.1375	3.3715	72.9690	0.0462	0.0137	14.7456	38
39	0.2873	21.9302	330.0537	3.4811	76.3405	0.0456	0.0131	15.0502	39
40	0.2782	22.2084	340.9045	3.5942	79.8216	0.0450	0.0125	15.3502	40
41	0.2695	22.4779	351.6832	3.7110	83.4158	0.0445	0.0120	15.6457	41
42	0.2610	22.7389	362.3837	3.8316	87.1268	0.0440	0.0115	15.9367	42
43	0.2528	22.9917	373.0001	3.9561	90.9584	0.0435	0.0110	16.2233	43
44	0.2448	23.2365	383.5271	4.0847	94.9146	0.0430	0.0105	16.5054	44
45	0.2371	23.4736	393.9599	4.2175	98.9993	0.0426	0.0101	16.7831	45
46	0.2296	23.7032	404.2939	4.3545	103.2168	0.0422	0.0097	17.0565	46
47	0.2224	23.9256	414.5251	4.4961	107.5713	0.0418	0.0093	17.3256	47
48	0.2154	24.1411	424.6496	4.6422	112.0674	0.0414	0.0089	17.5903	48
49	0.2086	24.3497	434.6641	4.7931	116.7096	0.0411	0.0086	17.8509	49
50	0.2021	24.5518	444.5654	4.9488	121.5026	0.0407	0.0082	18.1073	50
51	0.1957	24.7475	454.3507	5.1097	126.4515	0.0404	0.0079	18.3595	51
52	0.1895	24.9370	464.0176	5.2757	131.5611	0.0401	0.0076	18.6076	52
53	0.1836	25.1206	473.5638	5.4472	136.8369	0.0398	0.0073	18.8516	53
54	0.1778	25.2984	482.9873	5.6242	142.2841	0.0395	0.0070	19.0916	54
55	0.1722	25.4706	492.2864	5.8070	147.9083	0.0393	0.0068	19.3276	55
60	0.1468	26.2537	536.8703	6.8140	178.8930	0.0381	0.0056	20.4494	60
65	0.1251	26.9210	578.2021	7.9957	215.2509	0.0371	0.0046	21.4777	65
70	0.1066	27.4897	616.2692	9.3822	257.9135	0.0364	0.0039	22.4182	70
75	0.0908	27.9744	651.1340	11.0092	307.9744	0.0357	0.0032	23.2761	75
80	0.0774	28.3874	682.9114	12.9183	366.7164	0.0352	0.0027	24.0569	80
85	0.0660	28.7394	711.7527	15.1585	435.6450	0.0348	0.0023	24.7658	85
90	0.0562	29.0394	737.8316	17.7871	516.5265	0.0344	0.0019	25.4080	90
95	0.0479	29.2950	761.3346	20.8716	611.4338	0.0341	0.0016	25.9885	95
100	0.0408	29.5129	782.4537	24.4910	722.7992	0.0339	0.0014	26.5123	100

$I = 3.50\%$

n	(P/F)	(P/A)	(P/G)	(F/P)	(F/A)	(A/P)	(A/F)	(A/G)	n
1	0.9662	0.9662	0.0000	1.0350	1.0000	1.0350	1.0000	0.0000	1
2	0.9335	1.8997	0.9335	1.0712	2.0350	0.5264	0.4914	0.4914	2
3	0.9019	2.8016	2.7374	1.1087	3.1062	0.3569	0.3219	0.9771	3
4	0.8714	3.6731	5.3517	1.1475	4.2149	0.2723	0.2373	1.4570	4
5	0.8420	4.5151	8.7196	1.1877	5.3625	0.2215	0.1865	1.9312	5
6	0.8135	5.3286	12.7871	1.2293	6.5502	0.1877	0.1527	2.3997	6
7	0.7860	6.1145	17.5031	1.2723	7.7794	0.1635	0.1285	2.8625	7
8	0.7594	6.8740	22.8189	1.3168	9.0517	0.1455	0.1105	3.3196	8
9	0.7337	7.6077	28.6888	1.3629	10.3685	0.1314	0.0964	3.7710	9
10	0.7089	8.3166	35.0691	1.4106	11.7314	0.1202	0.0852	4.2168	10
11	0.6849	9.0016	41.9185	1.4600	13.1420	0.1111	0.0761	4.6568	11
12	0.6618	9.6633	49.1981	1.5111	14.6020	0.1035	0.0685	5.0912	12
13	0.6394	10.3027	56.8710	1.5640	16.1130	0.0971	0.0621	5.5200	13
14	0.6178	10.9205	64.9021	1.6187	17.6770	0.0916	0.0566	5.9431	14
15	0.5969	11.5174	73.2586	1.6753	19.2957	0.0868	0.0518	6.3607	15
16	0.5767	12.0941	81.9092	1.7340	20.9710	0.0827	0.0477	6.7726	16
17	0.5572	12.6513	90.8245	1.7947	22.7050	0.0790	0.0440	7.1791	17
18	0.5384	13.1897	99.9766	1.8575	24.4997	0.0758	0.0408	7.5799	18
19	0.5202	13.7098	109.3394	1.9225	26.3572	0.0729	0.0379	7.9753	19
20	0.5026	14.2124	118.8882	1.9898	28.2797	0.0704	0.0354	8.3651	20
21	0.4856	14.6980	128.5996	2.0594	30.2695	0.0680	0.0330	8.7495	21
22	0.4692	15.1671	138.4517	2.1315	32.3289	0.0659	0.0309	9.1284	22
23	0.4533	15.6204	148.4240	2.2061	34.4604	0.0640	0.0290	9.5019	23
24	0.4380	16.0584	158.4970	2.2833	36.6665	0.0623	0.0273	9.8701	24
25	0.4231	16.4815	168.6526	2.3632	38.9499	0.0607	0.0257	10.2328	25
26	0.4088	16.8904	178.8735	2.4460	41.3131	0.0592	0.0242	10.5903	26
27	0.3950	17.2854	189.1438	2.5316	43.7591	0.0579	0.0229	10.9424	27
28	0.3817	17.6670	199.4485	2.6202	46.2906	0.0566	0.0216	11.2893	28
29	0.3687	18.0358	209.7734	2.7119	48.9108	0.0554	0.0204	11.6310	29
30	0.3563	18.3920	220.1055	2.8068	51.6227	0.0544	0.0194	11.9674	30
31	0.3442	18.7363	230.4324	2.9050	54.4295	0.0534	0.0184	12.2987	31
32	0.3326	19.0689	240.7427	3.0067	57.3345	0.0524	0.0174	12.6249	32
33	0.3213	19.3902	251.0257	3.1119	60.3412	0.0516	0.0166	12.9460	33
34	0.3105	19.7007	261.2714	3.2209	63.4532	0.0508	0.0158	13.2620	34
35	0.3000	20.0007	271.4706	3.3336	66.6740	0.0500	0.0150	13.5731	35
36	0.2898	20.2905	281.6147	3.4503	70.0076	0.0493	0.0143	13.8791	36
37	0.2800	20.5705	291.6959	3.5710	73.4579	0.0486	0.0136	14.1803	37
38	0.2706	20.8411	301.7067	3.6960	77.0289	0.0480	0.0130	14.4765	38
39	0.2614	21.1025	311.6403	3.8254	80.7249	0.0474	0.0124	14.7679	39
40	0.2526	21.3551	321.4907	3.9593	84.5503	0.0468	0.0118	15.0545	40
41	0.2440	21.5991	331.2519	4.0978	88.5095	0.0463	0.0113	15.3364	41
42	0.2358	21.8349	340.9189	4.2413	92.6074	0.0458	0.0108	15.6135	42
43	0.2278	22.0627	350.4867	4.3897	96.8486	0.0453	0.0103	15.8859	43
44	0.2201	22.2828	359.9511	4.5433	101.2383	0.0449	0.0099	16.1538	44
45	0.2127	22.4955	369.3081	4.7024	105.7817	0.0445	0.0095	16.4170	45
46	0.2055	22.7009	378.5542	4.8669	110.4840	0.0441	0.0091	16.6757	46
47	0.1985	22.8994	387.6861	5.0373	115.3510	0.0437	0.0087	16.9299	47
48	0.1918	23.0912	396.7010	5.2136	120.3883	0.0433	0.0083	17.1797	48
49	0.1853	23.2766	405.5964	5.3961	125.6018	0.0430	0.0080	17.4251	49
50	0.1791	23.4556	414.3700	5.5849	130.9979	0.0426	0.0076	17.6661	50
51	0.1730	23.6286	423.0199	5.7804	136.5828	0.0423	0.0073	17.9029	51
52	0.1671	23.7958	431.5445	5.9827	142.3632	0.0420	0.0070	18.1353	52
53	0.1615	23.9573	439.9422	6.1921	148.3459	0.0417	0.0067	18.3636	53
54	0.1560	24.1133	448.2121	6.4088	154.5381	0.0415	0.0065	18.5878	54
55	0.1508	24.2641	456.3530	6.6331	160.9469	0.0412	0.0062	18.8078	55
60	0.1269	24.9447	495.1050	7.8781	196.5169	0.0401	0.0051	19.8481	60
65	0.1069	25.5178	530.5987	9.3567	238.7629	0.0392	0.0042	20.7932	65
70	0.0900	26.0004	562.8962	11.1128	288.9379	0.0385	0.0035	21.6495	70
75	0.0758	26.4067	592.1213	13.1986	348.5300	0.0379	0.0029	22.4232	75
80	0.0638	26.7488	618.4385	15.6757	419.3068	0.0374	0.0024	23.1203	80
85	0.0537	27.0368	642.0370	18.6179	503.3674	0.0370	0.0020	23.7468	85
90	0.0452	27.2793	663.1189	22.1122	603.2050	0.0367	0.0017	24.3085	90
95	0.0381	27.4835	681.8902	26.2623	721.7808	0.0364	0.0014	24.8109	95
100	0.0321	27.6554	698.5547	31.1914	862.6117	0.0362	0.0012	25.2592	100

$I = 3.75\%$

n	(P/F)	(P/A)	(P/G)	(F/P)	(F/A)	(A/P)	(A/F)	(A/G)	n
1	0.9639	0.9639	0.0000	1.0375	1.0000	1.0375	1.0000	0.0000	1
2	0.9290	1.8929	0.9290	1.0764	2.0375	0.5283	0.4908	0.4908	2
3	0.8954	2.7883	2.7199	1.1168	3.1139	0.3586	0.3211	0.9755	3
4	0.8631	3.6514	5.3091	1.1587	4.2307	0.2739	0.2364	1.4540	4
5	0.8319	4.4833	8.6366	1.2021	5.3893	0.2231	0.1856	1.9264	5
6	0.8018	5.2851	12.6457	1.2472	6.5914	0.1892	0.1517	2.3927	6
7	0.7728	6.0579	17.2826	1.2939	7.8386	0.1651	0.1276	2.8529	7
8	0.7449	6.8028	22.4969	1.3425	9.1326	0.1470	0.1095	3.3070	8
9	0.7180	7.5208	28.2407	1.3928	10.4750	0.1330	0.0955	3.7550	9
10	0.6920	8.2128	34.4689	1.4450	11.8678	0.1218	0.0843	4.1970	10
11	0.6670	8.8798	41.1389	1.4992	13.3129	0.1126	0.0751	4.6329	11
12	0.6429	9.5227	48.2108	1.5555	14.8121	0.1050	0.0675	5.0627	12
13	0.6197	10.1424	55.6468	1.6138	16.3676	0.0986	0.0611	5.4866	13
14	0.5973	10.7396	63.4112	1.6743	17.9814	0.0931	0.0556	5.9044	14
15	0.5757	11.3153	71.4707	1.7371	19.6557	0.0884	0.0509	6.3163	15
16	0.5549	11.8702	79.7937	1.8022	21.3927	0.0842	0.0467	6.7222	16
17	0.5348	12.4050	88.3507	1.8698	23.1950	0.0806	0.0431	7.1222	17
18	0.5155	12.9205	97.1139	1.9399	25.0648	0.0774	0.0399	7.5163	18
19	0.4969	13.4173	106.0572	2.0127	27.0047	0.0745	0.0370	7.9045	19
20	0.4789	13.8962	115.1562	2.0882	29.0174	0.0720	0.0345	8.2869	20
21	0.4616	14.3578	124.3879	2.1665	31.1055	0.0696	0.0321	8.6634	21
22	0.4449	14.8027	133.7307	2.2477	33.2720	0.0676	0.0301	9.0342	22
23	0.4288	15.2315	143.1647	2.3320	35.5197	0.0657	0.0282	9.3993	23
24	0.4133	15.6448	152.6711	2.4194	37.8517	0.0639	0.0264	9.7586	24
25	0.3984	16.0432	162.2322	2.5102	40.2711	0.0623	0.0248	10.1122	25
26	0.3840	16.4272	171.8317	2.6043	42.7813	0.0609	0.0234	10.4602	26
27	0.3701	16.7973	181.4544	2.7020	45.3856	0.0595	0.0220	10.8026	27
28	0.3567	17.1540	191.0859	2.8033	48.0875	0.0583	0.0208	11.1394	28
29	0.3438	17.4978	200.7132	2.9084	50.8908	0.0571	0.0196	11.4707	29
30	0.3314	17.8292	210.3239	3.0175	53.7992	0.0561	0.0186	11.7966	30
31	0.3194	18.1487	219.9066	3.1306	56.8167	0.0551	0.0176	12.1170	31
32	0.3079	18.4565	229.4509	3.2480	59.9473	0.0542	0.0167	12.4319	32
33	0.2968	18.7533	238.9469	3.3698	63.1954	0.0533	0.0158	12.7416	33
34	0.2860	19.0393	248.3858	3.4962	66.5652	0.0525	0.0150	13.0459	34
35	0.2757	19.3150	257.7591	3.6273	70.0614	0.0518	0.0143	13.3450	35
36	0.2657	19.5807	267.0594	3.7633	73.6887	0.0511	0.0136	13.6389	36
37	0.2561	19.8369	276.2797	3.9045	77.4520	0.0504	0.0129	13.9276	37
38	0.2469	20.0837	285.4135	4.0509	81.3565	0.0498	0.0123	14.2112	38
39	0.2379	20.3217	294.4552	4.2028	85.4073	0.0492	0.0117	14.4897	39
40	0.2293	20.5510	303.3993	4.3604	89.6101	0.0487	0.0112	14.7632	40
41	0.2210	20.7720	312.2413	4.5239	93.9705	0.0481	0.0106	15.0318	41
42	0.2131	20.9851	320.9767	4.6935	98.4944	0.0477	0.0102	15.2955	42
43	0.2054	21.1905	329.6017	4.8695	103.1879	0.0472	0.0097	15.5543	43
44	0.1979	21.3884	338.1129	5.0522	108.0575	0.0468	0.0093	15.8082	44
45	0.1908	21.5792	346.5073	5.2416	113.1096	0.0463	0.0088	16.0575	45
46	0.1839	21.7631	354.7821	5.4382	118.3512	0.0459	0.0084	16.3020	46
47	0.1772	21.9403	362.9351	5.6421	123.7894	0.0456	0.0081	16.5419	47
48	0.1708	22.1111	370.9643	5.8537	129.4315	0.0452	0.0077	16.7773	48
49	0.1647	22.2758	378.8679	6.0732	135.2852	0.0449	0.0074	17.0081	49
50	0.1587	22.4345	386.6445	6.3009	141.3584	0.0446	0.0071	17.2344	50
51	0.1530	22.5875	394.2930	6.5372	147.6593	0.0443	0.0068	17.4563	51
52	0.1474	22.7349	401.8125	6.7824	154.1965	0.0440	0.0065	17.6738	52
53	0.1421	22.8770	409.2023	7.0367	160.9789	0.0437	0.0062	17.8870	53
54	0.1370	23.0140	416.4620	7.3006	168.0156	0.0435	0.0060	18.0960	54
55	0.1320	23.1460	423.5913	7.5744	175.3162	0.0432	0.0057	18.3008	55
60	0.1098	23.7379	457.2860	9.1051	216.1369	0.0421	0.0046	19.2640	60
65	0.0914	24.2303	487.7779	10.9453	265.2074	0.0413	0.0038	20.1309	65
70	0.0760	24.6399	515.1914	13.1573	324.1952	0.0406	0.0031	20.9088	70
75	0.0632	24.9807	539.6998	15.8164	395.1043	0.0400	0.0025	21.6047	75
80	0.0526	25.2641	561.5051	19.0129	480.3441	0.0396	0.0021	22.2254	80
85	0.0438	25.4999	580.8234	22.8554	582.8109	0.0392	0.0017	22.7775	85
90	0.0364	25.6961	597.8747	27.4745	705.9861	0.0389	0.0014	23.2672	90
95	0.0303	25.8592	612.8752	33.0271	854.0551	0.0387	0.0012	23.7004	95
100	0.0252	25.9950	626.0325	39.7018	1032.0488	0.0385	0.0010	24.0828	100

$I = 4.00\%$

n	(P/F)	(P/A)	(P/G)	(F/P)	(F/A)	(A/P)	(A/F)	(A/G)	n
1	0.9615	0.9615	0.0000	1.0400	1.0000	1.0400	1.0000	0.0000	1
2	0.9246	1.8861	0.9246	1.0816	2.0400	0.5302	0.4902	0.4902	2
3	0.8890	2.7751	2.7025	1.1249	3.1216	0.3603	0.3203	0.9739	3
4	0.8548	3.6299	5.2670	1.1699	4.2465	0.2755	0.2355	1.4510	4
5	0.8219	4.4518	8.5547	1.2167	5.4163	0.2246	0.1846	1.9216	5
6	0.7903	5.2421	12.5062	1.2653	6.6330	0.1908	0.1508	2.3857	6
7	0.7599	6.0021	17.0657	1.3159	7.8983	0.1666	0.1266	2.8433	7
8	0.7307	6.7327	22.1806	1.3686	9.2142	0.1485	0.1085	3.2944	8
9	0.7026	7.4353	27.8013	1.4233	10.5828	0.1345	0.0945	3.7391	9
10	0.6756	8.1109	33.8814	1.4802	12.0061	0.1233	0.0833	4.1773	10
11	0.6496	8.7605	40.3772	1.5395	13.4864	0.1141	0.0741	4.6090	11
12	0.6246	9.3851	47.2477	1.6010	15.0258	0.1066	0.0666	5.0343	12
13	0.6006	9.9856	54.4546	1.6651	16.6268	0.1001	0.0601	5.4533	13
14	0.5775	10.5631	61.9618	1.7317	18.2919	0.0947	0.0547	5.8659	14
15	0.5553	11.1184	69.7355	1.8009	20.0236	0.0899	0.0499	6.2721	15
16	0.5339	11.6523	77.7441	1.8730	21.8245	0.0858	0.0458	6.6720	16
17	0.5134	12.1657	85.9581	1.9479	23.6975	0.0822	0.0422	7.0656	17
18	0.4936	12.6593	94.3498	2.0258	25.6454	0.0790	0.0390	7.4530	18
19	0.4746	13.1339	102.8933	2.1068	27.6712	0.0761	0.0361	7.8342	19
20	0.4564	13.5903	111.5647	2.1911	29.7781	0.0736	0.0336	8.2091	20
21	0.4388	14.0292	120.3414	2.2788	31.9692	0.0713	0.0313	8.5779	21
22	0.4220	14.4511	129.2024	2.3699	34.2480	0.0692	0.0292	8.9407	22
23	0.4057	14.8568	138.1284	2.4647	36.6179	0.0673	0.0273	9.2973	23
24	0.3901	15.2470	147.1012	2.5633	39.0826	0.0656	0.0256	9.6479	24
25	0.3751	15.6221	156.1040	2.6658	41.6459	0.0640	0.0240	9.9925	25
26	0.3607	15.9828	165.1212	2.7725	44.3117	0.0626	0.0226	10.3312	26
27	0.3468	16.3296	174.1385	2.8834	47.0842	0.0612	0.0212	10.6640	27
28	0.3335	16.6631	183.1424	2.9987	49.9676	0.0600	0.0200	10.9909	28
29	0.3207	16.9837	192.1206	3.1187	52.9663	0.0589	0.0189	11.3120	29
30	0.3083	17.2920	201.0618	3.2434	56.0849	0.0578	0.0178	11.6274	30
31	0.2965	17.5885	209.9556	3.3731	59.3283	0.0569	0.0169	11.9371	31
32	0.2851	17.8736	218.7924	3.5081	62.7015	0.0559	0.0159	12.2411	32
33	0.2741	18.1476	227.5634	3.6484	66.2095	0.0551	0.0151	12.5396	33
34	0.2636	18.4112	236.2607	3.7943	69.8579	0.0543	0.0143	12.8324	34
35	0.2534	18.6646	244.8768	3.9461	73.6522	0.0536	0.0136	13.1198	35
36	0.2437	18.9083	253.4052	4.1039	77.5983	0.0529	0.0129	13.4018	36
37	0.2343	19.1426	261.8399	4.2681	81.7022	0.0522	0.0122	13.6784	37
38	0.2253	19.3679	270.1754	4.4388	85.9703	0.0516	0.0116	13.9497	38
39	0.2166	19.5845	278.4070	4.6164	90.4091	0.0511	0.0111	14.2157	39
40	0.2083	19.7928	286.5303	4.8010	95.0255	0.0505	0.0105	14.4765	40
41	0.2003	19.9931	294.5414	4.9931	99.8265	0.0500	0.0100	14.7322	41
42	0.1926	20.1856	302.4370	5.1928	104.8196	0.0495	0.0095	14.9828	42
43	0.1852	20.3708	310.2141	5.4005	110.0124	0.0491	0.0091	15.2284	43
44	0.1780	20.5488	317.8700	5.6165	115.4129	0.0487	0.0087	15.4690	44
45	0.1712	20.7200	325.4028	5.8412	121.0294	0.0483	0.0083	15.7047	45
46	0.1646	20.8847	332.8104	6.0748	126.8706	0.0479	0.0079	15.9356	46
47	0.1583	21.0429	340.0914	6.3178	132.9454	0.0475	0.0075	16.1618	47
48	0.1522	21.1951	347.2446	6.5705	139.2632	0.0472	0.0072	16.3832	48
49	0.1463	21.3415	354.2689	6.8333	145.8337	0.0469	0.0069	16.6000	49
50	0.1407	21.4822	361.1638	7.1067	152.6671	0.0466	0.0066	16.8122	50
51	0.1353	21.6175	367.9289	7.3910	159.7738	0.0463	0.0063	17.0200	51
52	0.1301	21.7476	374.5638	7.6866	167.1647	0.0460	0.0060	17.2232	52
53	0.1251	21.8727	381.0686	7.9941	174.8513	0.0457	0.0057	17.4221	53
54	0.1203	21.9930	387.4436	8.3138	182.8454	0.0455	0.0055	17.6167	54
55	0.1157	22.1086	393.6890	8.6464	191.1592	0.0452	0.0052	17.8070	55
60	0.0951	22.6235	422.9966	10.5196	237.9907	0.0442	0.0042	18.6972	60
65	0.0781	23.0467	449.2014	12.7987	294.9684	0.0434	0.0034	19.4909	65
70	0.0642	23.3945	472.4789	15.5716	364.2905	0.0427	0.0027	20.1961	70
75	0.0528	23.6804	493.0408	18.9453	448.6314	0.0422	0.0022	20.8206	75
80	0.0434	23.9154	511.1161	23.0498	551.2450	0.0418	0.0018	21.3718	80
85	0.0357	24.1085	526.9384	28.0436	676.0901	0.0415	0.0015	21.8569	85
90	0.0293	24.2673	540.7369	34.1193	827.9833	0.0412	0.0012	22.2826	90
95	0.0241	24.3978	552.7307	41.5114	1012.7846	0.0410	0.0010	22.6550	95
100	0.0198	24.5050	563.1249	50.5049	1237.6237	0.0408	0.0008	22.9800	100

$I = 4.25\%$

n	(P/F)	(P/A)	(P/G)	(F/P)	(F/A)	(A/P)	(A/F)	(A/G)	n
1	0.9592	0.9592	0.0000	1.0425	1.0000	1.0425	1.0000	0.0000	1
2	0.9201	1.8794	0.9201	1.0868	2.0425	0.5321	0.4896	0.4896	2
3	0.8826	2.7620	2.6854	1.1330	3.1293	0.3621	0.3196	0.9723	3
4	0.8466	3.6086	5.2253	1.1811	4.2623	0.2771	0.2346	1.4480	4
5	0.8121	4.4207	8.4737	1.2313	5.4434	0.2262	0.1837	1.9168	5
6	0.7790	5.1997	12.3688	1.2837	6.6748	0.1923	0.1498	2.3787	6
7	0.7473	5.9470	16.8523	1.3382	7.9585	0.1682	0.1257	2.8338	7
8	0.7168	6.6638	21.8698	1.3951	9.2967	0.1501	0.1076	3.2819	8
9	0.6876	7.3513	27.3704	1.4544	10.6918	0.1360	0.0935	3.7232	9
10	0.6595	8.0109	33.3062	1.5162	12.1462	0.1248	0.0823	4.1576	10
11	0.6326	8.6435	39.6327	1.5807	13.6624	0.1157	0.0732	4.5852	11
12	0.6069	9.2504	46.3081	1.6478	15.2431	0.1081	0.0656	5.0061	12
13	0.5821	9.8325	53.2936	1.7179	16.8909	0.1017	0.0592	5.4201	13
14	0.5584	10.3909	60.5526	1.7909	18.6088	0.0962	0.0537	5.8275	14
15	0.5356	10.9265	68.0513	1.8670	20.3997	0.0915	0.0490	6.2281	15
16	0.5138	11.4403	75.7581	1.9463	22.2666	0.0874	0.0449	6.6220	16
17	0.4928	11.9332	83.6436	2.0291	24.2130	0.0838	0.0413	7.0093	17
18	0.4727	12.4059	91.6803	2.1153	26.2420	0.0806	0.0381	7.3901	18
19	0.4535	12.8594	99.8429	2.2052	28.3573	0.0778	0.0353	7.7642	19
20	0.4350	13.2944	108.1077	2.2989	30.5625	0.0752	0.0327	8.1318	20
21	0.4173	13.7116	116.4528	2.3966	32.8614	0.0729	0.0304	8.4930	21
22	0.4002	14.1119	124.8580	2.4985	35.2580	0.0709	0.0284	8.8477	22
23	0.3839	14.4958	133.3044	2.6047	37.7565	0.0690	0.0265	9.1961	23
24	0.3683	14.8641	141.7748	2.7153	40.3611	0.0673	0.0248	9.5381	24
25	0.3533	15.2173	150.2531	2.8308	43.0765	0.0657	0.0232	9.8738	25
26	0.3389	15.5562	158.7246	2.9511	45.9072	0.0643	0.0218	10.2033	26
27	0.3250	15.8812	167.1758	3.0765	48.8583	0.0630	0.0205	10.5266	27
28	0.3118	16.1930	175.5943	3.2072	51.9348	0.0618	0.0193	10.8438	28
29	0.2991	16.4921	183.9687	3.3435	55.1420	0.0606	0.0181	11.1549	29
30	0.2869	16.7790	192.2886	3.4856	58.4855	0.0596	0.0171	11.4601	30
31	0.2752	17.0542	200.5444	3.6338	61.9712	0.0586	0.0161	11.7592	31
32	0.2640	17.3182	208.7277	3.7882	65.6049	0.0577	0.0152	12.0525	32
33	0.2532	17.5714	216.8306	3.9492	69.3931	0.0569	0.0144	12.3400	33
34	0.2429	17.8143	224.8461	4.1171	73.3424	0.0561	0.0136	12.6217	34
35	0.2330	18.0473	232.7677	4.2920	77.4594	0.0554	0.0129	12.8977	35
36	0.2235	18.2708	240.5899	4.4744	81.7514	0.0547	0.0122	13.1680	36
37	0.2144	18.4852	248.3077	4.6646	86.2259	0.0541	0.0116	13.4328	37
38	0.2056	18.6908	255.9164	4.8628	90.8905	0.0535	0.0110	13.6921	38
39	0.1973	18.8881	263.4122	5.0695	95.7533	0.0529	0.0104	13.9460	39
40	0.1892	19.0773	270.7916	5.2850	100.8228	0.0524	0.0099	14.1945	40
41	0.1815	19.2588	278.0517	5.5096	106.1078	0.0519	0.0094	14.4377	41
42	0.1741	19.4329	285.1899	5.7437	111.6174	0.0515	0.0090	14.6756	42
43	0.1670	19.5999	292.2041	5.9878	117.3611	0.0510	0.0085	14.9085	43
44	0.1602	19.7601	299.0925	6.2423	123.3490	0.0506	0.0081	15.1362	44
45	0.1537	19.9137	305.8538	6.5076	129.5913	0.0502	0.0077	15.3589	45
46	0.1474	20.0611	312.4869	6.7842	136.0989	0.0498	0.0073	15.5767	46
47	0.1414	20.2025	318.9909	7.0725	142.8831	0.0495	0.0070	15.7896	47
48	0.1356	20.3382	325.3654	7.3731	149.9557	0.0492	0.0067	15.9978	48
49	0.1301	20.4683	331.6101	7.6865	157.3288	0.0489	0.0064	16.2012	49
50	0.1248	20.5931	337.7251	8.0131	165.0153	0.0486	0.0061	16.3999	50
51	0.1197	20.7128	343.7105	8.3537	173.0284	0.0483	0.0058	16.5941	51
52	0.1148	20.8276	349.5666	8.7087	181.3821	0.0480	0.0055	16.7838	52
53	0.1101	20.9377	355.2942	9.0789	190.0909	0.0478	0.0053	16.9691	53
54	0.1057	21.0434	360.8940	9.4647	199.1697	0.0475	0.0050	17.1500	54
55	0.1013	21.1447	366.3668	9.8670	208.6344	0.0473	0.0048	17.3266	55
60	0.0823	21.5928	391.8674	12.1497	262.3447	0.0463	0.0038	18.1481	60
65	0.0668	21.9566	414.3963	14.9604	328.4808	0.0455	0.0030	18.8734	65
70	0.0543	22.2521	434.1698	18.4215	409.9171	0.0449	0.0024	19.5114	70
75	0.0441	22.4921	451.4282	22.6832	510.1993	0.0445	0.0020	20.0705	75
80	0.0358	22.6870	466.4185	27.9309	633.6685	0.0441	0.0016	20.5588	80
85	0.0291	22.8453	479.3838	34.3926	785.7090	0.0438	0.0013	20.9839	85
90	0.0236	22.9738	490.5558	42.3493	972.9235	0.0435	0.0010	21.3528	90
95	0.0192	23.0782	500.1508	52.1466	1203.4496	0.0433	0.0008	21.6720	95
100	0.0156	23.1630	508.3669	64.2105	1487.3070	0.0432	0.0007	21.9474	100

ENGINEERING ECONOMIC ANALYSIS

$$I = 4.50\%$$

n	(P/F)	(P/A)	(P/G)	(F/P)	(F/A)	(A/P)	(A/F)	(A/G)	n
1	0.9569	0.9569	0.0000	1.0450	1.0000	1.0450	1.0000	0.0000	1
2	0.9157	1.8727	0.9157	1.0920	2.0450	0.5340	0.4890	0.4890	2
3	0.8763	2.7490	2.6683	1.1412	3.1370	0.3638	0.3188	0.9707	3
4	0.8386	3.5875	5.1840	1.1925	4.2782	0.2787	0.2337	1.4450	4
5	0.8025	4.3900	8.3938	1.2462	5.4707	0.2278	0.1828	1.9120	5
6	0.7679	5.1579	12.2333	1.3023	6.7169	0.1939	0.1489	2.3718	6
7	0.7348	5.8927	16.6423	1.3609	8.0192	0.1697	0.1247	2.8242	7
8	0.7032	6.5959	21.5646	1.4221	9.3800	0.1516	0.1066	3.2694	8
9	0.6729	7.2688	26.9478	1.4861	10.8021	0.1376	0.0926	3.7073	9
10	0.6439	7.9127	32.7431	1.5530	12.2882	0.1264	0.0814	4.1380	10
11	0.6162	8.5289	38.9051	1.6229	13.8412	0.1172	0.0722	4.5616	11
12	0.5897	9.1186	45.3914	1.6959	15.4640	0.1097	0.0647	4.9779	12
13	0.5643	9.6829	52.1627	1.7722	17.1599	0.1033	0.0583	5.3871	13
14	0.5400	10.2228	59.1823	1.8519	18.9321	0.0978	0.0528	5.7892	14
15	0.5167	10.7395	66.4164	1.9353	20.7841	0.0931	0.0481	6.1843	15
16	0.4945	11.2340	73.8335	2.0224	22.7193	0.0890	0.0440	6.5723	16
17	0.4732	11.7072	81.4043	2.1134	24.7417	0.0854	0.0404	6.9534	17
18	0.4528	12.1600	89.1019	2.2085	26.8551	0.0822	0.0372	7.3275	18
19	0.4333	12.5933	96.9013	2.3079	29.0636	0.0794	0.0344	7.6947	19
20	0.4146	13.0079	104.7795	2.4117	31.3714	0.0769	0.0319	8.0550	20
21	0.3968	13.4047	112.7153	2.5202	33.7831	0.0746	0.0296	8.4086	21
22	0.3797	13.7844	120.6890	2.6337	36.3034	0.0725	0.0275	8.7555	22
23	0.3634	14.1478	128.6827	2.7522	38.9370	0.0707	0.0257	9.0956	23
24	0.3477	14.4955	136.6799	2.8760	41.6892	0.0690	0.0240	9.4291	24
25	0.3327	14.8282	144.6654	3.0054	44.5652	0.0674	0.0224	9.7561	25
26	0.3184	15.1466	152.6255	3.1407	47.5706	0.0660	0.0210	10.0765	26
27	0.3047	15.4513	160.5475	3.2820	50.7113	0.0647	0.0197	10.3905	27
28	0.2916	15.7429	168.4199	3.4297	53.9933	0.0635	0.0185	10.6982	28
29	0.2790	16.0219	176.2323	3.5840	57.4230	0.0624	0.0174	10.9995	29
30	0.2670	16.2889	183.9753	3.7453	61.0071	0.0614	0.0164	11.2945	30
31	0.2555	16.5444	191.6404	3.9139	64.7524	0.0604	0.0154	11.5834	31
32	0.2445	16.7889	199.2199	4.0900	68.6662	0.0596	0.0146	11.8662	32
33	0.2340	17.0229	206.7069	4.2740	72.7562	0.0587	0.0137	12.1429	33
34	0.2239	17.2468	214.0955	4.4664	77.0303	0.0580	0.0130	12.4137	34
35	0.2143	17.4610	221.3802	4.6673	81.4966	0.0573	0.0123	12.6785	35
36	0.2050	17.6660	228.5561	4.8774	86.1640	0.0566	0.0116	12.9376	36
37	0.1962	17.8622	235.6193	5.0969	91.0413	0.0560	0.0110	13.1909	37
38	0.1878	18.0500	242.5661	5.3262	96.1382	0.0554	0.0104	13.4386	38
39	0.1797	18.2297	249.3934	5.5659	101.4644	0.0549	0.0099	13.6806	39
40	0.1719	18.4016	256.0986	5.8164	107.0303	0.0543	0.0093	13.9172	40
41	0.1645	18.5661	262.6796	6.0781	112.8467	0.0539	0.0089	14.1483	41
42	0.1574	18.7235	269.1346	6.3516	118.9248	0.0534	0.0084	14.3741	42
43	0.1507	18.8742	275.4624	6.6374	125.2764	0.0530	0.0080	14.5946	43
44	0.1442	19.0184	281.6618	6.9361	131.9138	0.0526	0.0076	14.8100	44
45	0.1380	19.1563	287.7322	7.2482	138.8500	0.0522	0.0072	15.0202	45
46	0.1320	19.2884	293.6733	7.5744	146.0982	0.0518	0.0068	15.2254	46
47	0.1263	19.4147	299.4848	7.9153	153.6726	0.0515	0.0065	15.4257	47
48	0.1209	19.5356	305.1670	8.2715	161.5879	0.0512	0.0062	15.6211	48
49	0.1157	19.6513	310.7202	8.6437	169.8594	0.0509	0.0059	15.8117	49
50	0.1107	19.7620	316.1450	9.0326	178.5030	0.0506	0.0056	15.9976	50
51	0.1059	19.8680	321.4421	9.4391	187.5357	0.0503	0.0053	16.1789	51
52	0.1014	19.9693	326.6125	9.8639	196.9748	0.0501	0.0051	16.3557	52
53	0.0970	20.0663	331.6573	10.3077	206.8386	0.0498	0.0048	16.5280	53
54	0.0928	20.1592	336.5776	10.7716	217.1464	0.0496	0.0046	16.6960	54
55	0.0888	20.2480	341.3749	11.2563	227.9180	0.0494	0.0044	16.8597	55
60	0.0713	20.6380	363.5707	14.0274	289.4980	0.0485	0.0035	17.6165	60
65	0.0572	20.9510	382.9465	17.4807	366.2378	0.0477	0.0027	18.2782	65
70	0.0459	21.2021	399.7503	21.7841	461.8697	0.0472	0.0022	18.8543	70
75	0.0368	21.4036	414.2422	27.1470	581.0444	0.0467	0.0017	19.3538	75
80	0.0296	21.5653	426.6797	33.8301	729.5577	0.0464	0.0014	19.7854	80
85	0.0237	21.6951	437.3091	42.1585	914.6323	0.0461	0.0011	20.1570	85
90	0.0190	21.7992	446.3592	52.5371	1145.2690	0.0459	0.0009	20.4759	90
95	0.0153	21.8828	454.0394	65.4708	1432.6843	0.0457	0.0007	20.7487	95
100	0.0123	21.9499	460.5376	81.5885	1790.8560	0.0456	0.0006	20.9814	100

$I = 4.75\%$

n	(P/F)	(P/A)	(P/G)	(F/P)	(F/A)	(A/P)	(A/F)	(A/G)	n
1	0.9547	0.9547	0.0000	1.0475	1.0000	1.0475	1.0000	0.0000	1
2	0.9114	1.8660	0.9114	1.0973	2.0475	0.5359	0.4884	0.4884	2
3	0.8700	2.7361	2.6514	1.1494	3.1448	0.3655	0.3180	0.9691	3
4	0.8306	3.5666	5.1432	1.2040	4.2941	0.2804	0.2329	1.4420	4
5	0.7929	4.3596	8.3149	1.2612	5.4981	0.2294	0.1819	1.9073	5
6	0.7570	5.1165	12.0997	1.3211	6.7593	0.1954	0.1479	2.3648	6
7	0.7226	5.8392	16.4355	1.3838	8.0803	0.1713	0.1238	2.8147	7
8	0.6899	6.5290	21.2646	1.4495	9.4641	0.1532	0.1057	3.2569	8
9	0.6586	7.1876	26.5333	1.5184	10.9137	0.1391	0.0916	3.6915	9
10	0.6287	7.8163	32.1918	1.5905	12.4321	0.1279	0.0804	4.1185	10
11	0.6002	8.4166	38.1940	1.6661	14.0226	0.1188	0.0713	4.5380	11
12	0.5730	8.9896	44.4969	1.7452	15.6887	0.1112	0.0637	4.9498	12
13	0.5470	9.5366	51.0611	1.8281	17.4339	0.1049	0.0574	5.3542	13
14	0.5222	10.0588	57.8498	1.9149	19.2620	0.0994	0.0519	5.7512	14
15	0.4985	10.5573	64.8292	2.0059	21.1770	0.0947	0.0472	6.1407	15
16	0.4759	11.0332	71.9680	2.1012	23.1829	0.0906	0.0431	6.5228	16
17	0.4543	11.4876	79.2375	2.2010	25.2840	0.0871	0.0396	6.8977	17
18	0.4337	11.9213	86.6110	2.3055	27.4850	0.0839	0.0364	7.2652	18
19	0.4141	12.3354	94.0643	2.4151	29.7906	0.0811	0.0336	7.6256	19
20	0.3953	12.7307	101.5748	2.5298	32.2056	0.0786	0.0311	7.9788	20
21	0.3774	13.1080	109.1222	2.6499	34.7354	0.0763	0.0288	8.3248	21
22	0.3603	13.4683	116.6876	2.7758	37.3853	0.0742	0.0267	8.6639	22
23	0.3439	13.8122	124.2538	2.9077	40.1611	0.0724	0.0249	8.9959	23
24	0.3283	14.1405	131.8053	3.0458	43.0688	0.0707	0.0232	9.3211	24
25	0.3134	14.4540	139.3277	3.1904	46.1146	0.0692	0.0217	9.6394	25
26	0.2992	14.7532	146.8083	3.3420	49.3050	0.0678	0.0203	9.9509	26
27	0.2857	15.0389	154.2353	3.5007	52.6470	0.0665	0.0190	10.2558	27
28	0.2727	15.3116	161.5983	3.6670	56.1477	0.0653	0.0178	10.5540	28
29	0.2603	15.5719	168.8877	3.8412	59.8147	0.0642	0.0167	10.8457	29
30	0.2485	15.8204	176.0950	4.0237	63.6559	0.0632	0.0157	11.1309	30
31	0.2373	16.0577	183.2128	4.2148	67.6796	0.0623	0.0148	11.4097	31
32	0.2265	16.2842	190.2344	4.4150	71.8944	0.0614	0.0139	11.6822	32
33	0.2162	16.5004	197.1538	4.6247	76.3094	0.0606	0.0131	11.9484	33
34	0.2064	16.7068	203.9658	4.8444	80.9341	0.0599	0.0124	12.2085	34
35	0.1971	16.9039	210.6660	5.0745	85.7784	0.0592	0.0117	12.4626	35
36	0.1881	17.0920	217.2505	5.3155	90.8529	0.0585	0.0110	12.7106	36
37	0.1796	17.2716	223.7160	5.5680	96.1684	0.0579	0.0104	12.9528	37
38	0.1715	17.4431	230.0598	5.8325	101.7364	0.0573	0.0098	13.1892	38
39	0.1637	17.6068	236.2796	6.1095	107.5689	0.0568	0.0093	13.4198	39
40	0.1563	17.7630	242.3736	6.3997	113.6784	0.0563	0.0088	13.6448	40
41	0.1492	17.9122	248.3405	6.7037	120.0781	0.0558	0.0083	13.8643	41
42	0.1424	18.0546	254.1791	7.0221	126.7818	0.0554	0.0079	14.0784	42
43	0.1359	18.1905	259.8890	7.3557	133.8040	0.0550	0.0075	14.2870	43
44	0.1298	18.3203	265.4697	7.7051	141.1597	0.0546	0.0071	14.4904	44
45	0.1239	18.4442	270.9213	8.0711	148.8648	0.0542	0.0067	14.6887	45
46	0.1183	18.5625	276.2439	8.4545	156.9358	0.0539	0.0064	14.8818	46
47	0.1129	18.6754	281.4381	8.8560	165.3903	0.0535	0.0060	15.0700	47
48	0.1078	18.7832	286.5046	9.2767	174.2463	0.0532	0.0057	15.2532	48
49	0.1029	18.8861	291.4442	9.7173	183.5230	0.0529	0.0054	15.4317	49
50	0.0982	18.9844	296.2581	10.1789	193.2404	0.0527	0.0052	15.6054	50
51	0.0938	19.0782	300.9475	10.6624	203.4193	0.0524	0.0049	15.7744	51
52	0.0895	19.1677	305.5137	11.1689	214.0817	0.0522	0.0047	15.9390	52
53	0.0855	19.2532	309.9584	11.6994	225.2506	0.0519	0.0044	16.0991	53
54	0.0816	19.3348	314.2831	12.2551	236.9500	0.0517	0.0042	16.2548	54
55	0.0779	19.4127	318.4896	12.8372	249.2051	0.0515	0.0040	16.4063	55
60	0.0618	19.7523	337.8155	16.1898	319.7856	0.0506	0.0031	17.1026	60
65	0.0490	20.0215	354.4857	20.4179	408.7989	0.0499	0.0024	17.7052	65
70	0.0388	20.2351	368.7715	25.7503	521.0588	0.0494	0.0019	18.2244	70
75	0.0308	20.4044	380.9455	32.4752	662.6366	0.0490	0.0015	18.6698	75
80	0.0244	20.5386	391.2698	40.9565	841.1889	0.0487	0.0012	19.0505	80
85	0.0194	20.6451	399.9883	51.6527	1066.3718	0.0484	0.0009	19.3745	85
90	0.0154	20.7295	407.3234	65.1423	1350.3635	0.0482	0.0007	19.6495	90
95	0.0122	20.7964	413.4742	82.1548	1708.5224	0.0481	0.0006	19.8820	95
100	0.0097	20.8494	418.6166	103.6104	2160.2180	0.0480	0.0005	20.0781	100

$I = 5.00\%$

n	(P/F)	(P/A)	(P/G)	(F/P)	(F/A)	(A/P)	(A/F)	(A/G)	n
1	0.9524	0.9524	0.0000	1.0500	1.0000	1.0500	1.0000	0.0000	1
2	0.9070	1.8594	0.9070	1.1025	2.0500	0.5378	0.4878	0.4878	2
3	0.8638	2.7232	2.6347	1.1576	3.1525	0.3672	0.3172	0.9675	3
4	0.8227	3.5460	5.1028	1.2155	4.3101	0.2820	0.2320	1.4391	4
5	0.7835	4.3295	8.2369	1.2763	5.5256	0.2310	0.1810	1.9025	5
6	0.7462	5.0757	11.9680	1.3401	6.8019	0.1970	0.1470	2.3579	6
7	0.7107	5.7864	16.2321	1.4071	8.1420	0.1728	0.1228	2.8052	7
8	0.6768	6.4632	20.9700	1.4775	9.5491	0.1547	0.1047	3.2445	8
9	0.6446	7.1078	26.1268	1.5513	11.0266	0.1407	0.0907	3.6758	9
10	0.6139	7.7217	31.6520	1.6289	12.5779	0.1295	0.0795	4.0991	10
11	0.5847	8.3064	37.4988	1.7103	14.2068	0.1204	0.0704	4.5144	11
12	0.5568	8.8633	43.6241	1.7959	15.9171	0.1128	0.0628	4.9219	12
13	0.5303	9.3936	49.9879	1.8856	17.7130	0.1065	0.0565	5.3215	13
14	0.5051	9.8986	56.5538	1.9799	19.5986	0.1010	0.0510	5.7133	14
15	0.4810	10.3797	63.2880	2.0789	21.5786	0.0963	0.0463	6.0973	15
16	0.4581	10.8378	70.1597	2.1829	23.6575	0.0923	0.0423	6.4736	16
17	0.4363	11.2741	77.1405	2.2920	25.8404	0.0887	0.0387	6.8423	17
18	0.4155	11.6896	84.2043	2.4066	28.1324	0.0855	0.0355	7.2034	18
19	0.3957	12.0853	91.3275	2.5270	30.5390	0.0827	0.0327	7.5569	19
20	0.3769	12.4622	98.4884	2.6533	33.0660	0.0802	0.0302	7.9030	20
21	0.3589	12.8212	105.6673	2.7860	35.7193	0.0780	0.0280	8.2416	21
22	0.3418	13.1630	112.8461	2.9253	38.5052	0.0760	0.0260	8.5730	22
23	0.3256	13.4886	120.0087	3.0715	41.4305	0.0741	0.0241	8.8971	23
24	0.3101	13.7986	127.1402	3.2251	44.5020	0.0725	0.0225	9.2140	24
25	0.2953	14.0939	134.2275	3.3864	47.7271	0.0710	0.0210	9.5238	25
26	0.2812	14.3752	141.2585	3.5557	51.1135	0.0696	0.0196	9.8266	26
27	0.2678	14.6430	148.2226	3.7335	54.6691	0.0683	0.0183	10.1224	27
28	0.2551	14.8981	155.1101	3.9201	58.4026	0.0671	0.0171	10.4114	28
29	0.2429	15.1411	161.9126	4.1161	62.3227	0.0660	0.0160	10.6936	29
30	0.2314	15.3725	168.6226	4.3219	66.4388	0.0651	0.0151	10.9691	30
31	0.2204	15.5928	175.2333	4.5380	70.7608	0.0641	0.0141	11.2381	31
32	0.2099	15.8027	181.7392	4.7649	75.2988	0.0633	0.0133	11.5005	32
33	0.1999	16.0025	188.1351	5.0032	80.0638	0.0625	0.0125	11.7566	33
34	0.1904	16.1929	194.4168	5.2533	85.0670	0.0618	0.0118	12.0063	34
35	0.1813	16.3742	200.5807	5.5160	90.3203	0.0611	0.0111	12.2498	35
36	0.1727	16.5469	206.6237	5.7918	95.8363	0.0604	0.0104	12.4872	36
37	0.1644	16.7113	212.5434	6.0814	101.6281	0.0598	0.0098	12.7186	37
38	0.1566	16.8679	218.3378	6.3855	107.7095	0.0593	0.0093	12.9440	38
39	0.1491	17.0170	224.0054	6.7048	114.0950	0.0588	0.0088	13.1636	39
40	0.1420	17.1591	229.5452	7.0400	120.7998	0.0583	0.0083	13.3775	40
41	0.1353	17.2944	234.9564	7.3920	127.8398	0.0578	0.0078	13.5857	41
42	0.1288	17.4232	240.2389	7.7616	135.2318	0.0574	0.0074	13.7884	42
43	0.1227	17.5459	245.3925	8.1497	142.9933	0.0570	0.0070	13.9857	43
44	0.1169	17.6628	250.4175	8.5572	151.1430	0.0566	0.0066	14.1777	44
45	0.1113	17.7741	255.3145	8.9850	159.7002	0.0563	0.0063	14.3644	45
46	0.1060	17.8801	260.0844	9.4343	168.6852	0.0559	0.0059	14.5461	46
47	0.1009	17.9810	264.7281	9.9060	178.1194	0.0556	0.0056	14.7226	47
48	0.0961	18.0772	269.2467	10.4013	188.0254	0.0553	0.0053	14.8943	48
49	0.0916	18.1687	273.6418	10.9213	198.4267	0.0550	0.0050	15.0611	49
50	0.0872	18.2559	277.9148	11.4674	209.3480	0.0548	0.0048	15.2233	50
51	0.0831	18.3390	282.0673	12.0408	220.8154	0.0545	0.0045	15.3808	51
52	0.0791	18.4181	286.1013	12.6428	232.8562	0.0543	0.0043	15.5337	52
53	0.0753	18.4934	290.0184	13.2749	245.4990	0.0541	0.0041	15.6823	53
54	0.0717	18.5651	293.8208	13.9387	258.7739	0.0539	0.0039	15.8265	54
55	0.0683	18.6335	297.5104	14.6356	272.7126	0.0537	0.0037	15.9664	55
60	0.0535	18.9293	314.3432	18.6792	353.5837	0.0528	0.0028	16.6062	60
65	0.0419	19.1611	328.6910	23.8399	456.7980	0.0522	0.0022	17.1541	65
70	0.0329	19.3427	340.8409	30.4264	588.5285	0.0517	0.0017	17.6212	70
75	0.0258	19.4850	351.0721	38.8327	756.6537	0.0513	0.0013	18.0176	75
80	0.0202	19.5965	359.6460	49.5614	971.2288	0.0510	0.0010	18.3526	80
85	0.0158	19.6838	366.8007	63.2544	1245.0871	0.0508	0.0008	18.6346	85
90	0.0124	19.7523	372.7488	80.7304	1594.6073	0.0506	0.0006	18.8712	90
95	0.0097	19.8059	377.6774	103.0347	2040.6935	0.0505	0.0005	19.0689	95
100	0.0076	19.8479	381.7492	131.5013	2610.0252	0.0504	0.0004	19.2337	100

$I = 5.25\%$

n	(P/F)	(P/A)	(P/G)	(F/P)	(F/A)	(A/P)	(A/F)	(A/G)	n
1	0.9501	0.9501	0.0000	1.0525	1.0000	1.0525	1.0000	0.0000	1
2	0.9027	1.8528	0.9027	1.1078	2.0525	0.5397	0.4872	0.4872	2
3	0.8577	2.7105	2.6181	1.1659	3.1603	0.3689	0.3164	0.9659	3
4	0.8149	3.5255	5.0629	1.2271	4.3262	0.2837	0.2312	1.4361	4
5	0.7743	4.2997	8.1599	1.2915	5.5533	0.2326	0.1801	1.8978	5
6	0.7356	5.0354	11.8381	1.3594	6.8448	0.1986	0.1461	2.3510	6
7	0.6989	5.7343	16.0318	1.4307	8.2042	0.1744	0.1219	2.7958	7
8	0.6641	6.3984	20.6804	1.5058	9.6349	0.1563	0.1038	3.2321	8
9	0.6310	7.0294	25.7281	1.5849	11.1407	0.1423	0.0898	3.6601	9
10	0.5995	7.6288	31.1235	1.6681	12.7256	0.1311	0.0786	4.0797	10
11	0.5696	8.1984	36.8193	1.7557	14.3937	0.1220	0.0695	4.4910	11
12	0.5412	8.7396	42.7722	1.8478	16.1494	0.1144	0.0619	4.8941	12
13	0.5142	9.2538	48.9423	1.9449	17.9972	0.1081	0.0556	5.2889	13
14	0.4885	9.7423	55.2932	2.0470	19.9421	0.1026	0.0501	5.6756	14
15	0.4642	10.2065	61.7914	2.1544	21.9891	0.0980	0.0455	6.0541	15
16	0.4410	10.6475	68.4065	2.2675	24.1435	0.0939	0.0414	6.4247	16
17	0.4190	11.0665	75.1107	2.3866	26.4110	0.0904	0.0379	6.7872	17
18	0.3981	11.4646	81.8786	2.5119	28.7976	0.0872	0.0347	7.1419	18
19	0.3783	11.8428	88.6871	2.6437	31.3095	0.0844	0.0319	7.4887	19
20	0.3594	12.2022	95.5154	2.7825	33.9532	0.0820	0.0295	7.8277	20
21	0.3415	12.5437	102.3445	2.9286	36.7358	0.0797	0.0272	8.1590	21
22	0.3244	12.8681	109.1574	3.0824	39.6644	0.0777	0.0252	8.4828	22
23	0.3082	13.1763	115.9387	3.2442	42.7468	0.0759	0.0234	8.7990	23
24	0.2929	13.4692	122.6747	3.4145	45.9910	0.0742	0.0217	9.1078	24
25	0.2783	13.7475	129.3528	3.5938	49.4055	0.0727	0.0202	9.4092	25
26	0.2644	14.0118	135.9623	3.7825	52.9993	0.0714	0.0189	9.7034	26
27	0.2512	14.2630	142.4932	3.9810	56.7818	0.0701	0.0176	9.9904	27
28	0.2387	14.5017	148.9371	4.1900	60.7628	0.0690	0.0165	10.2703	28
29	0.2268	14.7285	155.2863	4.4100	64.9529	0.0679	0.0154	10.5433	29
30	0.2154	14.9439	161.5342	4.6416	69.3629	0.0669	0.0144	10.8094	30
31	0.2047	15.1486	167.6751	4.8852	74.0044	0.0660	0.0135	11.0687	31
32	0.1945	15.3431	173.7043	5.1417	78.8897	0.0652	0.0127	11.3213	32
33	0.1848	15.5279	179.6174	5.4116	84.0314	0.0644	0.0119	11.5674	33
34	0.1756	15.7034	185.4112	5.6958	89.4430	0.0637	0.0112	11.8070	34
35	0.1668	15.8703	191.0828	5.9948	95.1388	0.0630	0.0105	12.0403	35
36	0.1585	16.0287	196.6300	6.3095	101.1336	0.0624	0.0099	12.2673	36
37	0.1506	16.1793	202.0511	6.6408	107.4431	0.0618	0.0093	12.4882	37
38	0.1431	16.3224	207.3448	6.9894	114.0838	0.0613	0.0088	12.7031	38
39	0.1359	16.4583	212.5104	7.3563	121.0732	0.0608	0.0083	12.9120	39
40	0.1292	16.5875	217.5475	7.7426	128.4296	0.0603	0.0078	13.1151	40
41	0.1227	16.7102	222.4560	8.1490	136.1721	0.0598	0.0073	13.3126	41
42	0.1166	16.8268	227.2363	8.5769	144.3212	0.0594	0.0069	13.5044	42
43	0.1108	16.9376	231.8890	9.0271	152.8980	0.0590	0.0065	13.6908	43
44	0.1053	17.0428	236.4148	9.5011	161.9252	0.0587	0.0062	13.8718	44
45	0.1000	17.1428	240.8148	9.9999	171.4262	0.0583	0.0058	14.0476	45
46	0.0950	17.2378	245.0904	10.5249	181.4261	0.0580	0.0055	14.2182	46
47	0.0903	17.3281	249.2430	11.0774	191.9510	0.0577	0.0052	14.3837	47
48	0.0858	17.4139	253.2742	11.6590	203.0284	0.0574	0.0049	14.5444	48
49	0.0815	17.4954	257.1859	12.2711	214.6874	0.0572	0.0047	14.7002	49
50	0.0774	17.5728	260.9798	12.9153	226.9585	0.0569	0.0044	14.8513	50
51	0.0736	17.6464	264.6581	13.5934	239.8738	0.0567	0.0042	14.9979	51
52	0.0699	17.7163	268.2228	14.3070	253.4672	0.0564	0.0039	15.1399	52
53	0.0664	17.7827	271.6760	15.0581	267.7742	0.0562	0.0037	15.2776	53
54	0.0631	17.8458	275.0202	15.8487	282.8324	0.0560	0.0035	15.4109	54
55	0.0599	17.9057	278.2574	16.6808	298.6811	0.0558	0.0033	15.5401	55
60	0.0464	18.1635	292.9237	21.5440	391.3142	0.0551	0.0026	16.1271	60
65	0.0359	18.3631	305.2772	27.8251	510.9544	0.0545	0.0020	16.6245	65
70	0.0278	18.5176	315.6147	35.9375	665.4753	0.0540	0.0015	17.0440	70
75	0.0215	18.6372	324.2168	46.4149	865.0466	0.0537	0.0012	17.3962	75
80	0.0167	18.7299	331.3404	59.9471	1122.8024	0.0534	0.0009	17.6905	80
85	0.0129	18.8016	337.2145	77.4246	1455.7064	0.0532	0.0007	17.9354	85
90	0.0100	18.8571	342.0403	99.9976	1885.6678	0.0530	0.0005	18.1385	90
95	0.0077	18.9001	345.9918	129.1516	2440.9836	0.0529	0.0004	18.3063	95
100	0.0060	18.9334	349.2177	166.8055	3158.2006	0.0528	0.0003	18.4445	100

ENGINEERING ECONOMIC ANALYSIS

$I = 5.50\%$

n	(P/F)	(P/A)	(P/G)	(F/P)	(F/A)	(A/P)	(A/F)	(A/G)	n
1	0.9479	0.9479	0.0000	1.0550	1.0000	1.0550	1.0000	0.0000	1
2	0.8985	1.8463	0.8985	1.1130	2.0550	0.5416	0.4866	0.4866	2
3	0.8516	2.6979	2.6017	1.1742	3.1680	0.3707	0.3157	0.9643	3
4	0.8072	3.5052	5.0233	1.2388	4.3423	0.2853	0.2303	1.4331	4
5	0.7651	4.2703	8.0839	1.3070	5.5811	0.2342	0.1792	1.8931	5
6	0.7252	4.9955	11.7101	1.3788	6.8881	0.2002	0.1452	2.3441	6
7	0.6874	5.6830	15.8347	1.4547	8.2669	0.1760	0.1210	2.7863	7
8	0.6516	6.3346	20.3959	1.5347	9.7216	0.1579	0.1029	3.2198	8
9	0.6176	6.9522	25.3369	1.6191	11.2563	0.1438	0.0888	3.6445	9
10	0.5854	7.5376	30.6058	1.7081	12.8754	0.1327	0.0777	4.0604	10
11	0.5549	8.0925	36.1549	1.8021	14.5835	0.1236	0.0686	4.4677	11
12	0.5260	8.6185	41.9407	1.9012	16.3856	0.1160	0.0610	4.8663	12
13	0.4986	9.1171	47.9234	2.0058	18.2868	0.1097	0.0547	5.2564	13
14	0.4726	9.5896	54.0669	2.1161	20.2926	0.1043	0.0493	5.6380	14
15	0.4479	10.0376	60.3379	2.2325	22.4087	0.0996	0.0446	6.0112	15
16	0.4246	10.4622	66.7066	2.3553	24.6411	0.0956	0.0406	6.3760	16
17	0.4024	10.8646	73.1458	2.4848	26.9964	0.0920	0.0370	6.7325	17
18	0.3815	11.2461	79.6307	2.6215	29.4812	0.0889	0.0339	7.0808	18
19	0.3616	11.6077	86.1391	2.7656	32.1027	0.0862	0.0312	7.4209	19
20	0.3427	11.9504	92.6510	2.9178	34.8683	0.0837	0.0287	7.7530	20
21	0.3249	12.2752	99.1482	3.0782	37.7861	0.0815	0.0265	8.0771	21
22	0.3079	12.5832	105.6146	3.2475	40.8643	0.0795	0.0245	8.3933	22
23	0.2919	12.8750	112.0358	3.4262	44.1118	0.0777	0.0227	8.7018	23
24	0.2767	13.1517	118.3989	3.6146	47.5380	0.0760	0.0210	9.0026	24
25	0.2622	13.4139	124.6925	3.8134	51.1526	0.0745	0.0195	9.2957	25
26	0.2486	13.6625	130.9066	4.0231	54.9660	0.0732	0.0182	9.5815	26
27	0.2356	13.8981	137.0323	4.2444	58.9891	0.0720	0.0170	9.8598	27
28	0.2233	14.1214	143.0620	4.4778	63.2335	0.0708	0.0158	10.1309	28
29	0.2117	14.3331	148.9890	4.7241	67.7114	0.0698	0.0148	10.3948	29
30	0.2006	14.5337	154.8077	4.9840	72.4355	0.0688	0.0138	10.6516	30
31	0.1902	14.7239	160.5132	5.2581	77.4194	0.0679	0.0129	10.9015	31
32	0.1803	14.9042	166.1016	5.5473	82.6775	0.0671	0.0121	11.1446	32
33	0.1709	15.0751	171.5695	5.8524	88.2248	0.0663	0.0113	11.3810	33
34	0.1620	15.2370	176.9142	6.1742	94.0771	0.0656	0.0106	11.6108	34
35	0.1535	15.3906	182.1339	6.5138	100.2514	0.0650	0.0100	11.8341	35
36	0.1455	15.5361	187.2270	6.8721	106.7652	0.0644	0.0094	12.0511	36
37	0.1379	15.6740	192.1925	7.2501	113.6373	0.0638	0.0088	12.2619	37
38	0.1307	15.8047	197.0298	7.6488	120.8873	0.0633	0.0083	12.4665	38
39	0.1239	15.9287	201.7389	8.0695	128.5361	0.0628	0.0078	12.6652	39
40	0.1175	16.0461	206.3200	8.5133	136.6056	0.0623	0.0073	12.8579	40
41	0.1113	16.1575	210.7736	8.9815	145.1189	0.0619	0.0069	13.0450	41
42	0.1055	16.2630	215.1005	9.4755	154.1005	0.0615	0.0065	13.2264	42
43	0.1000	16.3630	219.3019	9.9967	163.5760	0.0611	0.0061	13.4023	43
44	0.0948	16.4579	223.3791	10.5465	173.5727	0.0608	0.0058	13.5728	44
45	0.0899	16.5477	227.3336	11.1266	184.1192	0.0604	0.0054	13.7381	45
46	0.0852	16.6329	231.1671	11.7385	195.2457	0.0601	0.0051	13.8982	46
47	0.0807	16.7137	234.8815	12.3841	206.9842	0.0598	0.0048	14.0533	47
48	0.0765	16.7902	238.4789	13.0653	219.3684	0.0596	0.0046	14.2035	48
49	0.0725	16.8628	241.9612	13.7838	232.4336	0.0593	0.0043	14.3489	49
50	0.0688	16.9315	245.3308	14.5420	246.2175	0.0591	0.0041	14.4896	50
51	0.0652	16.9967	248.5898	15.3418	260.7594	0.0588	0.0038	14.6258	51
52	0.0618	17.0585	251.7408	16.1856	276.1012	0.0586	0.0036	14.7575	52
53	0.0586	17.1170	254.7860	17.0758	292.2868	0.0584	0.0034	14.8849	53
54	0.0555	17.1726	257.7281	18.0149	309.3625	0.0582	0.0032	15.0081	54
55	0.0526	17.2252	260.5693	19.0058	327.3775	0.0581	0.0031	15.1272	55
60	0.0403	17.4499	273.3522	24.8398	433.4504	0.0573	0.0023	15.6650	60
65	0.0308	17.6218	283.9925	32.4646	572.0834	0.0567	0.0017	16.1160	65
70	0.0236	17.7533	292.7914	42.4299	753.2712	0.0563	0.0013	16.4922	70
75	0.0180	17.8539	300.0269	55.4542	990.0764	0.0560	0.0010	16.8045	75
80	0.0138	17.9310	305.9481	72.4764	1299.5714	0.0558	0.0008	17.0626	80
85	0.0106	17.9899	310.7732	94.7238	1704.0689	0.0556	0.0006	17.2749	85
90	0.0081	18.0350	314.6905	123.8002	2232.7310	0.0554	0.0004	17.4489	90
95	0.0062	18.0694	317.8602	161.8019	2923.6712	0.0553	0.0003	17.5910	95
100	0.0047	18.0958	320.4174	211.4686	3826.7025	0.0553	0.0003	17.7067	100

$I = 5.75\%$

n	(P/F)	(P/A)	(P/G)	(F/P)	(F/A)	(A/P)	(A/F)	(A/G)	n
1	0.9456	0.9456	0.0000	1.0575	1.0000	1.0575	1.0000	0.0000	1
2	0.8942	1.8398	0.8942	1.1183	2.0575	0.5435	0.4860	0.4860	2
3	0.8456	2.6854	2.5854	1.1826	3.1758	0.3724	0.3149	0.9627	3
4	0.7996	3.4850	4.9842	1.2506	4.3584	0.2869	0.2294	1.4302	4
5	0.7561	4.2412	8.0087	1.3225	5.6090	0.2358	0.1783	1.8883	5
6	0.7150	4.9562	11.5838	1.3986	6.9315	0.2018	0.1443	2.3372	6
7	0.6761	5.6323	15.6407	1.4790	8.3301	0.1775	0.1200	2.7769	7
8	0.6394	6.2717	20.1163	1.5640	9.8091	0.1594	0.1019	3.2075	8
9	0.6046	6.8763	24.9532	1.6540	11.3731	0.1454	0.0879	3.6289	9
10	0.5717	7.4481	30.0989	1.7491	13.0271	0.1343	0.0768	4.0412	10
11	0.5406	7.9887	35.5054	1.8496	14.7761	0.1252	0.0677	4.4444	11
12	0.5113	8.5000	41.1291	1.9560	16.6257	0.1176	0.0601	4.8387	12
13	0.4835	8.9834	46.9306	2.0684	18.5817	0.1113	0.0538	5.2241	13
14	0.4572	9.4406	52.8737	2.1874	20.6502	0.1059	0.0484	5.6007	14
15	0.4323	9.8729	58.9261	2.3132	22.8376	0.1013	0.0438	5.9685	15
16	0.4088	10.2817	65.0581	2.4462	25.1507	0.0973	0.0398	6.3276	16
17	0.3866	10.6683	71.2433	2.5868	27.5969	0.0937	0.0362	6.6781	17
18	0.3656	11.0338	77.4578	2.7356	30.1837	0.0906	0.0331	7.0200	18
19	0.3457	11.3795	83.6800	2.8929	32.9193	0.0879	0.0304	7.3536	19
20	0.3269	11.7064	89.8908	3.0592	35.8121	0.0854	0.0279	7.6788	20
21	0.3091	12.0155	96.0729	3.2351	38.8713	0.0832	0.0257	7.9958	21
22	0.2923	12.3078	102.2113	3.4211	42.1064	0.0812	0.0237	8.3046	22
23	0.2764	12.5842	108.2923	3.6178	45.5275	0.0795	0.0220	8.6054	23
24	0.2614	12.8456	114.3040	3.8259	49.1454	0.0778	0.0203	8.8983	24
25	0.2472	13.0927	120.2360	4.0458	52.9712	0.0764	0.0189	9.1834	25
26	0.2337	13.3265	126.0792	4.2785	57.0171	0.0750	0.0175	9.4608	26
27	0.2210	13.5475	131.8257	4.5245	61.2956	0.0738	0.0163	9.7306	27
28	0.2090	13.7565	137.4687	4.7847	65.8201	0.0727	0.0152	9.9930	28
29	0.1976	13.9541	143.0026	5.0598	70.6047	0.0717	0.0142	10.2480	29
30	0.1869	14.1410	148.4224	5.3507	75.6645	0.0707	0.0132	10.4959	30
31	0.1767	14.3178	153.7243	5.6584	81.0152	0.0698	0.0123	10.7366	31
32	0.1671	14.4849	158.9050	5.9837	86.6736	0.0690	0.0115	10.9704	32
33	0.1580	14.6429	163.9621	6.3278	92.6573	0.0683	0.0108	11.1974	33
34	0.1494	14.7923	168.8936	6.6916	98.9851	0.0676	0.0101	11.4176	34
35	0.1413	14.9337	173.6983	7.0764	105.6767	0.0670	0.0095	11.6313	35
36	0.1336	15.0673	178.3754	7.4833	112.7532	0.0664	0.0089	11.8386	36
37	0.1264	15.1937	182.9245	7.9136	120.2365	0.0658	0.0083	12.0395	37
38	0.1195	15.3131	187.3458	8.3686	128.1501	0.0653	0.0078	12.2343	38
39	0.1130	15.4261	191.6396	8.8498	136.5187	0.0648	0.0073	12.4230	39
40	0.1069	15.5330	195.8069	9.3587	145.3685	0.0644	0.0069	12.6059	40
41	0.1010	15.6340	199.8486	9.8968	154.7272	0.0640	0.0065	12.7829	41
42	0.0955	15.7296	203.7661	10.4659	164.6240	0.0636	0.0061	12.9543	42
43	0.0904	15.8199	207.5609	11.0677	175.0899	0.0632	0.0057	13.1202	43
44	0.0854	15.9054	211.2349	11.7041	186.1576	0.0629	0.0054	13.2807	44
45	0.0808	15.9862	214.7898	12.3770	197.8616	0.0626	0.0051	13.4360	45
46	0.0764	16.0626	218.2279	13.0887	210.2387	0.0623	0.0048	13.5861	46
47	0.0722	16.1348	221.5513	13.8413	223.3274	0.0620	0.0045	13.7312	47
48	0.0683	16.2031	224.7623	14.6372	237.1687	0.0617	0.0042	13.8715	48
49	0.0646	16.2678	227.8633	15.4788	251.8059	0.0615	0.0040	14.0071	49
50	0.0611	16.3288	230.8568	16.3689	267.2848	0.0612	0.0037	14.1380	50
51	0.0578	16.3866	233.7453	17.3101	283.6536	0.0610	0.0035	14.2644	51
52	0.0546	16.4412	236.5313	18.3054	300.9637	0.0608	0.0033	14.3865	52
53	0.0517	16.4929	239.2176	19.3580	319.2691	0.0606	0.0031	14.5043	53
54	0.0488	16.5417	241.8066	20.4711	338.6271	0.0605	0.0030	14.6180	54
55	0.0462	16.5879	244.3010	21.6481	359.0982	0.0603	0.0028	14.7276	55
60	0.0349	16.7839	255.4462	28.6301	480.5231	0.0596	0.0021	15.2198	60
65	0.0264	16.9320	264.6142	37.8638	641.1099	0.0591	0.0016	15.6281	65
70	0.0200	17.0440	272.1064	50.0756	853.4890	0.0587	0.0012	15.9649	70
75	0.0151	17.1287	278.1950	66.2260	1134.3644	0.0584	0.0009	16.2415	75
80	0.0114	17.1927	283.1190	87.5851	1505.8273	0.0582	0.0007	16.4674	80
85	0.0086	17.2412	287.0843	115.8329	1997.0941	0.0580	0.0005	16.6511	85
90	0.0065	17.2778	290.2657	153.1912	2646.8036	0.0579	0.0004	16.7999	90
95	0.0049	17.3055	292.8096	202.5983	3506.0568	0.0578	0.0003	16.9201	95
100	0.0037	17.3264	294.8379	267.9400	4642.4353	0.0577	0.0002	17.0167	100

$I = 6.00\%$

n	(P/F)	(P/A)	(P/G)	(F/P)	(F/A)	(A/P)	(A/F)	(A/G)	n
1	0.9434	0.9434	0.0000	1.0600	1.0000	1.0600	1.0000	0.0000	1
2	0.8900	1.8334	0.8900	1.1236	2.0600	0.5454	0.4854	0.4854	2
3	0.8396	2.6730	2.5692	1.1910	3.1836	0.3741	0.3141	0.9612	3
4	0.7921	3.4651	4.9455	1.2625	4.3746	0.2886	0.2286	1.4272	4
5	0.7473	4.2124	7.9345	1.3382	5.6371	0.2374	0.1774	1.8836	5
6	0.7050	4.9173	11.4594	1.4185	6.9753	0.2034	0.1434	2.3304	6
7	0.6651	5.5824	15.4497	1.5036	8.3938	0.1791	0.1191	2.7676	7
8	0.6274	6.2098	19.8416	1.5938	9.8975	0.1610	0.1010	3.1952	8
9	0.5919	6.8017	24.5768	1.6895	11.4913	0.1470	0.0870	3.6133	9
10	0.5584	7.3601	29.6023	1.7908	13.1808	0.1359	0.0759	4.0220	10
11	0.5268	7.8869	34.8702	1.8983	14.9716	0.1268	0.0668	4.4213	11
12	0.4970	8.3838	40.3369	2.0122	16.8699	0.1193	0.0593	4.8113	12
13	0.4688	8.8527	45.9629	2.1329	18.8821	0.1130	0.0530	5.1920	13
14	0.4423	9.2950	51.7128	2.2609	21.0151	0.1076	0.0476	5.5635	14
15	0.4173	9.7122	57.5546	2.3966	23.2760	0.1030	0.0430	5.9260	15
16	0.3936	10.1059	63.4592	2.5404	25.6725	0.0990	0.0390	6.2794	16
17	0.3714	10.4773	69.4011	2.6928	28.2129	0.0954	0.0354	6.6240	17
18	0.3503	10.8276	75.3569	2.8543	30.9057	0.0924	0.0324	6.9597	18
19	0.3305	11.1581	81.3062	3.0256	33.7600	0.0896	0.0296	7.2867	19
20	0.3118	11.4699	87.2304	3.2071	36.7856	0.0872	0.0272	7.6051	20
21	0.2942	11.7641	93.1136	3.3996	39.9927	0.0850	0.0250	7.9151	21
22	0.2775	12.0416	98.9412	3.6035	43.3923	0.0830	0.0230	8.2166	22
23	0.2618	12.3034	104.7007	3.8197	46.9958	0.0813	0.0213	8.5099	23
24	0.2470	12.5504	110.3812	4.0489	50.8156	0.0797	0.0197	8.7951	24
25	0.2330	12.7834	115.9732	4.2919	54.8645	0.0782	0.0182	9.0722	25
26	0.2198	13.0032	121.4684	4.5494	59.1564	0.0769	0.0169	9.3414	26
27	0.2074	13.2105	126.8600	4.8223	63.7058	0.0757	0.0157	9.6029	27
28	0.1956	13.4062	132.1420	5.1117	68.5281	0.0746	0.0146	9.8568	28
29	0.1846	13.5907	137.3096	5.4184	73.6398	0.0736	0.0136	10.1032	29
30	0.1741	13.7648	142.3588	5.7435	79.0582	0.0726	0.0126	10.3422	30
31	0.1643	13.9291	147.2864	6.0881	84.8017	0.0718	0.0118	10.5740	31
32	0.1550	14.0840	152.0901	6.4534	90.8898	0.0710	0.0110	10.7988	32
33	0.1462	14.2302	156.7681	6.8406	97.3432	0.0703	0.0103	11.0166	33
34	0.1379	14.3681	161.3192	7.2510	104.1838	0.0696	0.0096	11.2276	34
35	0.1301	14.4982	165.7427	7.6861	111.4348	0.0690	0.0090	11.4319	35
36	0.1227	14.6210	170.0387	8.1473	119.1209	0.0684	0.0084	11.6298	36
37	0.1158	14.7368	174.2072	8.6361	127.2681	0.0679	0.0079	11.8213	37
38	0.1092	14.8460	178.2490	9.1543	135.9042	0.0674	0.0074	12.0065	38
39	0.1031	14.9491	182.1652	9.7035	145.0585	0.0669	0.0069	12.1857	39
40	0.0972	15.0463	185.9568	10.2857	154.7620	0.0665	0.0065	12.3590	40
41	0.0917	15.1380	189.6256	10.9029	165.0477	0.0661	0.0061	12.5264	41
42	0.0865	15.2245	193.1732	11.5570	175.9505	0.0657	0.0057	12.6883	42
43	0.0816	15.3062	196.6017	12.2505	187.5076	0.0653	0.0053	12.8446	43
44	0.0770	15.3832	199.9130	12.9855	199.7580	0.0650	0.0050	12.9956	44
45	0.0727	15.4558	203.1096	13.7646	212.7435	0.0647	0.0047	13.1413	45
46	0.0685	15.5244	206.1938	14.5905	226.5081	0.0644	0.0044	13.2819	46
47	0.0647	15.5890	209.1681	15.4659	241.0986	0.0641	0.0041	13.4177	47
48	0.0610	15.6500	212.0351	16.3939	256.5645	0.0639	0.0039	13.5485	48
49	0.0575	15.7076	214.7972	17.3775	272.9584	0.0637	0.0037	13.6748	49
50	0.0543	15.7619	217.4574	18.4202	290.3359	0.0634	0.0034	13.7964	50
51	0.0512	15.8131	220.0181	19.5254	308.7561	0.0632	0.0032	13.9137	51
52	0.0483	15.8614	222.4823	20.6969	328.2814	0.0630	0.0030	14.0267	52
53	0.0456	15.9070	224.8525	21.9387	348.9783	0.0629	0.0029	14.1355	53
54	0.0430	15.9500	227.1316	23.2550	370.9170	0.0627	0.0027	14.2402	54
55	0.0406	15.9905	229.3222	24.6503	394.1720	0.0625	0.0025	14.3411	55
60	0.0303	16.1614	239.0428	32.9877	533.1282	0.0619	0.0019	14.7909	60
65	0.0227	16.2891	246.9450	44.1450	719.0829	0.0614	0.0014	15.1601	65
70	0.0169	16.3845	253.3271	59.0759	967.9322	0.0610	0.0010	15.4613	70
75	0.0126	16.4558	258.4527	79.0569	1300.9487	0.0608	0.0008	15.7058	75
80	0.0095	16.5091	262.5493	105.7960	1746.5999	0.0606	0.0006	15.9033	80
85	0.0071	16.5489	265.8096	141.5789	2342.9817	0.0604	0.0004	16.0620	85
90	0.0053	16.5787	268.3946	189.4645	3141.0752	0.0603	0.0003	16.1891	90
95	0.0039	16.6009	270.4375	253.5463	4209.1042	0.0602	0.0002	16.2905	95
100	0.0029	16.6175	272.0471	339.3021	5638.3681	0.0602	0.0002	16.3711	100

$I = 6.25\%$

n	(P/F)	(P/A)	(P/G)	(F/P)	(F/A)	(A/P)	(A/F)	(A/G)	n
1	0.9412	0.9412	0.0000	1.0625	1.0000	1.0625	1.0000	0.0000	1
2	0.8858	1.8270	0.8858	1.1289	2.0625	0.5473	0.4848	0.4848	2
3	0.8337	2.6607	2.5532	1.1995	3.1914	0.3758	0.3133	0.9596	3
4	0.7847	3.4454	4.9072	1.2744	4.3909	0.2902	0.2277	1.4243	4
5	0.7385	4.1839	7.8613	1.3541	5.6653	0.2390	0.1765	1.8789	5
6	0.6951	4.8789	11.3366	1.4387	7.0194	0.2050	0.1425	2.3236	6
7	0.6542	5.5331	15.2617	1.5286	8.4581	0.1807	0.1182	2.7582	7
8	0.6157	6.1488	19.5716	1.6242	9.9867	0.1626	0.1001	3.1830	8
9	0.5795	6.7283	24.2074	1.7257	11.6109	0.1486	0.0861	3.5979	9
10	0.5454	7.2737	29.1160	1.8335	13.3366	0.1375	0.0750	4.0029	10
11	0.5133	7.7870	34.2491	1.9481	15.1701	0.1284	0.0659	4.3982	11
12	0.4831	8.2701	39.5634	2.0699	17.1182	0.1209	0.0584	4.7839	12
13	0.4547	8.7248	45.0198	2.1993	19.1881	0.1146	0.0521	5.1600	13
14	0.4280	9.1528	50.5831	2.3367	21.3874	0.1093	0.0468	5.5265	14
15	0.4028	9.5555	56.2220	2.4828	23.7241	0.1047	0.0422	5.8837	15
16	0.3791	9.9346	61.9083	2.6379	26.2069	0.1007	0.0382	6.2316	16
17	0.3568	10.2914	67.6169	2.8028	28.8448	0.0972	0.0347	6.5702	17
18	0.3358	10.6272	73.3255	2.9780	31.6476	0.0941	0.0316	6.8998	18
19	0.3160	10.9433	79.0143	3.1641	34.6256	0.0914	0.0289	7.2204	19
20	0.2975	11.2407	84.6659	3.3619	37.7897	0.0890	0.0265	7.5321	20
21	0.2800	11.5207	90.2651	3.5720	41.1515	0.0868	0.0243	7.8351	21
22	0.2635	11.7842	95.7984	3.7952	44.7235	0.0849	0.0224	8.1294	22
23	0.2480	12.0322	101.2542	4.0324	48.5187	0.0831	0.0206	8.4153	23
24	0.2334	12.2656	106.6224	4.2844	52.5511	0.0815	0.0190	8.6928	24
25	0.2197	12.4852	111.8946	4.5522	56.8356	0.0801	0.0176	8.9622	25
26	0.2068	12.6920	117.0633	4.8367	61.3878	0.0788	0.0163	9.2234	26
27	0.1946	12.8866	122.1227	5.1390	66.2245	0.0776	0.0151	9.4767	27
28	0.1831	13.0697	127.0675	5.4602	71.3635	0.0765	0.0140	9.7223	28
29	0.1724	13.2421	131.8939	5.8015	76.8238	0.0755	0.0130	9.9602	29
30	0.1622	13.4043	136.5985	6.1641	82.6253	0.0746	0.0121	10.1906	30
31	0.1527	13.5570	141.1792	6.5493	88.7893	0.0738	0.0113	10.4137	31
32	0.1437	13.7007	145.6340	6.9587	95.3387	0.0730	0.0105	10.6297	32
33	0.1353	13.8360	149.9621	7.3936	102.2973	0.0723	0.0098	10.8386	33
34	0.1273	13.9633	154.1629	7.8557	109.6909	0.0716	0.0091	11.0406	34
35	0.1198	14.0831	158.2364	8.3467	117.5466	0.0710	0.0085	11.2359	35
36	0.1128	14.1958	162.1830	8.8683	125.8933	0.0704	0.0079	11.4247	36
37	0.1061	14.3020	166.0036	9.4226	134.7616	0.0699	0.0074	11.6071	37
38	0.0999	14.4018	169.6993	10.0115	144.1842	0.0694	0.0069	11.7832	38
39	0.0940	14.4958	173.2717	10.6372	154.1957	0.0690	0.0065	11.9532	39
40	0.0885	14.5843	176.7224	11.3021	164.8329	0.0686	0.0061	12.1173	40
41	0.0833	14.6676	180.0534	12.0084	176.1350	0.0682	0.0057	12.2756	41
42	0.0784	14.7460	183.2668	12.7590	188.1434	0.0678	0.0053	12.4283	42
43	0.0738	14.8197	186.3650	13.5564	200.9024	0.0675	0.0050	12.5755	43
44	0.0694	14.8892	189.3503	14.4037	214.4588	0.0672	0.0047	12.7173	44
45	0.0653	14.9545	192.2254	15.3039	228.8625	0.0669	0.0044	12.8540	45
46	0.0615	15.0160	194.9929	16.2604	244.1664	0.0666	0.0041	12.9857	46
47	0.0579	15.0739	197.6554	17.2767	260.4268	0.0663	0.0038	13.1124	47
48	0.0545	15.1284	200.2158	18.3565	277.7034	0.0661	0.0036	13.2345	48
49	0.0513	15.1796	202.6769	19.5037	296.0599	0.0659	0.0034	13.3519	49
50	0.0483	15.2279	205.0415	20.7227	315.5637	0.0657	0.0032	13.4649	50
51	0.0454	15.2733	207.3123	22.0179	336.2864	0.0655	0.0030	13.5735	51
52	0.0427	15.3161	209.4924	23.3940	358.3043	0.0653	0.0028	13.6780	52
53	0.0402	15.3563	211.5844	24.8561	381.6983	0.0651	0.0026	13.7784	53
54	0.0379	15.3942	213.5913	26.4097	406.5544	0.0650	0.0025	13.8748	54
55	0.0356	15.4298	215.5157	28.0603	432.9641	0.0648	0.0023	13.9675	55
60	0.0263	15.5789	223.9965	37.9959	591.9338	0.0642	0.0017	14.3782	60
65	0.0194	15.6890	230.8102	51.4495	807.1917	0.0637	0.0012	14.7116	65
70	0.0144	15.7703	236.2488	69.6668	1098.6684	0.0634	0.0009	14.9806	70
75	0.0106	15.8304	240.5656	94.3345	1493.3515	0.0632	0.0007	15.1964	75
80	0.0078	15.8747	243.9752	127.7365	2027.7844	0.0630	0.0005	15.3688	80
85	0.0058	15.9075	246.6571	172.9656	2751.4499	0.0629	0.0004	15.5057	85
90	0.0043	15.9317	248.7586	234.2095	3731.3518	0.0628	0.0003	15.6141	90
95	0.0032	15.9495	250.3999	317.1387	5058.2185	0.0627	0.0002	15.6995	95
100	0.0023	15.9627	251.6780	429.4315	6854.9036	0.0626	0.0001	15.7666	100

$I = 6.50\%$

n	(P/F)	(P/A)	(P/G)	(F/P)	(F/A)	(A/P)	(A/F)	(A/G)	n
1	0.9390	0.9390	0.0000	1.0650	1.0000	1.0650	1.0000	0.0000	1
2	0.8817	1.8206	0.8817	1.1342	2.0650	0.5493	0.4843	0.4843	2
3	0.8278	2.6485	2.5374	1.2079	3.1992	0.3776	0.3126	0.9580	3
4	0.7773	3.4258	4.8693	1.2865	4.4072	0.2919	0.2269	1.4214	4
5	0.7299	4.1557	7.7889	1.3701	5.6936	0.2406	0.1756	1.8743	5
6	0.6853	4.8410	11.2155	1.4591	7.0637	0.2066	0.1416	2.3168	6
7	0.6435	5.4845	15.0766	1.5540	8.5229	0.1823	0.1173	2.7489	7
8	0.6042	6.0888	19.3062	1.6550	10.0769	0.1642	0.0992	3.1708	8
9	0.5674	6.6561	23.8450	1.7626	11.7319	0.1502	0.0852	3.5824	9
10	0.5327	7.1888	28.6395	1.8771	13.4944	0.1391	0.0741	3.9839	10
11	0.5002	7.6890	33.6417	1.9992	15.3716	0.1301	0.0651	4.3753	11
12	0.4697	8.1587	38.8082	2.1291	17.3707	0.1226	0.0576	4.7566	12
13	0.4410	8.5997	44.1004	2.2675	19.4998	0.1163	0.0513	5.1281	13
14	0.4141	9.0138	49.4837	2.4149	21.7673	0.1109	0.0459	5.4897	14
15	0.3888	9.4027	54.9272	2.5718	24.1822	0.1064	0.0414	5.8417	15
16	0.3651	9.7678	60.4037	2.7390	26.7540	0.1024	0.0374	6.1840	16
17	0.3428	10.1106	65.8887	2.9170	29.4930	0.0989	0.0339	6.5168	17
18	0.3219	10.4325	71.3608	3.1067	32.4101	0.0959	0.0309	6.8403	18
19	0.3022	10.7347	76.8012	3.3086	35.5167	0.0932	0.0282	7.1545	19
20	0.2838	11.0185	82.1933	3.5236	38.8253	0.0908	0.0258	7.4596	20
21	0.2665	11.2850	87.5229	3.7527	42.3490	0.0886	0.0236	7.7557	21
22	0.2502	11.5352	92.7773	3.9966	46.1016	0.0867	0.0217	8.0430	22
23	0.2349	11.7701	97.9460	4.2564	50.0982	0.0850	0.0200	8.3216	23
24	0.2206	11.9907	103.0199	4.5331	54.3546	0.0834	0.0184	8.5916	24
25	0.2071	12.1979	107.9912	4.8277	58.8877	0.0820	0.0170	8.8533	25
26	0.1945	12.3924	112.8536	5.1415	63.7154	0.0807	0.0157	9.1067	26
27	0.1826	12.5750	117.6018	5.4757	68.8569	0.0795	0.0145	9.3520	27
28	0.1715	12.7465	122.2318	5.8316	74.3326	0.0785	0.0135	9.5895	28
29	0.1610	12.9075	126.7401	6.2107	80.1642	0.0775	0.0125	9.8191	29
30	0.1512	13.0587	131.1245	6.6144	86.3749	0.0766	0.0116	10.0412	30
31	0.1420	13.2006	135.3833	7.0443	92.9892	0.0758	0.0108	10.2558	31
32	0.1333	13.3339	139.5154	7.5022	100.0335	0.0750	0.0100	10.4632	32
33	0.1252	13.4591	143.5205	7.9898	107.5357	0.0743	0.0093	10.6635	33
34	0.1175	13.5766	147.3987	8.5092	115.5255	0.0737	0.0087	10.8568	34
35	0.1103	13.6870	151.1505	9.0623	124.0347	0.0731	0.0081	11.0434	35
36	0.1036	13.7906	154.7770	9.6513	133.0969	0.0725	0.0075	11.2234	36
37	0.0973	13.8879	158.2794	10.2786	142.7482	0.0720	0.0070	11.3970	37
38	0.0914	13.9792	161.6594	10.9467	153.0269	0.0715	0.0065	11.5643	38
39	0.0858	14.0650	164.9189	11.6583	163.9736	0.0711	0.0061	11.7255	39
40	0.0805	14.1455	168.0599	12.4161	175.6319	0.0707	0.0057	11.8808	40
41	0.0756	14.2212	171.0850	13.2231	188.0480	0.0703	0.0053	12.0303	41
42	0.0710	14.2922	173.9963	14.0826	201.2711	0.0700	0.0050	12.1742	42
43	0.0667	14.3588	176.7967	14.9980	215.3537	0.0696	0.0046	12.3127	43
44	0.0626	14.4214	179.4888	15.9729	230.3517	0.0693	0.0043	12.4460	44
45	0.0588	14.4802	182.0753	17.0111	246.3246	0.0691	0.0041	12.5741	45
46	0.0552	14.5354	184.5592	18.1168	263.3357	0.0688	0.0038	12.6972	46
47	0.0518	14.5873	186.9433	19.2944	281.4525	0.0686	0.0036	12.8155	47
48	0.0487	14.6359	189.2306	20.5485	300.7469	0.0683	0.0033	12.9292	48
49	0.0457	14.6816	191.4239	21.8842	321.2955	0.0681	0.0031	13.0383	49
50	0.0429	14.7245	193.5264	23.3067	343.1797	0.0679	0.0029	13.1431	50
51	0.0403	14.7648	195.5407	24.8216	366.4864	0.0677	0.0027	13.2437	51
52	0.0378	14.8026	197.4700	26.4350	391.3080	0.0676	0.0026	13.3402	52
53	0.0355	14.8382	199.3170	28.1533	417.7430	0.0674	0.0024	13.4327	53
54	0.0334	14.8715	201.0847	29.9833	445.8963	0.0672	0.0022	13.5215	54
55	0.0313	14.9028	202.7758	31.9322	475.8795	0.0671	0.0021	13.6065	55
60	0.0229	15.0330	210.1774	43.7498	657.6898	0.0665	0.0015	13.9811	60
65	0.0167	15.1280	216.0547	59.9411	906.7857	0.0661	0.0011	14.2818	65
70	0.0122	15.1973	220.6910	82.1245	1248.0687	0.0658	0.0008	14.5217	70
75	0.0089	15.2479	224.3280	112.5176	1715.6559	0.0656	0.0006	14.7121	75
80	0.0065	15.2848	227.1673	154.1589	2356.2909	0.0654	0.0004	14.8623	80
85	0.0047	15.3118	229.3744	211.2111	3234.0163	0.0653	0.0003	14.9803	85
90	0.0035	15.3315	231.0837	289.3775	4436.5763	0.0652	0.0002	15.0725	90
95	0.0025	15.3458	232.4031	396.4722	6084.1877	0.0652	0.0002	15.1444	95
100	0.0018	15.3563	233.4185	543.2013	8341.5580	0.0651	0.0001	15.2002	100

$I = 6.75\%$

n	(P/F)	(P/A)	(P/G)	(F/P)	(F/A)	(A/P)	(A/F)	(A/G)	n
1	0.9368	0.9368	0.0000	1.0675	1.0000	1.0675	1.0000	0.0000	1
2	0.8775	1.8143	0.8775	1.1396	2.0675	0.5512	0.4837	0.4837	2
3	0.8220	2.6363	2.5216	1.2165	3.2071	0.3793	0.3118	0.9565	3
4	0.7701	3.4064	4.8318	1.2986	4.4235	0.2936	0.2261	1.4184	4
5	0.7214	4.1278	7.7173	1.3862	5.7221	0.2423	0.1748	1.8696	5
6	0.6758	4.8036	11.0961	1.4798	7.1084	0.2082	0.1407	2.3100	6
7	0.6330	5.4366	14.8943	1.5797	8.5882	0.1839	0.1164	2.7396	7
8	0.5930	6.0296	19.0453	1.6863	10.1679	0.1658	0.0983	3.1586	8
9	0.5555	6.5851	23.4894	1.8002	11.8542	0.1519	0.0844	3.5671	9
10	0.5204	7.1055	28.1728	1.9217	13.6544	0.1407	0.0732	3.9649	10
11	0.4875	7.5929	33.0476	2.0514	15.5760	0.1317	0.0642	4.3524	11
12	0.4567	8.0496	38.0707	2.1899	17.6274	0.1242	0.0567	4.7295	12
13	0.4278	8.4774	43.2041	2.3377	19.8173	0.1180	0.0505	5.0964	13
14	0.4007	8.8781	48.4135	2.4955	22.1549	0.1126	0.0451	5.4531	14
15	0.3754	9.2535	53.6690	2.6639	24.6504	0.1081	0.0406	5.7999	15
16	0.3517	9.6051	58.9438	2.8437	27.3143	0.1041	0.0366	6.1367	16
17	0.3294	9.9346	64.2144	3.0357	30.1580	0.1007	0.0332	6.4637	17
18	0.3086	10.2432	69.4604	3.2406	33.1937	0.0976	0.0301	6.7812	18
19	0.2891	10.5322	74.6638	3.4593	36.4343	0.0949	0.0274	7.0891	19
20	0.2708	10.8030	79.8089	3.6928	39.8936	0.0926	0.0251	7.3876	20
21	0.2537	11.0567	84.8824	3.9421	43.5864	0.0904	0.0229	7.6770	21
22	0.2376	11.2943	89.8726	4.2082	47.5285	0.0885	0.0210	7.9573	22
23	0.2226	11.5169	94.7700	4.4922	51.7366	0.0868	0.0193	8.2288	23
24	0.2085	11.7255	99.5662	4.7954	56.2289	0.0853	0.0178	8.4915	24
25	0.1953	11.9208	104.2545	5.1191	61.0243	0.0839	0.0164	8.7456	25
26	0.1830	12.1038	108.8293	5.4647	66.1435	0.0826	0.0151	8.9913	26
27	0.1714	12.2752	113.2863	5.8335	71.6081	0.0815	0.0140	9.2289	27
28	0.1606	12.4358	117.6220	6.2273	77.4417	0.0804	0.0129	9.4583	28
29	0.1504	12.5862	121.8341	6.6477	83.6690	0.0795	0.0120	9.6799	29
30	0.1409	12.7272	125.9207	7.0964	90.3167	0.0786	0.0111	9.8939	30
31	0.1320	12.8592	129.8809	7.5754	97.4130	0.0778	0.0103	10.1003	31
32	0.1237	12.9828	133.7143	8.0867	104.9884	0.0770	0.0095	10.2993	32
33	0.1158	13.0987	137.4212	8.6326	113.0751	0.0763	0.0088	10.4912	33
34	0.1085	13.2072	141.0022	9.2153	121.7077	0.0757	0.0082	10.6762	34
35	0.1017	13.3088	144.4584	9.8373	130.9230	0.0751	0.0076	10.8543	35
36	0.0952	13.4041	147.7913	10.5013	140.7603	0.0746	0.0071	11.0259	36
37	0.0892	13.4933	151.0027	11.2102	151.2616	0.0741	0.0066	11.1910	37
38	0.0836	13.5768	154.0946	11.9668	162.4717	0.0737	0.0062	11.3498	38
39	0.0783	13.6551	157.0693	12.7746	174.4386	0.0732	0.0057	11.5026	39
40	0.0733	13.7284	159.9291	13.6369	187.2132	0.0728	0.0053	11.6495	40
41	0.0687	13.7971	162.6769	14.5574	200.8501	0.0725	0.0050	11.7906	41
42	0.0644	13.8615	165.3152	15.5400	215.4075	0.0721	0.0046	11.9262	42
43	0.0603	13.9218	167.8470	16.5890	230.9475	0.0718	0.0043	12.0565	43
44	0.0565	13.9782	170.2752	17.7087	247.5364	0.0715	0.0040	12.1815	44
45	0.0529	14.0311	172.6028	18.9040	265.2451	0.0713	0.0038	12.3014	45
46	0.0496	14.0807	174.8327	20.1801	284.1492	0.0710	0.0035	12.4165	46
47	0.0464	14.1271	176.9680	21.5422	304.3292	0.0708	0.0033	12.5268	47
48	0.0435	14.1706	179.0118	22.9963	325.8715	0.0706	0.0031	12.6326	48
49	0.0407	14.2113	180.9671	24.5486	348.8678	0.0704	0.0029	12.7340	49
50	0.0382	14.2495	182.8370	26.2056	373.4164	0.0702	0.0027	12.8311	50
51	0.0357	14.2852	184.6243	27.9745	399.6220	0.0700	0.0025	12.9241	51
52	0.0335	14.3187	186.3321	29.8628	427.5965	0.0698	0.0023	13.0132	52
53	0.0314	14.3501	187.9633	31.8785	457.4592	0.0697	0.0022	13.0984	53
54	0.0294	14.3795	189.5208	34.0303	489.3377	0.0695	0.0020	13.1800	54
55	0.0275	14.4070	191.0072	36.3273	523.3680	0.0694	0.0019	13.2579	55
60	0.0199	14.5206	197.4692	50.3585	731.2374	0.0689	0.0014	13.5992	60
65	0.0143	14.6026	202.5405	69.8092	1019.3950	0.0685	0.0010	13.8702	65
70	0.0103	14.6617	206.4945	96.7725	1418.8515	0.0682	0.0007	14.0839	70
75	0.0075	14.7044	209.5601	134.1502	1972.5953	0.0680	0.0005	14.2515	75
80	0.0054	14.7352	211.9254	185.9648	2740.2189	0.0679	0.0004	14.3823	80
85	0.0039	14.7573	213.7426	257.7924	3804.3319	0.0678	0.0003	14.4838	85
90	0.0028	14.7734	215.1335	357.3630	5279.4512	0.0677	0.0002	14.5623	90
95	0.0020	14.7849	216.1947	495.3920	7324.3253	0.0676	0.0001	14.6227	95
100	0.0015	14.7932	217.0019	686.7337	10159.0180	0.0676	0.0001	14.6690	100

ENGINEERING ECONOMIC ANALYSIS

$I = 7.00\%$

n	(P/F)	(P/A)	(P/G)	(F/P)	(F/A)	(A/P)	(A/F)	(A/G)	n
1	0.9346	0.9346	0.0000	1.0700	1.0000	1.0700	1.0000	0.0000	1
2	0.8734	1.8080	0.8734	1.1449	2.0700	0.5531	0.4831	0.4831	2
3	0.8163	2.6243	2.5060	1.2250	3.2149	0.3811	0.3111	0.9549	3
4	0.7629	3.3872	4.7947	1.3108	4.4399	0.2952	0.2252	1.4155	4
5	0.7130	4.1002	7.6467	1.4026	5.7507	0.2439	0.1739	1.8650	5
6	0.6663	4.7665	10.9784	1.5007	7.1533	0.2098	0.1398	2.3032	6
7	0.6227	5.3893	14.7149	1.6058	8.6540	0.1856	0.1156	2.7304	7
8	0.5820	5.9713	18.7889	1.7182	10.2598	0.1675	0.0975	3.1465	8
9	0.5439	6.5152	23.1404	1.8385	11.9780	0.1535	0.0835	3.5517	9
10	0.5083	7.0236	27.7156	1.9672	13.8164	0.1424	0.0724	3.9461	10
11	0.4751	7.4987	32.4665	2.1049	15.7836	0.1334	0.0634	4.3296	11
12	0.4440	7.9427	37.3506	2.2522	17.8885	0.1259	0.0559	4.7025	12
13	0.4150	8.3577	42.3302	2.4098	20.1406	0.1197	0.0497	5.0648	13
14	0.3878	8.7455	47.3718	2.5785	22.5505	0.1143	0.0443	5.4167	14
15	0.3624	9.1079	52.4461	2.7590	25.1290	0.1098	0.0398	5.7583	15
16	0.3387	9.4466	57.5271	2.9522	27.8881	0.1059	0.0359	6.0897	16
17	0.3166	9.7632	62.5923	3.1588	30.8402	0.1024	0.0324	6.4110	17
18	0.2959	10.0591	67.6219	3.3799	33.9990	0.0994	0.0294	6.7225	18
19	0.2765	10.3356	72.5991	3.6165	37.3790	0.0968	0.0268	7.0242	19
20	0.2584	10.5940	77.5091	3.8697	40.9955	0.0944	0.0244	7.3163	20
21	0.2415	10.8355	82.3393	4.1406	44.8652	0.0923	0.0223	7.5990	21
22	0.2257	11.0612	87.0793	4.4304	49.0057	0.0904	0.0204	7.8725	22
23	0.2109	11.2722	91.7201	4.7405	53.4361	0.0887	0.0187	8.1369	23
24	0.1971	11.4693	96.2545	5.0724	58.1767	0.0872	0.0172	8.3923	24
25	0.1842	11.6536	100.6765	5.4274	63.2490	0.0858	0.0158	8.6391	25
26	0.1722	11.8258	104.9814	5.8074	68.6765	0.0846	0.0146	8.8773	26
27	0.1609	11.9867	109.1656	6.2139	74.4838	0.0834	0.0134	9.1072	27
28	0.1504	12.1371	113.2264	6.6488	80.6977	0.0824	0.0124	9.3289	28
29	0.1406	12.2777	117.1622	7.1143	87.3465	0.0814	0.0114	9.5427	29
30	0.1314	12.4090	120.9718	7.6123	94.4608	0.0806	0.0106	9.7487	30
31	0.1228	12.5318	124.6550	8.1451	102.0730	0.0798	0.0098	9.9471	31
32	0.1147	12.6466	128.2120	8.7153	110.2182	0.0791	0.0091	10.1381	32
33	0.1072	12.7538	131.6435	9.3253	118.9334	0.0784	0.0084	10.3219	33
34	0.1002	12.8540	134.9507	9.9781	128.2588	0.0778	0.0078	10.4987	34
35	0.0937	12.9477	138.1353	10.6766	138.2369	0.0772	0.0072	10.6687	35
36	0.0875	13.0352	141.1990	11.4239	148.9135	0.0767	0.0067	10.8321	36
37	0.0818	13.1170	144.1441	12.2236	160.3374	0.0762	0.0062	10.9891	37
38	0.0765	13.1935	146.9730	13.0793	172.5610	0.0758	0.0058	11.1398	38
39	0.0715	13.2649	149.6883	13.9948	185.6403	0.0754	0.0054	11.2845	39
40	0.0668	13.3317	152.2928	14.9745	199.6351	0.0750	0.0050	11.4233	40
41	0.0624	13.3941	154.7892	16.0227	214.6096	0.0747	0.0047	11.5565	41
42	0.0583	13.4524	157.1807	17.1443	230.6322	0.0743	0.0043	11.6842	42
43	0.0545	13.5070	159.4702	18.3444	247.7765	0.0740	0.0040	11.8065	43
44	0.0509	13.5579	161.6609	19.6285	266.1209	0.0738	0.0038	11.9237	44
45	0.0476	13.6055	163.7559	21.0025	285.7493	0.0735	0.0035	12.0360	45
46	0.0445	13.6500	165.7584	22.4726	306.7518	0.0733	0.0033	12.1435	46
47	0.0416	13.6916	167.6714	24.0457	329.2244	0.0730	0.0030	12.2463	47
48	0.0389	13.7305	169.4981	25.7289	353.2701	0.0728	0.0028	12.3447	48
49	0.0363	13.7668	171.2417	27.5299	378.9990	0.0726	0.0026	12.4387	49
50	0.0339	13.8007	172.9051	29.4570	406.5289	0.0725	0.0025	12.5287	50
51	0.0317	13.8325	174.4915	31.5190	435.9860	0.0723	0.0023	12.6146	51
52	0.0297	13.8621	176.0037	33.7253	467.5050	0.0721	0.0021	12.6967	52
53	0.0277	13.8898	177.4447	36.0861	501.2303	0.0720	0.0020	12.7751	53
54	0.0259	13.9157	178.8173	38.6122	537.3164	0.0719	0.0019	12.8500	54
55	0.0242	13.9399	180.1243	41.3150	575.9286	0.0717	0.0017	12.9215	55
60	0.0173	14.0392	185.7677	57.9464	813.5204	0.0712	0.0012	13.2321	60
65	0.0123	14.1099	190.1452	81.2729	1146.7552	0.0709	0.0009	13.4760	65
70	0.0088	14.1604	193.5185	113.9894	1614.1342	0.0706	0.0006	13.6662	70
75	0.0063	14.1964	196.1035	159.8760	2269.6574	0.0704	0.0004	13.8136	75
80	0.0045	14.2220	198.0748	224.2344	3189.0627	0.0703	0.0003	13.9273	80
85	0.0032	14.2403	199.5717	314.5003	4478.5761	0.0702	0.0002	14.0146	85
90	0.0023	14.2533	200.7042	441.1030	6287.1854	0.0702	0.0002	14.0812	90
95	0.0016	14.2626	201.5581	618.6697	8823.8535	0.0701	0.0001	14.1319	95
100	0.0012	14.2693	202.2001	867.7163	12381.6618	0.0701	0.0001	14.1703	100

$I = 7.25\%$

n	(P/F)	(P/A)	(P/G)	(F/P)	(F/A)	(A/P)	(A/F)	(A/G)	n
1	0.9324	0.9324	0.0000	1.0725	1.0000	1.0725	1.0000	0.0000	1
2	0.8694	1.8018	0.8694	1.1503	2.0725	0.5550	0.4825	0.4825	2
3	0.8106	2.6124	2.4906	1.2336	3.2228	0.3828	0.3103	0.9534	3
4	0.7558	3.3682	4.7580	1.3231	4.4564	0.2969	0.2244	1.4126	4
5	0.7047	4.0729	7.5769	1.4190	5.7795	0.2455	0.1730	1.8603	5
6	0.6571	4.7300	10.8622	1.5219	7.1985	0.2114	0.1389	2.2965	6
7	0.6127	5.3426	14.5382	1.6322	8.7204	0.1872	0.1147	2.7212	7
8	0.5712	5.9139	18.5369	1.7506	10.3526	0.1691	0.0966	3.1345	8
9	0.5326	6.4465	22.7979	1.8775	12.1032	0.1551	0.0826	3.5365	9
10	0.4966	6.9431	27.2675	2.0136	13.9807	0.1440	0.0715	3.9273	10
11	0.4631	7.4062	31.8981	2.1596	15.9943	0.1350	0.0625	4.3070	11
12	0.4318	7.8379	36.6473	2.3162	18.1539	0.1276	0.0551	4.6756	12
13	0.4026	8.2405	41.4781	2.4841	20.4700	0.1214	0.0489	5.0334	13
14	0.3754	8.6158	46.3576	2.6642	22.9541	0.1161	0.0436	5.3805	14
15	0.3500	8.9658	51.2573	2.8573	25.6183	0.1115	0.0390	5.7170	15
16	0.3263	9.2921	56.1521	3.0645	28.4756	0.1076	0.0351	6.0430	16
17	0.3043	9.5964	61.0203	3.2867	31.5401	0.1042	0.0317	6.3587	17
18	0.2837	9.8801	65.8431	3.5249	34.8267	0.1012	0.0287	6.6642	18
19	0.2645	10.1446	70.6044	3.7805	38.3517	0.0986	0.0261	6.9598	19
20	0.2466	10.3912	75.2904	4.0546	42.1322	0.0962	0.0237	7.2456	20
21	0.2300	10.6212	79.8897	4.3485	46.1867	0.0942	0.0217	7.5217	21
22	0.2144	10.8356	84.3924	4.6638	50.5353	0.0923	0.0198	7.7884	22
23	0.1999	11.0355	88.7907	5.0019	55.1991	0.0906	0.0181	8.0459	23
24	0.1864	11.2220	93.0781	5.3646	60.2010	0.0891	0.0166	8.2943	24
25	0.1738	11.3958	97.2495	5.7535	65.5656	0.0878	0.0153	8.5338	25
26	0.1621	11.5578	101.3009	6.1706	71.3191	0.0865	0.0140	8.7647	26
27	0.1511	11.7089	105.2296	6.6180	77.4897	0.0854	0.0129	8.9871	27
28	0.1409	11.8498	109.0336	7.0978	84.1077	0.0844	0.0119	9.2013	28
29	0.1314	11.9812	112.7118	7.6124	91.2055	0.0835	0.0110	9.4074	29
30	0.1225	12.1037	116.2638	8.1643	98.8179	0.0826	0.0101	9.6057	30
31	0.1142	12.2179	119.6900	8.7562	106.9823	0.0818	0.0093	9.7963	31
32	0.1065	12.3244	122.9910	9.3910	115.7385	0.0811	0.0086	9.9795	32
33	0.0993	12.4236	126.1682	10.0719	125.1295	0.0805	0.0080	10.1555	33
34	0.0926	12.5162	129.2231	10.8021	135.2014	0.0799	0.0074	10.3245	34
35	0.0863	12.6025	132.1579	11.5853	146.0035	0.0793	0.0068	10.4866	35
36	0.0805	12.6830	134.9748	12.4252	157.5887	0.0788	0.0063	10.6422	36
37	0.0750	12.7581	137.6762	13.3260	170.0139	0.0784	0.0059	10.7913	37
38	0.0700	12.8280	140.2651	14.2921	183.3399	0.0780	0.0055	10.9343	38
39	0.0652	12.8933	142.7441	15.3283	197.6321	0.0776	0.0051	11.0712	39
40	0.0608	12.9541	145.1165	16.4396	212.9604	0.0772	0.0047	11.2024	40
41	0.0567	13.0108	147.3851	17.6315	229.4000	0.0769	0.0044	11.3279	41
42	0.0529	13.0637	149.5533	18.9098	247.0315	0.0765	0.0040	11.4480	42
43	0.0493	13.1130	151.6242	20.2807	265.9413	0.0763	0.0038	11.5629	43
44	0.0460	13.1590	153.6012	21.7511	286.2221	0.0760	0.0035	11.6727	44
45	0.0429	13.2018	155.4873	23.3281	307.9732	0.0757	0.0032	11.7777	45
46	0.0400	13.2418	157.2859	25.0193	331.3012	0.0755	0.0030	11.8780	46
47	0.0373	13.2791	159.0002	26.8332	356.3206	0.0753	0.0028	11.9737	47
48	0.0347	13.3138	160.6334	28.7787	383.1538	0.0751	0.0026	12.0652	48
49	0.0324	13.3462	162.1885	30.8651	411.9325	0.0749	0.0024	12.1524	49
50	0.0302	13.3764	163.6687	33.1028	442.7976	0.0748	0.0023	12.2356	50
51	0.0282	13.4046	165.0771	35.5028	475.9004	0.0746	0.0021	12.3150	51
52	0.0263	13.4309	166.4165	38.0767	511.4032	0.0745	0.0020	12.3906	52
53	0.0245	13.4553	167.6898	40.8373	549.4799	0.0743	0.0018	12.4627	53
54	0.0228	13.4782	168.8999	43.7980	590.3172	0.0742	0.0017	12.5314	54
55	0.0213	13.4995	170.0495	46.9734	634.1152	0.0741	0.0016	12.5968	55
60	0.0150	13.5862	174.9797	66.6558	905.5975	0.0736	0.0011	12.8792	60
65	0.0106	13.6473	178.7596	94.5855	1290.8345	0.0733	0.0008	13.0986	65
70	0.0075	13.6903	181.6386	134.2181	1837.4909	0.0730	0.0005	13.2676	70
75	0.0053	13.7207	183.8192	190.4573	2613.2038	0.0729	0.0004	13.3972	75
80	0.0037	13.7421	185.4629	270.2614	3713.9507	0.0728	0.0003	13.4960	80
85	0.0026	13.7571	186.6965	383.5046	5275.9255	0.0727	0.0002	13.5709	85
90	0.0018	13.7678	187.6190	544.1982	7492.3886	0.0726	0.0001	13.6274	90
95	0.0013	13.7752	188.3065	772.2245	10637.5794	0.0726	0.0001	13.6699	95
100	0.0009	13.7805	188.8174	1095.7969	15100.6475	0.0726	0.0001	13.7018	100

ENGINEERING ECONOMIC ANALYSIS

$I = 7.50\%$

n	(P/F)	(P/A)	(P/G)	(F/P)	(F/A)	(A/P)	(A/F)	(A/G)	n
1	0.9302	0.9302	0.0000	1.0750	1.0000	1.0750	1.0000	0.0000	1
2	0.8653	1.7956	0.8653	1.1556	2.0750	0.5569	0.4819	0.4819	2
3	0.8050	2.6005	2.4753	1.2423	3.2306	0.3845	0.3095	0.9518	3
4	0.7488	3.3493	4.7217	1.3355	4.4729	0.2986	0.2236	1.4097	4
5	0.6966	4.0459	7.5079	1.4356	5.8084	0.2472	0.1722	1.8557	5
6	0.6480	4.6938	10.7477	1.5433	7.2440	0.2130	0.1380	2.2897	6
7	0.6028	5.2966	14.3642	1.6590	8.7873	0.1888	0.1138	2.7120	7
8	0.5607	5.8573	18.2891	1.7835	10.4464	0.1707	0.0957	3.1225	8
9	0.5216	6.3789	22.4618	1.9172	12.2298	0.1568	0.0818	3.5213	9
10	0.4852	6.8641	26.8286	2.0610	14.1471	0.1457	0.0707	3.9085	10
11	0.4513	7.3154	31.3420	2.2156	16.2081	0.1367	0.0617	4.2844	11
12	0.4199	7.7353	35.9604	2.3818	18.4237	0.1293	0.0543	4.6489	12
13	0.3906	8.1258	40.6471	2.5604	20.8055	0.1231	0.0481	5.0022	13
14	0.3633	8.4892	45.3702	2.7524	23.3659	0.1178	0.0428	5.3445	14
15	0.3380	8.8271	50.1017	2.9589	26.1184	0.1133	0.0383	5.6759	15
16	0.3144	9.1415	54.8175	3.1808	29.0772	0.1094	0.0344	5.9966	16
17	0.2925	9.4340	59.4968	3.4194	32.2580	0.1060	0.0310	6.3067	17
18	0.2720	9.7060	64.1216	3.6758	35.6774	0.1030	0.0280	6.6064	18
19	0.2531	9.9591	68.6769	3.9515	39.3532	0.1004	0.0254	6.8959	19
20	0.2354	10.1945	73.1497	4.2479	43.3047	0.0981	0.0231	7.1754	20
21	0.2190	10.4135	77.5295	4.5664	47.5525	0.0960	0.0210	7.4451	21
22	0.2037	10.6172	81.8074	4.9089	52.1190	0.0942	0.0192	7.7052	22
23	0.1895	10.8067	85.9764	5.2771	57.0279	0.0925	0.0175	7.9558	23
24	0.1763	10.9830	90.0308	5.6729	62.3050	0.0911	0.0161	8.1973	24
25	0.1640	11.1469	93.9663	6.0983	67.9779	0.0897	0.0147	8.4298	25
26	0.1525	11.2995	97.7797	6.5557	74.0762	0.0885	0.0135	8.6535	26
27	0.1419	11.4414	101.4690	7.0474	80.6319	0.0874	0.0124	8.8686	27
28	0.1320	11.5734	105.0329	7.5759	87.6793	0.0864	0.0114	9.0754	28
29	0.1228	11.6962	108.4710	8.1441	95.2553	0.0855	0.0105	9.2741	29
30	0.1142	11.8104	111.7834	8.7550	103.3994	0.0847	0.0097	9.4648	30
31	0.1063	11.9166	114.9710	9.4116	112.1544	0.0839	0.0089	9.6479	31
32	0.0988	12.0155	118.0350	10.1174	121.5659	0.0832	0.0082	9.8236	32
33	0.0919	12.1074	120.9772	10.8763	131.6834	0.0826	0.0076	9.9920	33
34	0.0855	12.1929	123.7996	11.6920	142.5596	0.0820	0.0070	10.1534	34
35	0.0796	12.2725	126.5047	12.5689	154.2516	0.0815	0.0065	10.3080	35
36	0.0740	12.3465	129.0951	13.5115	166.8205	0.0810	0.0060	10.4560	36
37	0.0688	12.4154	131.5736	14.5249	180.3320	0.0805	0.0055	10.5976	37
38	0.0640	12.4794	133.9432	15.6143	194.8569	0.0801	0.0051	10.7331	38
39	0.0596	12.5390	136.2071	16.7853	210.4712	0.0798	0.0048	10.8627	39
40	0.0554	12.5944	138.3685	18.0442	227.2565	0.0794	0.0044	10.9865	40
41	0.0516	12.6460	140.4306	19.3976	245.3008	0.0791	0.0041	11.1048	41
42	0.0480	12.6939	142.3968	20.8524	264.6983	0.0788	0.0038	11.2177	42
43	0.0446	12.7385	144.2704	22.4163	285.5507	0.0785	0.0035	11.3255	43
44	0.0415	12.7800	146.0548	24.0975	307.9670	0.0782	0.0032	11.4284	44
45	0.0386	12.8186	147.7534	25.9048	332.0645	0.0780	0.0030	11.5265	45
46	0.0359	12.8545	149.3693	27.8477	357.9694	0.0778	0.0028	11.6200	46
47	0.0334	12.8879	150.9059	29.9363	385.8171	0.0776	0.0026	11.7091	47
48	0.0311	12.9190	152.3664	32.1815	415.7533	0.0774	0.0024	11.7940	48
49	0.0289	12.9479	153.7538	34.5951	447.9348	0.0772	0.0022	11.8748	49
50	0.0269	12.9748	155.0714	37.1897	482.5299	0.0771	0.0021	11.9517	50
51	0.0250	12.9998	156.3221	39.9790	519.7197	0.0769	0.0019	12.0249	51
52	0.0233	13.0231	157.5087	42.9774	559.6987	0.0768	0.0018	12.0946	52
53	0.0216	13.0447	158.6343	46.2007	602.6761	0.0767	0.0017	12.1608	53
54	0.0201	13.0649	159.7014	49.6658	648.8768	0.0765	0.0015	12.2237	54
55	0.0187	13.0836	160.7128	53.3907	698.5425	0.0764	0.0014	12.2835	55
60	0.0130	13.1594	165.0213	76.6492	1008.6565	0.0760	0.0010	12.5402	60
65	0.0091	13.2122	168.2863	110.0399	1453.8653	0.0757	0.0007	12.7372	65
70	0.0063	13.2489	170.7444	157.9765	2093.0200	0.0755	0.0005	12.8874	70
75	0.0044	13.2745	172.5847	226.7957	3010.6094	0.0753	0.0003	13.0012	75
80	0.0031	13.2924	173.9557	325.5946	4327.9275	0.0752	0.0002	13.0869	80
85	0.0021	13.3048	174.9729	467.4331	6219.1080	0.0752	0.0002	13.1511	85
90	0.0015	13.3135	175.7246	671.0607	8934.1422	0.0751	0.0001	13.1990	90
95	0.0010	13.3195	176.2784	963.3944	12831.9249	0.0751	0.0001	13.2346	95
100	0.0007	13.3237	176.6852	1383.0772	18427.6961	0.0751	0.0001	13.2610	100

$I = 7.75\%$

n	(P/F)	(P/A)	(P/G)	(F/P)	(F/A)	(A/P)	(A/F)	(A/G)	n
1	0.9281	0.9281	0.0000	1.0775	1.0000	1.0775	1.0000	0.0000	1
2	0.8613	1.7894	0.8613	1.1610	2.0775	0.5588	0.4813	0.4813	2
3	0.7994	2.5888	2.4601	1.2510	3.2385	0.3863	0.3088	0.9503	3
4	0.7419	3.3306	4.6857	1.3479	4.4895	0.3002	0.2227	1.4068	4
5	0.6885	4.0192	7.4398	1.4524	5.8374	0.2488	0.1713	1.8511	5
6	0.6390	4.6582	10.6347	1.5650	7.2898	0.2147	0.1372	2.2830	6
7	0.5930	5.2512	14.1929	1.6862	8.8548	0.1904	0.1129	2.7028	7
8	0.5504	5.8016	18.0456	1.8169	10.5410	0.1724	0.0949	3.1105	8
9	0.5108	6.3124	22.1319	1.9577	12.3580	0.1584	0.0809	3.5061	9
10	0.4741	6.7864	26.3984	2.1095	14.3157	0.1474	0.0699	3.8899	10
11	0.4400	7.2264	30.7980	2.2730	16.4252	0.1384	0.0609	4.2619	11
12	0.4083	7.6347	35.2894	2.4491	18.6981	0.1310	0.0535	4.6223	12
13	0.3789	8.0136	39.8367	2.6389	21.1472	0.1248	0.0473	4.9711	13
14	0.3517	8.3653	44.4087	2.8434	23.7861	0.1195	0.0420	5.3087	14
15	0.3264	8.6917	48.9782	3.0638	26.6296	0.1151	0.0376	5.6350	15
16	0.3029	8.9946	53.5219	3.3012	29.6934	0.1112	0.0337	5.9504	16
17	0.2811	9.2757	58.0200	3.5571	32.9946	0.1078	0.0303	6.2550	17
18	0.2609	9.5367	62.4555	3.8328	36.5517	0.1049	0.0274	6.5490	18
19	0.2421	9.7788	66.8140	4.1298	40.3844	0.1023	0.0248	6.8325	19
20	0.2247	10.0035	71.0838	4.4499	44.5142	0.1000	0.0225	7.1059	20
21	0.2086	10.2121	75.2551	4.7947	48.9641	0.0979	0.0204	7.3692	21
22	0.1936	10.4057	79.3199	5.1663	53.7588	0.0961	0.0186	7.6228	22
23	0.1796	10.5853	83.2720	5.5667	58.9251	0.0945	0.0170	7.8668	23
24	0.1667	10.7520	87.1065	5.9981	64.4918	0.0930	0.0155	8.1014	24
25	0.1547	10.9067	90.8200	6.4630	70.4899	0.0917	0.0142	8.3270	25
26	0.1436	11.0503	94.4099	6.9638	76.9529	0.0905	0.0130	8.5436	26
27	0.1333	11.1836	97.8750	7.5035	83.9167	0.0894	0.0119	8.7516	27
28	0.1237	11.3073	101.2145	8.0851	91.4203	0.0884	0.0109	8.9513	28
29	0.1148	11.4221	104.4285	8.7117	99.5053	0.0875	0.0100	9.1427	29
30	0.1065	11.5286	107.5180	9.3868	108.2170	0.0867	0.0092	9.3262	30
31	0.0989	11.6275	110.4841	10.1143	117.6038	0.0860	0.0085	9.5020	31
32	0.0918	11.7192	113.3286	10.8982	127.7181	0.0853	0.0078	9.6703	32
33	0.0852	11.8044	116.0537	11.7428	138.6163	0.0847	0.0072	9.8314	33
34	0.0790	11.8834	118.6618	12.6528	150.3590	0.0842	0.0067	9.9855	34
35	0.0733	11.9568	121.1557	13.6334	163.0118	0.0836	0.0061	10.1328	35
36	0.0681	12.0249	123.5382	14.6900	176.6453	0.0832	0.0057	10.2736	36
37	0.0632	12.0880	125.8126	15.8285	191.3353	0.0827	0.0052	10.4080	37
38	0.0586	12.1467	127.9820	17.0552	207.1638	0.0823	0.0048	10.5364	38
39	0.0544	12.2011	130.0498	18.3770	224.2190	0.0820	0.0045	10.6589	39
40	0.0505	12.2516	132.0194	19.8012	242.5959	0.0816	0.0041	10.7757	40
41	0.0469	12.2985	133.8942	21.3358	262.3971	0.0813	0.0038	10.8871	41
42	0.0435	12.3420	135.6776	22.9893	283.7329	0.0810	0.0035	10.9932	42
43	0.0404	12.3823	137.3732	24.7710	306.7222	0.0808	0.0033	11.0943	43
44	0.0375	12.4198	138.9842	26.6907	331.4931	0.0805	0.0030	11.1905	44
45	0.0348	12.4546	140.5142	28.7592	358.1839	0.0803	0.0028	11.2821	45
46	0.0323	12.4868	141.9663	30.9881	386.9431	0.0801	0.0026	11.3693	46
47	0.0299	12.5168	143.3440	33.3897	417.9312	0.0799	0.0024	11.4521	47
48	0.0278	12.5446	144.6504	35.9774	451.3209	0.0797	0.0022	11.5309	48
49	0.0258	12.5704	145.8886	38.7656	487.2982	0.0796	0.0021	11.6057	49
50	0.0239	12.5943	147.0617	41.7699	526.0639	0.0794	0.0019	11.6768	50
51	0.0222	12.6165	148.1726	45.0071	567.8338	0.0793	0.0018	11.7443	51
52	0.0206	12.6372	149.2243	48.4952	612.8409	0.0791	0.0016	11.8084	52
53	0.0191	12.6563	150.2194	52.2535	661.3361	0.0790	0.0015	11.8692	53
54	0.0178	12.6741	151.1608	56.3032	713.5896	0.0789	0.0014	11.9268	54
55	0.0165	12.6905	152.0509	60.6667	769.8928	0.0788	0.0013	11.9814	55
60	0.0113	12.7568	155.8172	88.1123	1124.0302	0.0784	0.0009	12.2145	60
65	0.0078	12.8024	158.6385	127.9744	1638.3794	0.0781	0.0006	12.3913	65
70	0.0054	12.8338	160.7380	185.8701	2385.4206	0.0779	0.0004	12.5246	70
75	0.0037	12.8554	162.2917	269.9578	3470.4235	0.0778	0.0003	12.6244	75
80	0.0026	12.8703	163.4359	392.0869	5046.2823	0.0777	0.0002	12.6987	80
85	0.0018	12.8806	164.2749	569.4672	7335.0604	0.0776	0.0001	12.7537	85
90	0.0012	12.8876	164.8879	827.0944	10659.2829	0.0776	0.0001	12.7943	90
95	0.0008	12.8925	165.3342	1201.2724	15487.3855	0.0776	0.0001	12.8241	95
100	0.0006	12.8958	165.6583	1744.7286	22499.7240	0.0775	0.0000	12.8459	100

$I = 8.00\%$

n	(P/F)	(P/A)	(P/G)	(F/P)	(F/A)	(A/P)	(A/F)	(A/G)	n
1	0.9259	0.9259	0.0000	1.0800	1.0000	1.0800	1.0000	0.0000	1
2	0.8573	1.7833	0.8573	1.1664	2.0800	0.5608	0.4808	0.4808	2
3	0.7938	2.5771	2.4450	1.2597	3.2464	0.3880	0.3080	0.9487	3
4	0.7350	3.3121	4.6501	1.3605	4.5061	0.3019	0.2219	1.4040	4
5	0.6806	3.9927	7.3724	1.4693	5.8666	0.2505	0.1705	1.8465	5
6	0.6302	4.6229	10.5233	1.5869	7.3359	0.2163	0.1363	2.2763	6
7	0.5835	5.2064	14.0242	1.7138	8.9228	0.1921	0.1121	2.6937	7
8	0.5403	5.7466	17.8061	1.8509	10.6366	0.1740	0.0940	3.0985	8
9	0.5002	6.2469	21.8081	1.9990	12.4876	0.1601	0.0801	3.4910	9
10	0.4632	6.7101	25.9768	2.1589	14.4866	0.1490	0.0690	3.8713	10
11	0.4289	7.1390	30.2657	2.3316	16.6455	0.1401	0.0601	4.2395	11
12	0.3971	7.5361	34.6339	2.5182	18.9771	0.1327	0.0527	4.5957	12
13	0.3677	7.9038	39.0463	2.7196	21.4953	0.1265	0.0465	4.9402	13
14	0.3405	8.2442	43.4723	2.9372	24.2149	0.1213	0.0413	5.2731	14
15	0.3152	8.5595	47.8857	3.1722	27.1521	0.1168	0.0368	5.5945	15
16	0.2919	8.8514	52.2640	3.4259	30.3243	0.1130	0.0330	5.9046	16
17	0.2703	9.1216	56.5883	3.7000	33.7502	0.1096	0.0296	6.2037	17
18	0.2502	9.3719	60.8426	3.9960	37.4502	0.1067	0.0267	6.4920	18
19	0.2317	9.6036	65.0134	4.3157	41.4463	0.1041	0.0241	6.7697	19
20	0.2145	9.8181	69.0898	4.6610	45.7620	0.1019	0.0219	7.0369	20
21	0.1987	10.0168	73.0629	5.0338	50.4229	0.0998	0.0198	7.2940	21
22	0.1839	10.2007	76.9257	5.4365	55.4568	0.0980	0.0180	7.5412	22
23	0.1703	10.3711	80.6726	5.8715	60.8933	0.0964	0.0164	7.7786	23
24	0.1577	10.5288	84.2997	6.3412	66.7648	0.0950	0.0150	8.0066	24
25	0.1460	10.6748	87.8041	6.8485	73.1059	0.0937	0.0137	8.2254	25
26	0.1352	10.8100	91.1842	7.3964	79.9544	0.0925	0.0125	8.4352	26
27	0.1252	10.9352	94.4390	7.9881	87.3508	0.0914	0.0114	8.6363	27
28	0.1159	11.0511	97.5687	8.6271	95.3388	0.0905	0.0105	8.8289	28
29	0.1073	11.1584	100.5738	9.3173	103.9659	0.0896	0.0096	9.0133	29
30	0.0994	11.2578	103.4558	10.0627	113.2832	0.0888	0.0088	9.1897	30
31	0.0920	11.3498	106.2163	10.8677	123.3459	0.0881	0.0081	9.3584	31
32	0.0852	11.4350	108.8575	11.7371	134.2135	0.0875	0.0075	9.5197	32
33	0.0789	11.5139	111.3819	12.6760	145.9506	0.0869	0.0069	9.6737	33
34	0.0730	11.5869	113.7924	13.6901	158.6267	0.0863	0.0063	9.8208	34
35	0.0676	11.6546	116.0920	14.7853	172.3168	0.0858	0.0058	9.9611	35
36	0.0626	11.7172	118.2839	15.9682	187.1021	0.0853	0.0053	10.0949	36
37	0.0580	11.7752	120.3713	17.2456	203.0703	0.0849	0.0049	10.2225	37
38	0.0537	11.8289	122.3579	18.6253	220.3159	0.0845	0.0045	10.3440	38
39	0.0497	11.8786	124.2470	20.1153	238.9412	0.0842	0.0042	10.4597	39
40	0.0460	11.9246	126.0422	21.7245	259.0565	0.0839	0.0039	10.5699	40
41	0.0426	11.9672	127.7470	23.4625	280.7810	0.0836	0.0036	10.6747	41
42	0.0395	12.0067	129.3651	25.3395	304.2435	0.0833	0.0033	10.7744	42
43	0.0365	12.0432	130.8998	27.3666	329.5830	0.0830	0.0030	10.8692	43
44	0.0338	12.0771	132.3547	29.5560	356.9496	0.0828	0.0028	10.9592	44
45	0.0313	12.1084	133.7331	31.9204	386.5056	0.0826	0.0026	11.0447	45
46	0.0290	12.1374	135.0384	34.4741	418.4261	0.0824	0.0024	11.1258	46
47	0.0269	12.1643	136.2739	37.2320	452.9002	0.0822	0.0022	11.2028	47
48	0.0249	12.1891	137.4428	40.2106	490.1322	0.0820	0.0020	11.2758	48
49	0.0230	12.2122	138.5480	43.4274	530.3427	0.0819	0.0019	11.3451	49
50	0.0213	12.2335	139.5928	46.9016	573.7702	0.0817	0.0017	11.4107	50
51	0.0197	12.2532	140.5799	50.6537	620.6718	0.0816	0.0016	11.4729	51
52	0.0183	12.2715	141.5121	54.7060	671.3255	0.0815	0.0015	11.5318	52
53	0.0169	12.2884	142.3923	59.0825	726.0316	0.0814	0.0014	11.5875	53
54	0.0157	12.3041	143.2229	63.8091	785.1141	0.0813	0.0013	11.6403	54
55	0.0145	12.3186	144.0065	68.9139	848.9232	0.0812	0.0012	11.6902	55
60	0.0099	12.3766	147.3000	101.2571	1253.2133	0.0808	0.0008	11.9015	60
65	0.0067	12.4160	149.7387	148.7798	1847.2481	0.0805	0.0005	12.0602	65
70	0.0046	12.4428	151.5326	218.6064	2720.0801	0.0804	0.0004	12.1783	70
75	0.0031	12.4611	152.8448	321.2045	4002.5566	0.0802	0.0002	12.2658	75
80	0.0021	12.4735	153.8001	471.9548	5886.9354	0.0802	0.0002	12.3301	80
85	0.0014	12.4820	154.4925	693.4565	8655.7061	0.0801	0.0001	12.3772	85
90	0.0010	12.4877	154.9925	1018.9151	12723.9386	0.0801	0.0001	12.4116	90
95	0.0007	12.4917	155.3524	1497.1205	18701.5069	0.0801	0.0001	12.4365	95
100	0.0005	12.4943	155.6107	2199.7613	27484.5157	0.0800	0.0000	12.4545	100

$I = 8.25\%$

n	(P/F)	(P/A)	(P/G)	(F/P)	(F/A)	(A/P)	(A/F)	(A/G)	n
1	0.9238	0.9238	0.0000	1.0825	1.0000	1.0825	1.0000	0.0000	1
2	0.8534	1.7772	0.8534	1.1718	2.0825	0.5627	0.4802	0.4802	2
3	0.7883	2.5655	2.4301	1.2685	3.2543	0.3898	0.3073	0.9472	3
4	0.7283	3.2938	4.6149	1.3731	4.5228	0.3036	0.2211	1.4011	4
5	0.6728	3.9665	7.3059	1.4864	5.8959	0.2521	0.1696	1.8419	5
6	0.6215	4.5880	10.4133	1.6090	7.3823	0.2180	0.1355	2.2697	6
7	0.5741	5.1621	13.8581	1.7418	8.9914	0.1937	0.1112	2.6846	7
8	0.5304	5.6925	17.5707	1.8855	10.7332	0.1757	0.0932	3.0866	8
9	0.4899	6.1825	21.4902	2.0410	12.6186	0.1617	0.0792	3.4760	9
10	0.4526	6.6351	25.5637	2.2094	14.6597	0.1507	0.0682	3.8528	10
11	0.4181	7.0532	29.7448	2.3917	16.8691	0.1418	0.0593	4.2172	11
12	0.3862	7.4394	33.9935	2.5890	19.2608	0.1344	0.0519	4.5694	12
13	0.3568	7.7962	38.2752	2.8026	21.8498	0.1283	0.0458	4.9094	13
14	0.3296	8.1259	42.5603	3.0338	24.6524	0.1231	0.0406	5.2376	14
15	0.3045	8.4304	46.8232	3.2841	27.6863	0.1186	0.0361	5.5541	15
16	0.2813	8.7116	51.0426	3.5551	30.9704	0.1148	0.0323	5.8591	16
17	0.2599	8.9715	55.2002	3.8483	34.5254	0.1115	0.0290	6.1528	17
18	0.2400	9.2115	59.2810	4.1658	38.3738	0.1086	0.0261	6.4355	18
19	0.2218	9.4333	63.2725	4.5095	42.5396	0.1060	0.0235	6.7074	19
20	0.2049	9.6381	67.1648	4.8816	47.0491	0.1038	0.0213	6.9686	20
21	0.1892	9.8274	70.9496	5.2843	51.9307	0.1018	0.0193	7.2196	21
22	0.1748	10.0022	74.6207	5.7202	57.2150	0.1000	0.0175	7.4604	22
23	0.1615	10.1637	78.1736	6.1922	62.9352	0.0984	0.0159	7.6915	23
24	0.1492	10.3129	81.6049	6.7030	69.1274	0.0970	0.0145	7.9129	24
25	0.1378	10.4507	84.9125	7.2560	75.8304	0.0957	0.0132	8.1251	25
26	0.1273	10.5780	88.0954	7.8546	83.0864	0.0945	0.0120	8.3282	26
27	0.1176	10.6956	91.1532	8.5026	90.9410	0.0935	0.0110	8.5225	27
28	0.1086	10.8043	94.0867	9.2041	99.4436	0.0926	0.0101	8.7083	28
29	0.1004	10.9046	96.8970	9.9634	108.6477	0.0917	0.0092	8.8858	29
30	0.0927	10.9974	99.5858	10.7854	118.6112	0.0909	0.0084	9.0554	30
31	0.0857	11.0830	102.1553	11.6752	129.3966	0.0902	0.0077	9.2173	31
32	0.0791	11.1621	104.6082	12.6384	141.0718	0.0896	0.0071	9.3717	32
33	0.0731	11.2352	106.9472	13.6811	153.7102	0.0890	0.0065	9.5189	33
34	0.0675	11.3028	109.1754	14.8098	167.3913	0.0885	0.0060	9.6592	34
35	0.0624	11.3651	111.2962	16.0316	182.2011	0.0880	0.0055	9.7928	35
36	0.0576	11.4228	113.3130	17.3542	198.2327	0.0875	0.0050	9.9199	36
37	0.0532	11.4760	115.2294	18.7859	215.5869	0.0871	0.0046	10.0409	37
38	0.0492	11.5252	117.0488	20.3358	234.3728	0.0868	0.0043	10.1559	38
39	0.0454	11.5706	118.7750	22.0135	254.7086	0.0864	0.0039	10.2653	39
40	0.0420	11.6125	120.4117	23.8296	276.7221	0.0861	0.0036	10.3691	40
41	0.0388	11.6513	121.9623	25.7955	300.5516	0.0858	0.0033	10.4677	41
42	0.0358	11.6871	123.4306	27.9236	326.3471	0.0856	0.0031	10.5612	42
43	0.0331	11.7202	124.8201	30.2273	354.2708	0.0853	0.0028	10.6500	43
44	0.0306	11.7508	126.1342	32.7211	384.4981	0.0851	0.0026	10.7341	44
45	0.0282	11.7790	127.3764	35.4206	417.2192	0.0849	0.0024	10.8139	45
46	0.0261	11.8051	128.5501	38.3428	452.6398	0.0847	0.0022	10.8894	46
47	0.0241	11.8292	129.6583	41.5061	490.9826	0.0845	0.0020	10.9609	47
48	0.0223	11.8514	130.7044	44.9303	532.4886	0.0844	0.0019	11.0286	48
49	0.0206	11.8720	131.6913	48.6371	577.4190	0.0842	0.0017	11.0926	49
50	0.0190	11.8910	132.6220	52.6496	626.0560	0.0841	0.0016	11.1532	50
51	0.0175	11.9085	133.4993	56.9932	678.7056	0.0840	0.0015	11.2104	51
52	0.0162	11.9247	134.3259	61.6952	735.6989	0.0839	0.0014	11.2645	52
53	0.0150	11.9397	135.1045	66.7850	797.3940	0.0838	0.0013	11.3156	53
54	0.0138	11.9535	135.8377	72.2948	864.1790	0.0837	0.0012	11.3638	54
55	0.0128	11.9663	136.5277	78.2591	936.4738	0.0836	0.0011	11.4093	55
60	0.0086	12.0170	139.4087	116.3253	1397.8828	0.0832	0.0007	11.6009	60
65	0.0058	12.0511	141.5174	172.9075	2083.7272	0.0830	0.0005	11.7431	65
70	0.0039	12.0741	143.0508	257.0120	3103.1754	0.0828	0.0003	11.8478	70
75	0.0026	12.0895	144.1595	382.0260	4618.4965	0.0827	0.0002	11.9244	75
80	0.0018	12.0999	144.9574	567.8484	6870.8896	0.0826	0.0001	11.9801	80
85	0.0012	12.1069	145.5291	844.0573	10218.8763	0.0826	0.0001	12.0204	85
90	0.0008	12.1116	145.9372	1254.6178	15195.3675	0.0826	0.0001	12.0494	90
95	0.0005	12.1147	146.2275	1864.8804	22592.4893	0.0825	0.0000	12.0702	95
100	0.0004	12.1168	146.4335	2771.9826	33587.6681	0.0825	0.0000	12.0851	100

$I = 8.50\%$

n	(P/F)	(P/A)	(P/G)	(F/P)	(F/A)	(A/P)	(A/F)	(A/G)	n
1	0.9217	0.9217	0.0000	1.0850	1.0000	1.0850	1.0000	0.0000	1
2	0.8495	1.7711	0.8495	1.1772	2.0850	0.5646	0.4796	0.4796	2
3	0.7829	2.5540	2.4153	1.2773	3.2622	0.3915	0.3065	0.9457	3
4	0.7216	3.2756	4.5800	1.3859	4.5395	0.3053	0.2203	1.3982	4
5	0.6650	3.9406	7.2402	1.5037	5.9254	0.2538	0.1688	1.8373	5
6	0.6129	4.5536	10.3049	1.6315	7.4290	0.2196	0.1346	2.2630	6
7	0.5649	5.1185	13.6945	1.7701	9.0605	0.1954	0.1104	2.6755	7
8	0.5207	5.6392	17.3391	1.9206	10.8306	0.1773	0.0923	3.0748	8
9	0.4799	6.1191	21.1782	2.0839	12.7512	0.1634	0.0784	3.4610	9
10	0.4423	6.5613	25.1588	2.2610	14.8351	0.1524	0.0674	3.8344	10
11	0.4076	6.9690	29.2351	2.4532	17.0961	0.1435	0.0585	4.1950	11
12	0.3757	7.3447	33.3678	2.6617	19.5492	0.1362	0.0512	4.5431	12
13	0.3463	7.6910	37.5231	2.8879	22.2109	0.1300	0.0450	4.8789	13
14	0.3191	8.0101	41.6719	3.1334	25.0989	0.1248	0.0398	5.2024	14
15	0.2941	8.3042	45.7899	3.3997	28.2323	0.1204	0.0354	5.5140	15
16	0.2711	8.5753	49.8563	3.6887	31.6320	0.1166	0.0316	5.8139	16
17	0.2499	8.8252	53.8541	4.0023	35.3207	0.1133	0.0283	6.1023	17
18	0.2303	9.0555	57.7689	4.3425	39.3230	0.1104	0.0254	6.3794	18
19	0.2122	9.2677	61.5893	4.7116	43.6654	0.1079	0.0229	6.6456	19
20	0.1956	9.4633	65.3060	5.1120	48.3770	0.1057	0.0207	6.9009	20
21	0.1803	9.6436	68.9118	5.5466	53.4891	0.1037	0.0187	7.1458	21
22	0.1662	9.8098	72.4013	6.0180	59.0356	0.1019	0.0169	7.3805	22
23	0.1531	9.9629	75.7706	6.5296	65.0537	0.1004	0.0154	7.6052	23
24	0.1412	10.1041	79.0171	7.0846	71.5832	0.0990	0.0140	7.8203	24
25	0.1301	10.2342	82.1394	7.6868	78.6678	0.0977	0.0127	8.0260	25
26	0.1199	10.3541	85.1369	8.3401	86.3546	0.0966	0.0116	8.2225	26
27	0.1105	10.4646	88.0101	9.0490	94.6947	0.0956	0.0106	8.4103	27
28	0.1019	10.5665	90.7601	9.8182	103.7437	0.0946	0.0096	8.5895	28
29	0.0939	10.6603	93.3886	10.6528	113.5620	0.0938	0.0088	8.7604	29
30	0.0865	10.7468	95.8976	11.5583	124.2147	0.0931	0.0081	8.9233	30
31	0.0797	10.8266	98.2898	12.5407	135.7730	0.0924	0.0074	9.0786	31
32	0.0735	10.9001	100.5681	13.6067	148.3137	0.0917	0.0067	9.2264	32
33	0.0677	10.9678	102.7356	14.7632	161.9203	0.0912	0.0062	9.3670	33
34	0.0624	11.0302	104.7958	16.0181	176.6836	0.0907	0.0057	9.5008	34
35	0.0575	11.0878	106.7521	17.3796	192.7017	0.0902	0.0052	9.6279	35
36	0.0530	11.1408	108.6082	18.8569	210.0813	0.0898	0.0048	9.7487	36
37	0.0489	11.1897	110.3678	20.4597	228.9382	0.0894	0.0044	9.8633	37
38	0.0450	11.2347	112.0345	22.1988	249.3980	0.0890	0.0040	9.9722	38
39	0.0415	11.2763	113.6122	24.0857	271.5968	0.0887	0.0037	10.0754	39
40	0.0383	11.3145	115.1046	26.1330	295.6825	0.0884	0.0034	10.1732	40
41	0.0353	11.3498	116.5153	28.3543	321.8156	0.0881	0.0031	10.2659	41
42	0.0325	11.3823	117.8480	30.7644	350.1699	0.0879	0.0029	10.3536	42
43	0.0300	11.4123	119.1063	33.3794	380.9343	0.0876	0.0026	10.4367	43
44	0.0276	11.4399	120.2936	36.2167	414.3137	0.0874	0.0024	10.5153	44
45	0.0254	11.4653	121.4133	39.2951	450.5304	0.0872	0.0022	10.5896	45
46	0.0235	11.4888	122.4688	42.6352	489.8255	0.0870	0.0020	10.6599	46
47	0.0216	11.5104	123.4632	46.2592	532.4606	0.0869	0.0019	10.7262	47
48	0.0199	11.5303	124.3996	50.1912	578.7198	0.0867	0.0017	10.7889	48
49	0.0184	11.5487	125.2810	54.4574	628.9110	0.0866	0.0016	10.8481	49
50	0.0169	11.5656	126.1103	59.0863	683.3684	0.0865	0.0015	10.9039	50
51	0.0156	11.5812	126.8902	64.1087	742.4547	0.0863	0.0013	10.9566	51
52	0.0144	11.5956	127.6234	69.5579	806.5634	0.0862	0.0012	11.0062	52
53	0.0133	11.6088	128.3124	75.4703	876.1213	0.0861	0.0011	11.0530	53
54	0.0122	11.6210	128.9597	81.8853	951.5916	0.0861	0.0011	11.0971	54
55	0.0113	11.6323	129.5675	88.8455	1033.4769	0.0860	0.0010	11.1386	55
60	0.0075	11.6766	132.0884	133.5932	1559.9198	0.0856	0.0006	11.3122	60
65	0.0050	11.7061	133.9125	200.8783	2351.5092	0.0854	0.0004	11.4395	65
70	0.0033	11.7258	135.2236	302.0520	3541.7879	0.0853	0.0003	11.5322	70
75	0.0022	11.7388	136.1608	454.1825	5331.5584	0.0852	0.0002	11.5992	75
80	0.0015	11.7475	136.8275	682.9345	8022.7589	0.0851	0.0001	11.6474	80
85	0.0010	11.7532	137.2997	1026.8990	12069.4004	0.0851	0.0001	11.6819	85
90	0.0006	11.7571	137.6329	1544.1036	18154.1600	0.0851	0.0001	11.7064	90
95	0.0004	11.7596	137.8673	2321.8017	27303.5496	0.0850	0.0000	11.7238	95
100	0.0003	11.7613	138.0317	3491.1927	41061.0904	0.0850	0.0000	11.7361	100

$I = 8.75\%$

n	(P/F)	(P/A)	(P/G)	(F/P)	(F/A)	(A/P)	(A/F)	(A/G)	n
1	0.9195	0.9195	0.0000	1.0875	1.0000	1.0875	1.0000	0.0000	1
2	0.8456	1.7651	0.8456	1.1827	2.0875	0.5665	0.4790	0.4790	2
3	0.7775	2.5426	2.4006	1.2861	3.2702	0.3933	0.3058	0.9441	3
4	0.7150	3.2576	4.5455	1.3987	4.5563	0.3070	0.2195	1.3954	4
5	0.6574	3.9150	7.1752	1.5211	5.9550	0.2554	0.1679	1.8327	5
6	0.6045	4.5196	10.1979	1.6542	7.4760	0.2213	0.1338	2.2564	6
7	0.5559	5.0755	13.5333	1.7989	9.1302	0.1970	0.1095	2.6664	7
8	0.5112	5.5866	17.1115	1.9563	10.9291	0.1790	0.0915	3.0629	8
9	0.4700	6.0567	20.8718	2.1275	12.8854	0.1651	0.0776	3.4461	9
10	0.4322	6.4889	24.7618	2.3136	15.0128	0.1541	0.0666	3.8160	10
11	0.3974	6.8863	28.7363	2.5161	17.3265	0.1452	0.0577	4.1729	11
12	0.3655	7.2518	32.7564	2.7362	19.8425	0.1379	0.0504	4.5170	12
13	0.3361	7.5879	36.7892	2.9756	22.5787	0.1318	0.0443	4.8484	13
14	0.3090	7.8969	40.8065	3.2360	25.5544	0.1266	0.0391	5.1674	14
15	0.2842	8.1810	44.7847	3.5192	28.7904	0.1222	0.0347	5.4742	15
16	0.2613	8.4423	48.7041	3.8271	32.3096	0.1185	0.0310	5.7690	16
17	0.2403	8.6826	52.5485	4.1620	36.1366	0.1152	0.0277	6.0522	17
18	0.2209	8.9035	56.3045	4.5261	40.2986	0.1123	0.0248	6.3238	18
19	0.2032	9.1067	59.9614	4.9222	44.8247	0.1098	0.0223	6.5843	19
20	0.1868	9.2935	63.5109	5.3529	49.7469	0.1076	0.0201	6.8339	20
21	0.1718	9.4653	66.9466	5.8212	55.0997	0.1056	0.0181	7.0728	21
22	0.1580	9.6233	70.2638	6.3306	60.9210	0.1039	0.0164	7.3014	22
23	0.1453	9.7685	73.4594	6.8845	67.2516	0.1024	0.0149	7.5200	23
24	0.1336	9.9021	76.5314	7.4869	74.1361	0.1010	0.0135	7.7288	24
25	0.1228	10.0249	79.4791	8.1420	81.6230	0.0998	0.0123	7.9282	25
26	0.1129	10.1379	82.3026	8.8544	89.7650	0.0986	0.0111	8.1183	26
27	0.1039	10.2417	85.0027	9.6292	98.6194	0.0976	0.0101	8.2997	27
28	0.0955	10.3372	87.5810	10.4718	108.2486	0.0967	0.0092	8.4724	28
29	0.0878	10.4250	90.0398	11.3880	118.7204	0.0959	0.0084	8.6369	29
30	0.0807	10.5058	92.3814	12.3845	130.1084	0.0952	0.0077	8.7934	30
31	0.0742	10.5800	94.6089	13.4681	142.4929	0.0945	0.0070	8.9422	31
32	0.0683	10.6483	96.7254	14.6466	155.9610	0.0939	0.0064	9.0837	32
33	0.0628	10.7111	98.7344	15.9282	170.6076	0.0934	0.0059	9.2180	33
34	0.0577	10.7688	100.6395	17.3219	186.5358	0.0929	0.0054	9.3455	34
35	0.0531	10.8219	102.4444	18.8375	203.8577	0.0924	0.0049	9.4664	35
36	0.0488	10.8707	104.1529	20.4858	222.6952	0.0920	0.0045	9.5811	36
37	0.0449	10.9156	105.7689	22.2783	243.1810	0.0916	0.0041	9.6897	37
38	0.0413	10.9569	107.2960	24.2277	265.4594	0.0913	0.0038	9.7926	38
39	0.0380	10.9948	108.7383	26.3476	289.6871	0.0910	0.0035	9.8900	39
40	0.0349	11.0297	110.0994	28.6530	316.0347	0.0907	0.0032	9.9821	40
41	0.0321	11.0618	111.3831	31.1602	344.6877	0.0904	0.0029	10.0692	41
42	0.0295	11.0913	112.5930	33.8867	375.8479	0.0902	0.0027	10.1515	42
43	0.0271	11.1184	113.7327	36.8518	409.7346	0.0899	0.0024	10.2292	43
44	0.0250	11.1434	114.8057	40.0763	446.5864	0.0897	0.0022	10.3026	44
45	0.0229	11.1663	115.8152	43.5830	486.6627	0.0896	0.0021	10.3718	45
46	0.0211	11.1874	116.7647	47.3965	530.2456	0.0894	0.0019	10.4371	46
47	0.0194	11.2068	117.6571	51.5437	577.6421	0.0892	0.0017	10.4987	47
48	0.0178	11.2247	118.4956	56.0538	629.1858	0.0891	0.0016	10.5567	48
49	0.0164	11.2411	119.2830	60.9585	685.2396	0.0890	0.0015	10.6113	49
50	0.0151	11.2562	120.0222	66.2923	746.1981	0.0888	0.0013	10.6628	50
51	0.0139	11.2700	120.7157	72.0929	812.4904	0.0887	0.0012	10.7112	51
52	0.0128	11.2828	121.3662	78.4010	884.5833	0.0886	0.0011	10.7567	52
53	0.0117	11.2945	121.9761	85.2611	962.9843	0.0885	0.0010	10.7996	53
54	0.0108	11.3053	122.5477	92.7215	1048.2455	0.0885	0.0010	10.8398	54
55	0.0099	11.3152	123.0832	100.8346	1140.9669	0.0884	0.0009	10.8777	55
60	0.0065	11.3541	125.2898	153.3755	1741.4341	0.0881	0.0006	11.0348	60
65	0.0043	11.3796	126.8682	233.2933	2654.7806	0.0879	0.0004	11.1488	65
70	0.0028	11.3964	127.9897	354.8531	4044.0353	0.0877	0.0002	11.2307	70
75	0.0019	11.4074	128.7822	539.7528	6157.1751	0.0877	0.0002	11.2894	75
80	0.0012	11.4147	129.3395	820.9964	9371.3874	0.0876	0.0001	11.3310	80
85	0.0008	11.4194	129.7298	1248.7847	14260.3969	0.0876	0.0001	11.3605	85
90	0.0005	11.4226	130.0020	1899.4764	21696.8734	0.0875	0.0000	11.3812	90
95	0.0003	11.4246	130.1913	2889.2175	33008.1998	0.0875	0.0000	11.3957	95
100	0.0002	11.4260	130.3225	4394.6730	50213.4054	0.0875	0.0000	11.4058	100

$$I = 9.00\%$$

n	(P/F)	(P/A)	(P/G)	(F/P)	(F/A)	(A/P)	(A/F)	(A/G)	n
1	0.9174	0.9174	0.0000	1.0900	1.0000	1.0900	1.0000	0.0000	1
2	0.8417	1.7591	0.8417	1.1881	2.0900	0.5685	0.4785	0.4785	2
3	0.7722	2.5313	2.3860	1.2950	3.2781	0.3951	0.3051	0.9426	3
4	0.7084	3.2397	4.5113	1.4116	4.5731	0.3087	0.2187	1.3925	4
5	0.6499	3.8897	7.1110	1.5386	5.9847	0.2571	0.1671	1.8282	5
6	0.5963	4.4859	10.0924	1.6771	7.5233	0.2229	0.1329	2.2498	6
7	0.5470	5.0330	13.3746	1.8280	9.2004	0.1987	0.1087	2.6574	7
8	0.5019	5.5348	16.8877	1.9926	11.0285	0.1807	0.0907	3.0512	8
9	0.4604	5.9952	20.5711	2.1719	13.0210	0.1668	0.0768	3.4312	9
10	0.4224	6.4177	24.3728	2.3674	15.1929	0.1558	0.0658	3.7978	10
11	0.3875	6.8052	28.2481	2.5804	17.5603	0.1469	0.0569	4.1510	11
12	0.3555	7.1607	32.1590	2.8127	20.1407	0.1397	0.0497	4.4910	12
13	0.3262	7.4869	36.0731	3.0658	22.9534	0.1336	0.0436	4.8182	13
14	0.2992	7.7862	39.9633	3.3417	26.0192	0.1284	0.0384	5.1326	14
15	0.2745	8.0607	43.8069	3.6425	29.3609	0.1241	0.0341	5.4346	15
16	0.2519	8.3126	47.5849	3.9703	33.0034	0.1203	0.0303	5.7245	16
17	0.2311	8.5436	51.2821	4.3276	36.9737	0.1170	0.0270	6.0024	17
18	0.2120	8.7556	54.8860	4.7171	41.3013	0.1142	0.0242	6.2687	18
19	0.1945	8.9501	58.3868	5.1417	46.0185	0.1117	0.0217	6.5236	19
20	0.1784	9.1285	61.7770	5.6044	51.1601	0.1095	0.0195	6.7674	20
21	0.1637	9.2922	65.0509	6.1088	56.7645	0.1076	0.0176	7.0006	21
22	0.1502	9.4424	68.2048	6.6586	62.8733	0.1059	0.0159	7.2232	22
23	0.1378	9.5802	71.2359	7.2579	69.5319	0.1044	0.0144	7.4357	23
24	0.1264	9.7066	74.1433	7.9111	76.7898	0.1030	0.0130	7.6384	24
25	0.1160	9.8226	76.9265	8.6231	84.7009	0.1018	0.0118	7.8316	25
26	0.1064	9.9290	79.5863	9.3992	93.3240	0.1007	0.0107	8.0156	26
27	0.0976	10.0266	82.1241	10.2451	102.7231	0.0997	0.0097	8.1906	27
28	0.0895	10.1161	84.5419	11.1671	112.9682	0.0989	0.0089	8.3571	28
29	0.0822	10.1983	86.8422	12.1722	124.1354	0.0981	0.0081	8.5154	29
30	0.0754	10.2737	89.0280	13.2677	136.3075	0.0973	0.0073	8.6657	30
31	0.0691	10.3428	91.1024	14.4618	149.5752	0.0967	0.0067	8.8083	31
32	0.0634	10.4062	93.0690	15.7633	164.0370	0.0961	0.0061	8.9436	32
33	0.0582	10.4644	94.9314	17.1820	179.8003	0.0956	0.0056	9.0718	33
34	0.0534	10.5178	96.6935	18.7284	196.9823	0.0951	0.0051	9.1933	34
35	0.0490	10.5668	98.3590	20.4140	215.7108	0.0946	0.0046	9.3083	35
36	0.0449	10.6118	99.9319	22.2512	236.1247	0.0942	0.0042	9.4171	36
37	0.0412	10.6530	101.4162	24.2538	258.3759	0.0939	0.0039	9.5200	37
38	0.0378	10.6908	102.8158	26.4367	282.6298	0.0935	0.0035	9.6172	38
39	0.0347	10.7255	104.1345	28.8160	309.0665	0.0932	0.0032	9.7090	39
40	0.0318	10.7574	105.3762	31.4094	337.8824	0.0930	0.0030	9.7957	40
41	0.0292	10.7866	106.5445	34.2363	369.2919	0.0927	0.0027	9.8775	41
42	0.0268	10.8134	107.6432	37.3175	403.5281	0.0925	0.0025	9.9546	42
43	0.0246	10.8380	108.6758	40.6761	440.8457	0.0923	0.0023	10.0273	43
44	0.0226	10.8605	109.6456	44.3370	481.5218	0.0921	0.0021	10.0958	44
45	0.0207	10.8812	110.5561	48.3273	525.8587	0.0919	0.0019	10.1603	45
46	0.0190	10.9002	111.4103	52.6767	574.1860	0.0917	0.0017	10.2210	46
47	0.0174	10.9176	112.2115	57.4176	626.8628	0.0916	0.0016	10.2780	47
48	0.0160	10.9336	112.9625	62.5852	684.2804	0.0915	0.0015	10.3317	48
49	0.0147	10.9482	113.6661	68.2179	746.8656	0.0913	0.0013	10.3821	49
50	0.0134	10.9617	114.3251	74.3575	815.0836	0.0912	0.0012	10.4295	50
51	0.0123	10.9740	114.9420	81.0497	889.4411	0.0911	0.0011	10.4740	51
52	0.0113	10.9853	115.5193	88.3442	970.4908	0.0910	0.0010	10.5158	52
53	0.0104	10.9957	116.0593	96.2951	1058.8349	0.0909	0.0009	10.5549	53
54	0.0095	11.0053	116.5642	104.9617	1155.1301	0.0909	0.0009	10.5917	54
55	0.0087	11.0140	117.0362	114.4083	1260.0918	0.0908	0.0008	10.6261	55
60	0.0057	11.0480	118.9683	176.0313	1944.7921	0.0905	0.0005	10.7683	60
65	0.0037	11.0701	120.3344	270.8460	2998.2885	0.0903	0.0003	10.8702	65
70	0.0024	11.0844	121.2942	416.7301	4619.2232	0.0902	0.0002	10.9427	70
75	0.0016	11.0938	121.9646	641.1909	7113.2321	0.0901	0.0001	10.9940	75
80	0.0010	11.0998	122.4306	986.5517	10950.5741	0.0901	0.0001	11.0299	80
85	0.0007	11.1038	122.7533	1517.9320	16854.8003	0.0901	0.0001	11.0551	85
90	0.0004	11.1064	122.9758	2335.5266	25939.1842	0.0900	0.0000	11.0726	90
95	0.0003	11.1080	123.1287	3593.4971	39916.6350	0.0900	0.0000	11.0847	95
100	0.0002	11.1091	123.2335	5529.0408	61422.6755	0.0900	0.0000	11.0930	100

$I = 9.25\%$

n	(P/F)	(P/A)	(P/G)	(F/P)	(F/A)	(A/P)	(A/F)	(A/G)	n
1	0.9153	0.9153	0.0000	1.0925	1.0000	1.0925	1.0000	0.0000	1
2	0.8378	1.7532	0.8378	1.1936	2.0925	0.5704	0.4779	0.4779	2
3	0.7669	2.5201	2.3716	1.3040	3.2861	0.3968	0.3043	0.9411	3
4	0.7020	3.2220	4.4775	1.4246	4.5900	0.3104	0.2179	1.3897	4
5	0.6425	3.8646	7.0476	1.5563	6.0146	0.2588	0.1663	1.8237	5
6	0.5881	4.4527	9.9883	1.7003	7.5709	0.2246	0.1321	2.2432	6
7	0.5383	4.9910	13.2183	1.8576	9.2713	0.2004	0.1079	2.6484	7
8	0.4928	5.4838	16.6675	2.0294	11.1288	0.1824	0.0899	3.0394	8
9	0.4510	5.9348	20.2758	2.2171	13.1583	0.1685	0.0760	3.4164	9
10	0.4128	6.3476	23.9914	2.4222	15.3754	0.1575	0.0650	3.7796	10
11	0.3779	6.7255	27.7703	2.6463	17.7976	0.1487	0.0562	4.1291	11
12	0.3459	7.0714	31.5751	2.8911	20.4439	0.1414	0.0489	4.4652	12
13	0.3166	7.3880	35.3744	3.1585	23.3350	0.1354	0.0429	4.7881	13
14	0.2898	7.6778	39.1418	3.4506	26.4935	0.1302	0.0377	5.0980	14
15	0.2653	7.9431	42.8555	3.7698	29.9441	0.1259	0.0334	5.3953	15
16	0.2428	8.1859	46.4975	4.1185	33.7139	0.1222	0.0297	5.6802	16
17	0.2222	8.4081	50.0535	4.4995	37.8325	0.1189	0.0264	5.9530	17
18	0.2034	8.6116	53.5118	4.9157	42.3320	0.1161	0.0236	6.2139	18
19	0.1862	8.7978	56.8635	5.3704	47.2477	0.1137	0.0212	6.4634	19
20	0.1704	8.9682	60.1019	5.8672	52.6181	0.1115	0.0190	6.7016	20
21	0.1560	9.1242	63.2220	6.4099	58.4853	0.1096	0.0171	6.9290	21
22	0.1428	9.2670	66.2208	7.0028	64.8951	0.1079	0.0154	7.1459	22
23	0.1307	9.3977	69.0964	7.6506	71.8980	0.1064	0.0139	7.3525	23
24	0.1196	9.5174	71.8482	8.3582	79.5485	0.1051	0.0126	7.5492	24
25	0.1095	9.6269	74.4765	9.1314	87.9067	0.1039	0.0114	7.7363	25
26	0.1002	9.7271	76.9825	9.9760	97.0381	0.1028	0.0103	7.9142	26
27	0.0918	9.8189	79.3681	10.8988	107.0141	0.1018	0.0093	8.0832	27
28	0.0840	9.9029	81.6357	11.9069	117.9130	0.1010	0.0085	8.2436	28
29	0.0769	9.9797	83.7882	13.0083	129.8199	0.1002	0.0077	8.3958	29
30	0.0704	10.0501	85.8287	14.2116	142.8282	0.0995	0.0070	8.5401	30
31	0.0644	10.1145	87.7610	15.5262	157.0399	0.0989	0.0064	8.6767	31
32	0.0590	10.1735	89.5885	16.9624	172.5660	0.0983	0.0058	8.8061	32
33	0.0540	10.2274	91.3153	18.5314	189.5284	0.0978	0.0053	8.9285	33
34	0.0494	10.2768	92.9453	20.2455	208.0598	0.0973	0.0048	9.0442	34
35	0.0452	10.3220	94.4825	22.1182	228.3053	0.0969	0.0044	9.1535	35
36	0.0414	10.3634	95.9309	24.1642	250.4236	0.0965	0.0040	9.2567	36
37	0.0379	10.4013	97.2946	26.3994	274.5877	0.0961	0.0036	9.3541	37
38	0.0347	10.4360	98.5775	28.8413	300.9871	0.0958	0.0033	9.4459	38
39	0.0317	10.4677	99.7835	31.5091	329.8284	0.0955	0.0030	9.5325	39
40	0.0290	10.4968	100.9164	34.4237	361.3375	0.0953	0.0028	9.6141	40
41	0.0266	10.5233	101.9800	37.6079	395.7613	0.0950	0.0025	9.6908	41
42	0.0243	10.5477	102.9779	41.0866	433.3692	0.0948	0.0023	9.7631	42
43	0.0223	10.5700	103.9136	44.8872	474.4558	0.0946	0.0021	9.8310	43
44	0.0204	10.5904	104.7905	49.0392	519.3430	0.0944	0.0019	9.8949	44
45	0.0187	10.6090	105.6117	53.5754	568.3822	0.0943	0.0018	9.9549	45
46	0.0171	10.6261	106.3806	58.5311	621.9576	0.0941	0.0016	10.0112	46
47	0.0156	10.6417	107.0999	63.9452	680.4886	0.0940	0.0015	10.0641	47
48	0.0143	10.6561	107.7727	69.8601	744.4338	0.0938	0.0013	10.1137	48
49	0.0131	10.6692	108.4016	76.3222	814.2940	0.0937	0.0012	10.1603	49
50	0.0120	10.6812	108.9893	83.3820	890.6162	0.0936	0.0011	10.2039	50
51	0.0110	10.6921	109.5381	91.0948	973.9982	0.0935	0.0010	10.2447	51
52	0.0100	10.7022	110.0506	99.5211	1065.0930	0.0934	0.0009	10.2830	52
53	0.0092	10.7114	110.5289	108.7268	1164.6141	0.0934	0.0009	10.3188	53
54	0.0084	10.7198	110.9750	118.7840	1273.3409	0.0933	0.0008	10.3523	54
55	0.0077	10.7275	111.3912	129.7716	1392.1249	0.0932	0.0007	10.3837	55
60	0.0050	10.7573	113.0834	201.9699	2172.6480	0.0930	0.0005	10.5123	60
65	0.0032	10.7764	114.2663	314.3359	3387.4150	0.0928	0.0003	10.6034	65
70	0.0020	10.7887	115.0879	489.2166	5278.0174	0.0927	0.0002	10.6674	70
75	0.0013	10.7966	115.6552	761.3922	8220.4562	0.0926	0.0001	10.7122	75
80	0.0008	10.8017	116.0452	1184.9926	12799.9204	0.0926	0.0001	10.7432	80
85	0.0005	10.8049	116.3120	1844.2631	19927.1688	0.0926	0.0001	10.7647	85
90	0.0003	10.8070	116.4939	2870.3186	31019.6607	0.0925	0.0000	10.7794	90
95	0.0002	10.8084	116.6176	4467.2199	48283.4589	0.0925	0.0000	10.7895	95
100	0.0001	10.8093	116.7013	6952.5571	75151.9686	0.0925	0.0000	10.7964	100

$$I = 9.50\%$$

n	(P/F)	(P/A)	(P/G)	(F/P)	(F/A)	(A/P)	(A/F)	(A/G)	n
1	0.9132	0.9132	0.0000	1.0950	1.0000	1.0950	1.0000	0.0000	1
2	0.8340	1.7473	0.8340	1.1990	2.0950	0.5723	0.4773	0.4773	2
3	0.7617	2.5089	2.3573	1.3129	3.2940	0.3986	0.3036	0.9396	3
4	0.6956	3.2045	4.4440	1.4377	4.6070	0.3121	0.2171	1.3868	4
5	0.6352	3.8397	6.9850	1.5742	6.0446	0.2604	0.1654	1.8191	5
6	0.5801	4.4198	9.8855	1.7238	7.6189	0.2263	0.1313	2.2366	6
7	0.5298	4.9496	13.0643	1.8876	9.3426	0.2020	0.1070	2.6395	7
8	0.4838	5.4334	16.4510	2.0669	11.2302	0.1840	0.0890	3.0277	8
9	0.4418	5.8753	19.9858	2.2632	13.2971	0.1702	0.0752	3.4017	9
10	0.4035	6.2788	23.6174	2.4782	15.5603	0.1593	0.0643	3.7615	10
11	0.3685	6.6473	27.3025	2.7137	18.0385	0.1504	0.0554	4.1073	11
12	0.3365	6.9838	31.0044	2.9715	20.7522	0.1432	0.0482	4.4394	12
13	0.3073	7.2912	34.6924	3.2537	23.7236	0.1372	0.0422	4.7581	13
14	0.2807	7.5719	38.3412	3.5629	26.9774	0.1321	0.0371	5.0636	14
15	0.2563	7.8282	41.9297	3.9013	30.5402	0.1277	0.0327	5.3563	15
16	0.2341	8.0623	45.4410	4.2719	34.4416	0.1240	0.0290	5.6363	16
17	0.2138	8.2760	48.8614	4.6778	38.7135	0.1208	0.0258	5.9040	17
18	0.1952	8.4713	52.1803	5.1222	43.3913	0.1180	0.0230	6.1597	18
19	0.1783	8.6496	55.3896	5.6088	48.5135	0.1156	0.0206	6.4037	19
20	0.1628	8.8124	58.4832	6.1416	54.1222	0.1135	0.0185	6.6365	20
21	0.1487	8.9611	61.4572	6.7251	60.2638	0.1116	0.0166	6.8582	21
22	0.1358	9.0969	64.3089	7.3639	66.9889	0.1099	0.0149	7.0693	22
23	0.1240	9.2209	67.0373	8.0635	74.3529	0.1084	0.0134	7.2701	23
24	0.1133	9.3341	69.6421	8.8296	82.4164	0.1071	0.0121	7.4610	24
25	0.1034	9.4376	72.1245	9.6684	91.2459	0.1060	0.0110	7.6423	25
26	0.0945	9.5320	74.4859	10.5869	100.9143	0.1049	0.0099	7.8143	26
27	0.0863	9.6183	76.7287	11.5926	111.5012	0.1040	0.0090	7.9774	27
28	0.0788	9.6971	78.8557	12.6939	123.0938	0.1031	0.0081	8.1319	28
29	0.0719	9.7690	80.8701	13.8998	135.7877	0.1024	0.0074	8.2782	29
30	0.0657	9.8347	82.7755	15.2203	149.6875	0.1017	0.0067	8.4167	30
31	0.0600	9.8947	84.5755	16.6662	164.9078	0.1011	0.0061	8.5475	31
32	0.0548	9.9495	86.2742	18.2495	181.5741	0.1005	0.0055	8.6712	32
33	0.0500	9.9996	87.8755	19.9832	199.8236	0.1000	0.0050	8.7879	33
34	0.0457	10.0453	89.3836	21.8816	219.8068	0.0995	0.0045	8.8981	34
35	0.0417	10.0870	90.8026	23.9604	241.6885	0.0991	0.0041	9.0020	35
36	0.0381	10.1251	92.1367	26.2366	265.6489	0.0988	0.0038	9.0998	36
37	0.0348	10.1599	93.3897	28.7291	291.8855	0.0984	0.0034	9.1920	37
38	0.0318	10.1917	94.5659	31.4584	320.6147	0.0981	0.0031	9.2787	38
39	0.0290	10.2207	95.6690	34.4469	352.0731	0.0978	0.0028	9.3603	39
40	0.0265	10.2472	96.7030	37.7194	386.5200	0.0976	0.0026	9.4370	40
41	0.0242	10.2715	97.6715	41.3027	424.2394	0.0974	0.0024	9.5090	41
42	0.0221	10.2936	98.5720	45.2265	465.5421	0.0971	0.0021	9.5767	42
43	0.0202	10.3138	99.4261	49.5230	510.7686	0.0970	0.0020	9.6401	43
44	0.0184	10.3322	100.2190	54.2277	560.2917	0.0968	0.0018	9.6997	44
45	0.0168	10.3490	100.9600	59.3793	614.5194	0.0966	0.0016	9.7555	45
46	0.0154	10.3644	101.6521	65.0204	673.8987	0.0965	0.0015	9.8078	46
47	0.0140	10.3785	102.2982	71.1973	738.9191	0.0964	0.0014	9.8568	47
48	0.0128	10.3913	102.9011	77.9611	810.1164	0.0962	0.0012	9.9026	48
49	0.0117	10.4030	103.4634	85.3674	888.0775	0.0961	0.0011	9.9455	49
50	0.0107	10.4137	103.9876	93.4773	973.4448	0.0960	0.0010	9.9856	50
51	0.0098	10.4235	104.4760	102.3576	1066.9221	0.0959	0.0009	10.0231	51
52	0.0089	10.4324	104.9311	112.0816	1169.2797	0.0959	0.0009	10.0582	52
53	0.0081	10.4405	105.3548	122.7293	1281.3612	0.0958	0.0008	10.0909	53
54	0.0074	10.4480	105.7491	134.3886	1404.0905	0.0957	0.0007	10.1215	54
55	0.0068	10.4548	106.1161	147.1555	1538.4791	0.0956	0.0006	10.1500	55
60	0.0043	10.4809	107.5987	231.6579	2427.9781	0.0954	0.0004	10.2662	60
65	0.0027	10.4975	108.6233	364.6849	3828.2618	0.0953	0.0003	10.3476	65
70	0.0017	10.5080	109.3268	574.1011	6032.6426	0.0952	0.0002	10.4042	70
75	0.0011	10.5147	109.8072	903.7721	9502.8644	0.0951	0.0001	10.4432	75
80	0.0007	10.5189	110.1336	1422.7531	14965.8219	0.0951	0.0001	10.4700	80
85	0.0004	10.5216	110.3544	2239.7530	23565.8212	0.0950	0.0000	10.4883	85
90	0.0003	10.5233	110.5032	3525.9060	37104.2733	0.0950	0.0000	10.5008	90
95	0.0002	10.5244	110.6032	5550.6178	58417.0292	0.0950	0.0000	10.5092	95
100	0.0001	10.5251	110.6702	8737.9975	91968.3951	0.0950	0.0000	10.5149	100

$I = 9.75\%$

n	(P/F)	(P/A)	(P/G)	(F/P)	(F/A)	(A/P)	(A/F)	(A/G)	n
1	0.9112	0.9112	0.0000	1.0975	1.0000	1.0975	1.0000	0.0000	1
2	0.8302	1.7414	0.8302	1.2045	2.0975	0.5743	0.4768	0.4768	2
3	0.7565	2.4978	2.3431	1.3219	3.3020	0.4003	0.3028	0.9381	3
4	0.6893	3.1871	4.4109	1.4508	4.6240	0.3138	0.2163	1.3840	4
5	0.6280	3.8151	6.9230	1.5923	6.0748	0.2621	0.1646	1.8146	5
6	0.5722	4.3874	9.7842	1.7475	7.6671	0.2279	0.1304	2.2301	6
7	0.5214	4.9088	12.9126	1.9179	9.4146	0.2037	0.1062	2.6305	7
8	0.4751	5.3838	16.2381	2.1049	11.3325	0.1857	0.0882	3.0161	8
9	0.4329	5.8167	19.7011	2.3102	13.4375	0.1719	0.0744	3.3870	9
10	0.3944	6.2111	23.2508	2.5354	15.7476	0.1610	0.0635	3.7434	10
11	0.3594	6.5705	26.8446	2.7826	18.2830	0.1522	0.0547	4.0856	11
12	0.3275	6.8979	30.4465	3.0539	21.0656	0.1450	0.0475	4.4139	12
13	0.2984	7.1963	34.0269	3.3517	24.1195	0.1390	0.0415	4.7284	13
14	0.2719	7.4682	37.5610	3.6784	27.4712	0.1339	0.0364	5.0295	14
15	0.2477	7.7159	41.0288	4.0371	31.1496	0.1296	0.0321	5.3175	15
16	0.2257	7.9416	44.4143	4.4307	35.1867	0.1259	0.0284	5.5926	16
17	0.2056	8.1472	47.7046	4.8627	39.6174	0.1227	0.0252	5.8553	17
18	0.1874	8.3346	50.8901	5.3368	44.4801	0.1200	0.0225	6.1059	18
19	0.1707	8.5053	53.9632	5.8571	49.8169	0.1176	0.0201	6.3446	19
20	0.1556	8.6609	56.9190	6.4282	55.6740	0.1155	0.0180	6.5720	20
21	0.1417	8.8026	59.7538	7.0550	62.1022	0.1136	0.0161	6.7882	21
22	0.1292	8.9318	62.4660	7.7428	69.1572	0.1120	0.0145	6.9937	22
23	0.1177	9.0495	65.0549	8.4978	76.9000	0.1105	0.0130	7.1888	23
24	0.1072	9.1567	67.5211	9.3263	85.3978	0.1092	0.0117	7.3740	24
25	0.0977	9.2544	69.8659	10.2356	94.7241	0.1081	0.0106	7.5495	25
26	0.0890	9.3434	72.0913	11.2336	104.9597	0.1070	0.0095	7.7158	26
27	0.0811	9.4245	74.2002	12.3288	116.1932	0.1061	0.0086	7.8731	27
28	0.0739	9.4984	76.1956	13.5309	128.5221	0.1053	0.0078	8.0219	28
29	0.0673	9.5658	78.0811	14.8502	142.0530	0.1045	0.0070	8.1626	29
30	0.0614	9.6271	79.8605	16.2981	156.9032	0.1039	0.0064	8.2954	30
31	0.0559	9.6830	81.5377	17.8871	173.2012	0.1033	0.0058	8.4207	31
32	0.0509	9.7340	83.1168	19.6311	191.0883	0.1027	0.0052	8.5389	32
33	0.0464	9.7804	84.6021	21.5451	210.7195	0.1022	0.0047	8.6502	33
34	0.0423	9.8227	85.9976	23.6458	232.2646	0.1018	0.0043	8.7550	34
35	0.0385	9.8612	87.3078	25.9513	255.9104	0.1014	0.0039	8.8537	35
36	0.0351	9.8963	88.5367	28.4815	281.8617	0.1010	0.0035	8.9464	36
37	0.0320	9.9283	89.6884	31.2585	310.3432	0.1007	0.0032	9.0336	37
38	0.0291	9.9574	90.7669	34.3062	341.6016	0.1004	0.0029	9.1155	38
39	0.0266	9.9840	91.7761	37.6510	375.9078	0.1002	0.0027	9.1923	39
40	0.0242	10.0082	92.7200	41.3220	413.5588	0.0999	0.0024	9.2644	40
41	0.0221	10.0303	93.6020	45.3509	454.8808	0.0997	0.0022	9.3320	41
42	0.0201	10.0503	94.4257	49.7726	500.2317	0.0995	0.0020	9.3953	42
43	0.0183	10.0687	95.1946	54.6254	550.0042	0.0993	0.0018	9.4546	43
44	0.0167	10.0853	95.9118	59.9514	604.6297	0.0992	0.0017	9.5100	44
45	0.0152	10.1005	96.5806	65.7967	664.5811	0.0990	0.0015	9.5619	45
46	0.0138	10.1144	97.2037	72.2118	730.3777	0.0989	0.0014	9.6105	46
47	0.0126	10.1270	97.7841	79.2525	802.5895	0.0987	0.0012	9.6558	47
48	0.0115	10.1385	98.3245	86.9796	881.8420	0.0986	0.0011	9.6981	48
49	0.0105	10.1490	98.8273	95.4601	968.8216	0.0985	0.0010	9.7377	49
50	0.0095	10.1585	99.2950	104.7675	1064.2817	0.0984	0.0009	9.7746	50
51	0.0087	10.1672	99.7299	114.9823	1169.0492	0.0984	0.0009	9.8090	51
52	0.0079	10.1751	100.1340	126.1931	1284.0315	0.0983	0.0008	9.8411	52
53	0.0072	10.1824	100.5095	138.4969	1410.2245	0.0982	0.0007	9.8709	53
54	0.0066	10.1889	100.8582	152.0003	1548.7214	0.0981	0.0006	9.8988	54
55	0.0060	10.1949	101.1819	166.8204	1700.7218	0.0981	0.0006	9.9247	55
60	0.0038	10.2178	102.4812	265.6267	2714.1201	0.0979	0.0004	10.0297	60
65	0.0024	10.2322	103.3690	422.9552	4327.7457	0.0977	0.0002	10.1024	65
70	0.0015	10.2412	103.9717	673.4681	6897.1086	0.0976	0.0001	10.1523	70
75	0.0009	10.2468	104.3785	1072.3577	10988.2840	0.0976	0.0001	10.1864	75
80	0.0006	10.2504	104.6518	1707.5063	17502.6287	0.0976	0.0001	10.2095	80
85	0.0004	10.2526	104.8346	2718.8482	27875.3660	0.0975	0.0000	10.2251	85
90	0.0002	10.2540	104.9564	4329.1995	44391.7900	0.0975	0.0000	10.2356	90
95	0.0001	10.2549	105.0373	6893.3487	70690.7557	0.0975	0.0000	10.2426	95
100	0.0001	10.2555	105.0909	10976.2222	112566.3819	0.0975	0.0000	10.2473	100

$I = 10.00\%$

n	(P/F)	(P/A)	(P/G)	(F/P)	(F/A)	(A/P)	(A/F)	(A/G)	n
1	0.9091	0.9091	0.0000	1.1000	1.0000	1.1000	1.0000	0.0000	1
2	0.8264	1.7355	0.8264	1.2100	2.1000	0.5762	0.4762	0.4762	2
3	0.7513	2.4869	2.3291	1.3310	3.3100	0.4021	0.3021	0.9366	3
4	0.6830	3.1699	4.3781	1.4641	4.6410	0.3155	0.2155	1.3812	4
5	0.6209	3.7908	6.8618	1.6105	6.1051	0.2638	0.1638	1.8101	5
6	0.5645	4.3553	9.6842	1.7716	7.7156	0.2296	0.1296	2.2236	6
7	0.5132	4.8684	12.7631	1.9487	9.4872	0.2054	0.1054	2.6216	7
8	0.4665	5.3349	16.0287	2.1436	11.4359	0.1874	0.0874	3.0045	8
9	0.4241	5.7590	19.4215	2.3579	13.5795	0.1736	0.0736	3.3724	9
10	0.3855	6.1446	22.8913	2.5937	15.9374	0.1627	0.0627	3.7255	10
11	0.3505	6.4951	26.3963	2.8531	18.5312	0.1540	0.0540	4.0641	11
12	0.3186	6.8137	29.9012	3.1384	21.3843	0.1468	0.0468	4.3884	12
13	0.2897	7.1034	33.3772	3.4523	24.5227	0.1408	0.0408	4.6988	13
14	0.2633	7.3667	36.8005	3.7975	27.9750	0.1357	0.0357	4.9955	14
15	0.2394	7.6061	40.1520	4.1772	31.7725	0.1315	0.0315	5.2789	15
16	0.2176	7.8237	43.4164	4.5950	35.9497	0.1278	0.0278	5.5493	16
17	0.1978	8.0216	46.5819	5.0545	40.5447	0.1247	0.0247	5.8071	17
18	0.1799	8.2014	49.6395	5.5599	45.5992	0.1219	0.0219	6.0526	18
19	0.1635	8.3649	52.5827	6.1159	51.1591	0.1195	0.0195	6.2861	19
20	0.1486	8.5136	55.4069	6.7275	57.2750	0.1175	0.0175	6.5081	20
21	0.1351	8.6487	58.1095	7.4002	64.0025	0.1156	0.0156	6.7189	21
22	0.1228	8.7715	60.6893	8.1403	71.4027	0.1140	0.0140	6.9189	22
23	0.1117	8.8832	63.1462	8.9543	79.5430	0.1126	0.0126	7.1085	23
24	0.1015	8.9847	65.4813	9.8497	88.4973	0.1113	0.0113	7.2881	24
25	0.0923	9.0770	67.6964	10.8347	98.3471	0.1102	0.0102	7.4580	25
26	0.0839	9.1609	69.7940	11.9182	109.1818	0.1092	0.0092	7.6186	26
27	0.0763	9.2372	71.7773	13.1100	121.0999	0.1083	0.0083	7.7704	27
28	0.0693	9.3066	73.6495	14.4210	134.2099	0.1075	0.0075	7.9137	28
29	0.0630	9.3696	75.4146	15.8631	148.6309	0.1067	0.0067	8.0489	29
30	0.0573	9.4269	77.0766	17.4494	164.4940	0.1061	0.0061	8.1762	30
31	0.0521	9.4790	78.6395	19.1943	181.9434	0.1055	0.0055	8.2962	31
32	0.0474	9.5264	80.1078	21.1138	201.1378	0.1050	0.0050	8.4091	32
33	0.0431	9.5694	81.4856	23.2252	222.2515	0.1045	0.0045	8.5152	33
34	0.0391	9.6086	82.7773	25.5477	245.4767	0.1041	0.0041	8.6149	34
35	0.0356	9.6442	83.9872	28.1024	271.0244	0.1037	0.0037	8.7086	35
36	0.0323	9.6765	85.1194	30.9127	299.1268	0.1033	0.0033	8.7965	36
37	0.0294	9.7059	86.1781	34.0039	330.0395	0.1030	0.0030	8.8789	37
38	0.0267	9.7327	87.1673	37.4043	364.0434	0.1027	0.0027	8.9562	38
39	0.0243	9.7570	88.0908	41.1448	401.4478	0.1025	0.0025	9.0285	39
40	0.0221	9.7791	88.9525	45.2593	442.5926	0.1023	0.0023	9.0962	40
41	0.0201	9.7991	89.7560	49.7852	487.8518	0.1020	0.0020	9.1596	41
42	0.0183	9.8174	90.5047	54.7637	537.6370	0.1019	0.0019	9.2188	42
43	0.0166	9.8340	91.2019	60.2401	592.4007	0.1017	0.0017	9.2741	43
44	0.0151	9.8491	91.8508	66.2641	652.6408	0.1015	0.0015	9.3258	44
45	0.0137	9.8628	92.4544	72.8905	718.9048	0.1014	0.0014	9.3740	45
46	0.0125	9.8753	93.0157	80.1795	791.7953	0.1013	0.0013	9.4190	46
47	0.0113	9.8866	93.5372	88.1975	871.9749	0.1011	0.0011	9.4610	47
48	0.0103	9.8969	94.0217	97.0172	960.1723	0.1010	0.0010	9.5001	48
49	0.0094	9.9063	94.4715	106.7190	1057.1896	0.1009	0.0009	9.5365	49
50	0.0085	9.9148	94.8889	117.3909	1163.9085	0.1009	0.0009	9.5704	50
51	0.0077	9.9226	95.2761	129.1299	1281.2994	0.1008	0.0008	9.6020	51
52	0.0070	9.9296	95.6351	142.0429	1410.4293	0.1007	0.0007	9.6313	52
53	0.0064	9.9360	95.9679	156.2472	1552.4723	0.1006	0.0006	9.6586	53
54	0.0058	9.9418	96.2763	171.8719	1708.7195	0.1006	0.0006	9.6840	54
55	0.0053	9.9471	96.5619	189.0591	1880.5914	0.1005	0.0005	9.7075	55
60	0.0033	9.9672	97.7010	304.4816	3034.8164	0.1003	0.0003	9.8023	60
65	0.0020	9.9796	98.4705	490.3707	4893.7073	0.1002	0.0002	9.8672	65
70	0.0013	9.9873	98.9870	789.7470	7887.4696	0.1001	0.0001	9.9113	70
75	0.0008	9.9921	99.3317	1271.8954	12708.9537	0.1001	0.0001	9.9410	75
80	0.0005	9.9951	99.5606	2048.4002	20474.0021	0.1000	0.0000	9.9609	80
85	0.0003	9.9970	99.7120	3298.9690	32979.6903	0.1000	0.0000	9.9742	85
90	0.0002	9.9981	99.8118	5313.0226	53120.2261	0.1000	0.0000	9.9831	90
95	0.0001	9.9988	99.8773	8556.6760	85556.7605	0.1000	0.0000	9.9889	95
100	0.0001	9.9993	99.9202	13780.6123	137796.1234	0.1000	0.0000	9.9927	100

$I = 10.25\%$

n	(P/F)	(P/A)	(P/G)	(F/P)	(F/A)	(A/P)	(A/F)	(A/G)	n
1	0.9070	0.9070	0.0000	1.1025	1.0000	1.1025	1.0000	0.0000	1
2	0.8227	1.7297	0.8227	1.2155	2.1025	0.5781	0.4756	0.4756	2
3	0.7462	2.4759	2.3151	1.3401	3.3180	0.4039	0.3014	0.9350	3
4	0.6768	3.1528	4.3457	1.4775	4.6581	0.3172	0.2147	1.3784	4
5	0.6139	3.7667	6.8013	1.6289	6.1356	0.2655	0.1630	1.8056	5
6	0.5568	4.3235	9.5855	1.7959	7.7645	0.2313	0.1288	2.2170	6
7	0.5051	4.8286	12.6159	1.9799	9.5603	0.2071	0.1046	2.6127	7
8	0.4581	5.2867	15.8227	2.1829	11.5402	0.1892	0.0867	2.9929	8
9	0.4155	5.7022	19.1468	2.4066	13.7231	0.1754	0.0729	3.3578	9
10	0.3769	6.0791	22.5389	2.6533	16.1297	0.1645	0.0620	3.7076	10
11	0.3418	6.4210	25.9573	2.9253	18.7830	0.1557	0.0532	4.0426	11
12	0.3101	6.7310	29.3681	3.2251	21.7083	0.1486	0.0461	4.3631	12
13	0.2812	7.0123	32.7430	3.5557	24.9334	0.1426	0.0401	4.6694	13
14	0.2551	7.2674	36.0592	3.9201	28.4891	0.1376	0.0351	4.9618	14
15	0.2314	7.4988	39.2985	4.3219	32.4092	0.1334	0.0309	5.2407	15
16	0.2099	7.7086	42.4465	4.7649	36.7311	0.1297	0.0272	5.5064	16
17	0.1904	7.8990	45.4922	5.2533	41.4961	0.1266	0.0241	5.7592	17
18	0.1727	8.0716	48.4273	5.7918	46.7494	0.1239	0.0214	5.9997	18
19	0.1566	8.2282	51.2462	6.3855	52.5412	0.1215	0.0190	6.2281	19
20	0.1420	8.3703	53.9451	7.0400	58.9267	0.1195	0.0170	6.4448	20
21	0.1288	8.4991	56.5219	7.7616	65.9667	0.1177	0.0152	6.6503	21
22	0.1169	8.6160	58.9760	8.5572	73.7283	0.1161	0.0136	6.8449	22
23	0.1060	8.7220	61.3079	9.4343	82.2854	0.1147	0.0122	7.0291	23
24	0.0961	8.8181	63.5192	10.4013	91.7197	0.1134	0.0109	7.2033	24
25	0.0872	8.9053	65.6121	11.4674	102.1210	0.1123	0.0098	7.3677	25
26	0.0791	8.9844	67.5895	12.6428	113.5884	0.1113	0.0088	7.5230	26
27	0.0717	9.0562	69.4548	13.9387	126.2312	0.1104	0.0079	7.6693	27
28	0.0651	9.1212	71.2117	15.3674	140.1699	0.1096	0.0071	7.8072	28
29	0.0590	9.1803	72.8644	16.9426	155.5373	0.1089	0.0064	7.9371	29
30	0.0535	9.2338	74.4169	18.6792	172.4799	0.1083	0.0058	8.0592	30
31	0.0486	9.2824	75.8737	20.5938	191.1590	0.1077	0.0052	8.1740	31
32	0.0440	9.3264	77.2390	22.7047	211.7529	0.1072	0.0047	8.2818	32
33	0.0399	9.3664	78.5174	25.0319	234.4575	0.1068	0.0043	8.3829	33
34	0.0362	9.4026	79.7131	27.5977	259.4894	0.1064	0.0039	8.4778	34
35	0.0329	9.4355	80.8306	30.4264	287.0871	0.1060	0.0035	8.5667	35
36	0.0298	9.4653	81.8740	33.5451	317.5135	0.1056	0.0031	8.6499	36
37	0.0270	9.4923	82.8474	36.9835	351.0586	0.1053	0.0028	8.7278	37
38	0.0245	9.5168	83.7548	40.7743	388.0421	0.1051	0.0026	8.8007	38
39	0.0222	9.5391	84.6001	44.9537	428.8165	0.1048	0.0023	8.8688	39
40	0.0202	9.5592	85.3870	49.5614	473.7702	0.1046	0.0021	8.9324	40
41	0.0183	9.5776	86.1191	54.6415	523.3316	0.1044	0.0019	8.9918	41
42	0.0166	9.5941	86.7997	60.2422	577.9731	0.1042	0.0017	9.0471	42
43	0.0151	9.6092	87.4320	66.4171	638.2153	0.1041	0.0016	9.0988	43
44	0.0137	9.6229	88.0193	73.2248	704.6324	0.1039	0.0014	9.1469	44
45	0.0124	9.6352	88.5643	80.7304	777.8572	0.1038	0.0013	9.1917	45
46	0.0112	9.6465	89.0699	89.0052	858.5876	0.1037	0.0012	9.2334	46
47	0.0102	9.6567	89.5386	98.1283	947.5928	0.1036	0.0011	9.2722	47
48	0.0092	9.6659	89.9731	108.1864	1045.7211	0.1035	0.0010	9.3083	48
49	0.0084	9.6743	90.3755	119.2755	1153.9075	0.1034	0.0009	9.3418	49
50	0.0076	9.6819	90.7481	131.5013	1273.1830	0.1033	0.0008	9.3730	50
51	0.0069	9.6888	91.0930	144.9801	1404.6843	0.1032	0.0007	9.4019	51
52	0.0063	9.6951	91.4121	159.8406	1549.6644	0.1031	0.0006	9.4287	52
53	0.0057	9.7007	91.7071	176.2243	1709.5050	0.1031	0.0006	9.4536	53
54	0.0051	9.7059	91.9799	194.2872	1885.7293	0.1030	0.0005	9.4767	54
55	0.0047	9.7106	92.2320	214.2017	2080.0165	0.1030	0.0005	9.4981	55
60	0.0029	9.7281	93.2310	348.9120	3394.2633	0.1028	0.0003	9.5836	60
65	0.0018	9.7389	93.8982	568.3409	5535.0328	0.1027	0.0002	9.6415	65
70	0.0011	9.7456	94.3409	925.7674	9022.1207	0.1026	0.0001	9.6804	70
75	0.0007	9.7496	94.6331	1507.9775	14702.2195	0.1026	0.0001	9.7063	75
80	0.0004	9.7521	94.8249	2456.3364	23954.5019	0.1025	0.0000	9.7235	80
85	0.0002	9.7537	94.9504	4001.1132	39025.4949	0.1025	0.0000	9.7348	85
90	0.0002	9.7546	95.0321	6517.3918	63574.5545	0.1025	0.0000	9.7423	90
95	0.0001	9.7552	95.0852	10616.1446	103562.3859	0.1025	0.0000	9.7471	95
100	0.0001	9.7555	95.1195	17292.5808	168698.3494	0.1025	0.0000	9.7503	100

ENGINEERING ECONOMIC ANALYSIS

$I = 10.50\%$

n	(P/F)	(P/A)	(P/G)	(F/P)	(F/A)	(A/P)	(A/F)	(A/G)	n
1	0.9050	0.9050	0.0000	1.1050	1.0000	1.1050	1.0000	0.0000	1
2	0.8190	1.7240	0.8190	1.2210	2.1050	0.5801	0.4751	0.4751	2
3	0.7412	2.4651	2.3013	1.3492	3.3260	0.4057	0.3007	0.9335	3
4	0.6707	3.1359	4.3135	1.4909	4.6753	0.3189	0.2139	1.3755	4
5	0.6070	3.7429	6.7415	1.6474	6.1662	0.2672	0.1622	1.8012	5
6	0.5493	4.2922	9.4881	1.8204	7.8136	0.2330	0.1280	2.2106	6
7	0.4971	4.7893	12.4709	2.0116	9.6340	0.2088	0.1038	2.6039	7
8	0.4499	5.2392	15.6201	2.2228	11.6456	0.1909	0.0859	2.9814	8
9	0.4071	5.6463	18.8771	2.4562	13.8684	0.1771	0.0721	3.3433	9
10	0.3684	6.0148	22.1932	2.7141	16.3246	0.1663	0.0613	3.6898	10
11	0.3334	6.3482	25.5276	2.9991	19.0387	0.1575	0.0525	4.0212	11
12	0.3018	6.6500	28.8469	3.3140	22.0377	0.1504	0.0454	4.3379	12
13	0.2731	6.9230	32.1238	3.6619	25.3517	0.1444	0.0394	4.6401	13
14	0.2471	7.1702	35.3365	4.0464	29.0136	0.1395	0.0345	4.9283	14
15	0.2236	7.3938	38.4676	4.4713	33.0600	0.1352	0.0302	5.2027	15
16	0.2024	7.5962	41.5036	4.9408	37.5313	0.1316	0.0266	5.4637	16
17	0.1832	7.7794	44.4342	5.4596	42.4721	0.1285	0.0235	5.7118	17
18	0.1658	7.9451	47.2521	6.0328	47.9317	0.1259	0.0209	5.9473	18
19	0.1500	8.0952	49.9523	6.6663	53.9645	0.1235	0.0185	6.1706	19
20	0.1358	8.2309	52.5316	7.3662	60.6308	0.1215	0.0165	6.3822	20
21	0.1229	8.3538	54.9887	8.1397	67.9970	0.1197	0.0147	6.5825	21
22	0.1112	8.4649	57.3235	8.9944	76.1367	0.1181	0.0131	6.7719	22
23	0.1006	8.5656	59.5370	9.9388	85.1311	0.1167	0.0117	6.9507	23
24	0.0911	8.6566	61.6313	10.9823	95.0699	0.1155	0.0105	7.1196	24
25	0.0824	8.7390	63.6090	12.1355	106.0522	0.1144	0.0094	7.2787	25
26	0.0746	8.8136	65.4733	13.4097	118.1877	0.1135	0.0085	7.4287	26
27	0.0675	8.8811	67.2280	14.8177	131.5974	0.1126	0.0076	7.5698	27
28	0.0611	8.9422	68.8770	16.3736	146.4151	0.1118	0.0068	7.7025	28
29	0.0553	8.9974	70.4245	18.0928	162.7887	0.1111	0.0061	7.8272	29
30	0.0500	9.0474	71.8751	19.9926	180.8815	0.1105	0.0055	7.9442	30
31	0.0453	9.0927	73.2331	22.0918	200.8741	0.1100	0.0050	8.0540	31
32	0.0410	9.1337	74.5029	24.4114	222.9658	0.1095	0.0045	8.1570	32
33	0.0371	9.1707	75.6892	26.9746	247.3772	0.1090	0.0040	8.2533	33
34	0.0335	9.2043	76.7964	29.8069	274.3518	0.1086	0.0036	8.3435	34
35	0.0304	9.2347	77.8287	32.9367	304.1588	0.1083	0.0033	8.4279	35
36	0.0275	9.2621	78.7903	36.3950	337.0955	0.1080	0.0030	8.5067	36
37	0.0249	9.2870	79.6855	40.2165	373.4905	0.1077	0.0027	8.5803	37
38	0.0225	9.3095	80.5181	44.4392	413.7070	0.1074	0.0024	8.6490	38
39	0.0204	9.3299	81.2919	49.1054	458.1462	0.1072	0.0022	8.7131	39
40	0.0184	9.3483	82.0107	54.2614	507.2516	0.1070	0.0020	8.7728	40
41	0.0167	9.3650	82.6778	59.9589	561.5130	0.1068	0.0018	8.8284	41
42	0.0151	9.3801	83.2966	66.2545	621.4719	0.1066	0.0016	8.8802	42
43	0.0137	9.3937	83.8703	73.2113	687.7264	0.1065	0.0015	8.9283	43
44	0.0124	9.4061	84.4018	80.8985	760.9377	0.1063	0.0013	8.9731	44
45	0.0112	9.4173	84.8940	89.3928	841.8361	0.1062	0.0012	9.0147	45
46	0.0101	9.4274	85.3496	98.7790	931.2289	0.1061	0.0011	9.0534	46
47	0.0092	9.4366	85.7710	109.1508	1030.0080	0.1060	0.0010	9.0892	47
48	0.0083	9.4448	86.1607	120.6117	1139.1588	0.1059	0.0009	9.1225	48
49	0.0075	9.4524	86.5209	133.2759	1259.7705	0.1058	0.0008	9.1534	49
50	0.0068	9.4591	86.8536	147.2699	1393.0464	0.1057	0.0007	9.1820	50
51	0.0061	9.4653	87.1608	162.7332	1540.3162	0.1056	0.0006	9.2085	51
52	0.0056	9.4708	87.4445	179.8202	1703.0494	0.1056	0.0006	9.2330	52
53	0.0050	9.4759	87.7062	198.7013	1882.8696	0.1055	0.0005	9.2557	53
54	0.0046	9.4804	87.9476	219.5649	2081.5710	0.1055	0.0005	9.2767	54
55	0.0041	9.4846	88.1701	242.6193	2301.1359	0.1054	0.0004	9.2962	55
60	0.0025	9.5000	89.0464	399.7023	3797.1651	0.1053	0.0003	9.3733	60
65	0.0015	9.5093	89.6251	658.4883	6261.7935	0.1052	0.0002	9.4249	65
70	0.0009	9.5150	90.0048	1084.8244	10322.1375	0.1051	0.0001	9.4592	70
75	0.0006	9.5185	90.2525	1787.1905	17011.3383	0.1051	0.0001	9.4818	75
80	0.0003	9.5206	90.4134	2944.3012	28031.4404	0.1050	0.0000	9.4966	80
85	0.0002	9.5218	90.5174	4850.5796	46186.4719	0.1050	0.0000	9.5063	85
90	0.0001	9.5226	90.5843	7991.0716	76095.9200	0.1050	0.0000	9.5125	90
95	0.0001	9.5231	90.6273	13164.8651	125370.1434	0.1050	0.0000	9.5166	95
100	0.0000	9.5234	90.6549	21688.4144	206546.8035	0.1050	0.0000	9.5192	100

$I = 10.75\%$

n	(P/F)	(P/A)	(P/G)	(F/P)	(F/A)	(A/P)	(A/F)	(A/G)	n
1	0.9029	0.9029	0.0000	1.1075	1.0000	1.1075	1.0000	0.0000	1
2	0.8153	1.7182	0.8153	1.2266	2.1075	0.5820	0.4745	0.4745	2
3	0.7362	2.4544	2.2876	1.3584	3.3341	0.4074	0.2999	0.9320	3
4	0.6647	3.1191	4.2817	1.5044	4.6925	0.3206	0.2131	1.3727	4
5	0.6002	3.7193	6.6824	1.6662	6.1969	0.2689	0.1614	1.7967	5
6	0.5419	4.2612	9.3920	1.8453	7.8631	0.2347	0.1272	2.2041	6
7	0.4893	4.7505	12.3280	2.0436	9.7084	0.2105	0.1030	2.5951	7
8	0.4418	5.1923	15.4207	2.2633	11.7520	0.1926	0.0851	2.9699	8
9	0.3989	5.5913	18.6122	2.5066	14.0153	0.1789	0.0714	3.3288	9
10	0.3602	5.9515	21.8542	2.7761	16.5220	0.1680	0.0605	3.6721	10
11	0.3253	6.2767	25.1067	3.0745	19.2981	0.1593	0.0518	4.0000	11
12	0.2937	6.5704	28.3372	3.4051	22.3727	0.1522	0.0447	4.3128	12
13	0.2652	6.8356	31.5193	3.7711	25.7777	0.1463	0.0388	4.6111	13
14	0.2394	7.0750	34.6319	4.1765	29.5488	0.1413	0.0338	4.8950	14
15	0.2162	7.2912	37.6586	4.6255	33.7253	0.1372	0.0297	5.1649	15
16	0.1952	7.4864	40.5868	5.1227	38.3508	0.1336	0.0261	5.4214	16
17	0.1763	7.6627	43.4070	5.6734	43.4735	0.1305	0.0230	5.6647	17
18	0.1592	7.8218	46.1126	6.2833	49.1469	0.1278	0.0203	5.8954	18
19	0.1437	7.9655	48.6992	6.9587	55.4302	0.1255	0.0180	6.1137	19
20	0.1298	8.0953	51.1646	7.7068	62.3889	0.1235	0.0160	6.3203	20
21	0.1172	8.2125	53.5078	8.5353	70.0957	0.1218	0.0143	6.5154	21
22	0.1058	8.3182	55.7293	9.4528	78.6310	0.1202	0.0127	6.6996	22
23	0.0955	8.4138	57.8308	10.4690	88.0839	0.1189	0.0114	6.8734	23
24	0.0862	8.5000	59.8145	11.5944	98.5529	0.1176	0.0101	7.0370	24
25	0.0779	8.5779	61.6835	12.8408	110.1473	0.1166	0.0091	7.1910	25
26	0.0703	8.6482	63.4415	14.2212	122.9882	0.1156	0.0081	7.3358	26
27	0.0635	8.7117	65.0923	15.7500	137.2094	0.1148	0.0073	7.4718	27
28	0.0573	8.7690	66.6401	17.4431	152.9594	0.1140	0.0065	7.5995	28
29	0.0518	8.8208	68.0895	19.3183	170.4025	0.1134	0.0059	7.7192	29
30	0.0467	8.8675	69.4450	21.3950	189.7208	0.1128	0.0053	7.8314	30
31	0.0422	8.9097	70.7111	23.6949	211.1158	0.1122	0.0047	7.9364	31
32	0.0381	8.9478	71.8924	26.2422	234.8108	0.1118	0.0043	8.0346	32
33	0.0344	8.9823	72.9935	29.0632	261.0529	0.1113	0.0038	8.1264	33
34	0.0311	9.0133	74.0187	32.1875	290.1161	0.1109	0.0034	8.2121	34
35	0.0281	9.0414	74.9725	35.6476	322.3036	0.1106	0.0031	8.2922	35
36	0.0253	9.0667	75.8590	39.4798	357.9512	0.1103	0.0028	8.3668	36
37	0.0229	9.0896	76.6824	43.7238	397.4310	0.1100	0.0025	8.4363	37
38	0.0207	9.1102	77.4464	48.4241	441.1548	0.1098	0.0023	8.5010	38
39	0.0186	9.1289	78.1550	53.6297	489.5789	0.1095	0.0020	8.5613	39
40	0.0168	9.1457	78.8116	59.3949	543.2087	0.1093	0.0018	8.6173	40
41	0.0152	9.1609	79.4197	65.7799	602.6036	0.1092	0.0017	8.6694	41
42	0.0137	9.1746	79.9825	72.8512	668.3835	0.1090	0.0015	8.7178	42
43	0.0124	9.1870	80.5031	80.6827	741.2347	0.1088	0.0013	8.7627	43
44	0.0112	9.1982	80.9843	89.3561	821.9175	0.1087	0.0012	8.8043	44
45	0.0101	9.2083	81.4289	98.9619	911.2736	0.1086	0.0011	8.8430	45
46	0.0091	9.2175	81.8395	109.6003	1010.2355	0.1085	0.0010	8.8788	46
47	0.0082	9.2257	82.2184	121.3823	1119.8358	0.1084	0.0009	8.9119	47
48	0.0074	9.2331	82.5681	134.4310	1241.2182	0.1083	0.0008	8.9426	48
49	0.0067	9.2398	82.8905	148.8823	1375.6491	0.1082	0.0007	8.9710	49
50	0.0061	9.2459	83.1876	164.8871	1524.5314	0.1082	0.0007	8.9972	50
51	0.0055	9.2514	83.4614	182.6125	1689.4185	0.1081	0.0006	9.0215	51
52	0.0049	9.2563	83.7136	202.2433	1872.0310	0.1080	0.0005	9.0439	52
53	0.0045	9.2608	83.9458	223.9845	2074.2743	0.1080	0.0005	9.0646	53
54	0.0040	9.2648	84.1594	248.0628	2298.2588	0.1079	0.0004	9.0838	54
55	0.0036	9.2685	84.3560	274.7296	2546.3217	0.1079	0.0004	9.1014	55
60	0.0022	9.2820	85.1249	457.7455	4248.7954	0.1077	0.0002	9.1710	60
65	0.0013	9.2901	85.6270	762.6807	7085.4021	0.1076	0.0001	9.2170	65
70	0.0008	9.2950	85.9927	1270.7539	11811.6640	0.1076	0.0001	9.2472	70
75	0.0005	9.2979	86.1629	2117.2889	19686.4082	0.1076	0.0001	9.2669	75
80	0.0003	9.2997	86.2978	3527.7581	32807.0516	0.1075	0.0000	9.2796	80
85	0.0002	9.3007	86.3840	5877.8360	54668.2421	0.1075	0.0000	9.2879	85
90	0.0001	9.3014	86.4389	9793.4597	91092.6484	0.1075	0.0000	9.2931	90
95	0.0001	9.3018	86.4738	16317.5448	151781.8125	0.1075	0.0000	9.2965	95
100	0.0000	9.3020	86.4959	27187.7639	252900.1291	0.1075	0.0000	9.2986	100

$I = 11.00\%$

n	(P/F)	(P/A)	(P/G)	(F/P)	(F/A)	(A/P)	(A/F)	(A/G)	n
1	0.9009	0.9009	0.0000	1.1100	1.0000	1.1100	1.0000	0.0000	1
2	0.8116	1.7125	0.8116	1.2321	2.1100	0.5839	0.4739	0.4739	2
3	0.7312	2.4437	2.2740	1.3676	3.3421	0.4092	0.2992	0.9306	3
4	0.6587	3.1024	4.2502	1.5181	4.7097	0.3223	0.2123	1.3700	4
5	0.5935	3.6959	6.6240	1.6851	6.2278	0.2706	0.1606	1.7923	5
6	0.5346	4.2305	9.2972	1.8704	7.9129	0.2364	0.1264	2.1976	6
7	0.4817	4.7122	12.1872	2.0762	9.7833	0.2122	0.1022	2.5863	7
8	0.4339	5.1461	15.2246	2.3045	11.8594	0.1943	0.0843	2.9585	8
9	0.3909	5.5370	18.3520	2.5580	14.1640	0.1806	0.0706	3.3144	9
10	0.3522	5.8892	21.5217	2.8394	16.7220	0.1698	0.0598	3.6544	10
11	0.3173	6.2065	24.6945	3.1518	19.5614	0.1611	0.0511	3.9788	11
12	0.2858	6.4924	27.8388	3.4985	22.7132	0.1540	0.0440	4.2879	12
13	0.2575	6.7499	30.9290	3.8833	26.2116	0.1482	0.0382	4.5822	13
14	0.2320	6.9819	33.9449	4.3104	30.0949	0.1432	0.0332	4.8619	14
15	0.2090	7.1909	36.8709	4.7846	34.4054	0.1391	0.0291	5.1275	15
16	0.1883	7.3792	39.6953	5.3109	39.1899	0.1355	0.0255	5.3794	16
17	0.1696	7.5488	42.4095	5.8951	44.5008	0.1325	0.0225	5.6180	17
18	0.1528	7.7016	45.0074	6.5436	50.3959	0.1298	0.0198	5.8439	18
19	0.1377	7.8393	47.4856	7.2633	56.9395	0.1276	0.0176	6.0574	19
20	0.1240	7.9633	49.8423	8.0623	64.2028	0.1256	0.0156	6.2590	20
21	0.1117	8.0751	52.0771	8.9492	72.2651	0.1238	0.0138	6.4491	21
22	0.1007	8.1757	54.1912	9.9336	81.2143	0.1223	0.0123	6.6283	22
23	0.0907	8.2664	56.1864	11.0263	91.1479	0.1210	0.0110	6.7969	23
24	0.0817	8.3481	58.0656	12.2392	102.1742	0.1198	0.0098	6.9555	24
25	0.0736	8.4217	59.8322	13.5855	114.4133	0.1187	0.0087	7.1045	25
26	0.0663	8.4881	61.4900	15.0799	127.9988	0.1178	0.0078	7.2443	26
27	0.0597	8.5478	63.0433	16.7386	143.0786	0.1170	0.0070	7.3754	27
28	0.0538	8.6016	64.4965	18.5799	159.8173	0.1163	0.0063	7.4982	28
29	0.0485	8.6501	65.8542	20.6237	178.3972	0.1156	0.0056	7.6131	29
30	0.0437	8.6938	67.1210	22.8923	199.0209	0.1150	0.0050	7.7206	30
31	0.0394	8.7331	68.3016	25.4104	221.9132	0.1145	0.0045	7.8210	31
32	0.0355	8.7686	69.4007	28.2056	247.3236	0.1140	0.0040	7.9147	32
33	0.0319	8.8005	70.4228	31.3082	275.5292	0.1136	0.0036	8.0021	33
34	0.0288	8.8293	71.3724	34.7521	306.8374	0.1133	0.0033	8.0836	34
35	0.0259	8.8552	72.2538	38.5749	341.5896	0.1129	0.0029	8.1594	35
36	0.0234	8.8786	73.0712	42.8181	380.1644	0.1126	0.0026	8.2300	36
37	0.0210	8.8996	73.8286	47.5281	422.9825	0.1124	0.0024	8.2957	37
38	0.0190	8.9186	74.5300	52.7562	470.5106	0.1121	0.0021	8.3567	38
39	0.0171	8.9357	75.1789	58.5593	523.2667	0.1119	0.0019	8.4133	39
40	0.0154	8.9511	75.7789	65.0009	581.8261	0.1117	0.0017	8.4659	40
41	0.0139	8.9649	76.3333	72.1510	646.8269	0.1115	0.0015	8.5147	41
42	0.0125	8.9774	76.8452	80.0876	718.9779	0.1114	0.0014	8.5599	42
43	0.0112	8.9886	77.3176	88.8972	799.0655	0.1113	0.0013	8.6017	43
44	0.0101	8.9988	77.7534	98.6759	887.9627	0.1111	0.0011	8.6404	44
45	0.0091	9.0079	78.1551	109.5302	986.6386	0.1110	0.0010	8.6763	45
46	0.0082	9.0161	78.5253	121.5786	1096.1688	0.1109	0.0009	8.7094	46
47	0.0074	9.0235	78.8661	134.9522	1217.7474	0.1108	0.0008	8.7400	47
48	0.0067	9.0302	79.1799	149.7970	1352.6996	0.1107	0.0007	8.7683	48
49	0.0060	9.0362	79.4686	166.2746	1502.4965	0.1107	0.0007	8.7944	49
50	0.0054	9.0417	79.7341	184.5648	1668.7712	0.1106	0.0006	8.8185	50
51	0.0049	9.0465	79.9781	204.8670	1853.3360	0.1105	0.0005	8.8407	51
52	0.0044	9.0509	80.2024	227.4023	2058.2029	0.1105	0.0005	8.8612	52
53	0.0040	9.0549	80.4084	252.4166	2285.6053	0.1104	0.0004	8.8801	53
54	0.0036	9.0585	80.5976	280.1824	2538.0218	0.1104	0.0004	8.8975	54
55	0.0032	9.0617	80.7712	311.0025	2818.2042	0.1104	0.0004	8.9135	55
60	0.0019	9.0736	81.4461	524.0572	4755.0658	0.1102	0.0002	8.9762	60
65	0.0011	9.0806	81.8819	883.0669	8018.7903	0.1101	0.0001	9.0172	65
70	0.0007	9.0848	82.1614	1488.0191	13518.3557	0.1101	0.0001	9.0438	70
75	0.0004	9.0873	82.3397	2507.3988	22785.4434	0.1100	0.0000	9.0610	75
80	0.0002	9.0888	82.4529	4225.1128	38401.0250	0.1100	0.0000	9.0720	80
85	0.0001	9.0896	82.5245	7119.5607	64714.1881	0.1100	0.0000	9.0790	85
90	0.0001	9.0902	82.5695	11996.8738	109053.3983	0.1100	0.0000	9.0834	90
95	0.0000	9.0905	82.5978	20215.4301	183767.5459	0.1100	0.0000	9.0862	95
100	0.0000	9.0906	82.6155	34064.1753	309665.2297	0.1100	0.0000	9.0880	100

$I = 11.25\%$

n	(P/F)	(P/A)	(P/G)	(F/P)	(F/A)	(A/P)	(A/F)	(A/G)	n
1	0.8989	0.8989	0.0000	1.1125	1.0000	1.1125	1.0000	0.0000	1
2	0.8080	1.7069	0.8080	1.2377	2.1125	0.5859	0.4734	0.4734	2
3	0.7263	2.4331	2.2605	1.3769	3.3502	0.4110	0.2985	0.9291	3
4	0.6528	3.0860	4.2190	1.5318	4.7270	0.3240	0.2115	1.3672	4
5	0.5868	3.6728	6.5663	1.7041	6.2588	0.2723	0.1598	1.7878	5
6	0.5275	4.2002	9.2036	1.8958	7.9630	0.2381	0.1256	2.1912	6
7	0.4741	4.6744	12.0484	2.1091	9.8588	0.2139	0.1014	2.5775	7
8	0.4262	5.1006	15.0317	2.3464	11.9679	0.1961	0.0836	2.9471	8
9	0.3831	5.4837	18.0964	2.6104	14.3143	0.1824	0.0699	3.3001	9
10	0.3443	5.8280	21.1956	2.9040	16.9247	0.1716	0.0591	3.6369	10
11	0.3095	6.1375	24.2909	3.2307	19.8287	0.1629	0.0504	3.9578	11
12	0.2782	6.4158	27.3514	3.5942	23.0594	0.1559	0.0434	4.2632	12
13	0.2501	6.6658	30.3525	3.9985	26.6536	0.1500	0.0375	4.5534	13
14	0.2248	6.8907	33.2749	4.4484	30.6521	0.1451	0.0326	4.8290	14
15	0.2021	7.0927	36.1039	4.9488	35.1005	0.1410	0.0285	5.0903	15
16	0.1816	7.2744	38.8284	5.5055	40.0493	0.1375	0.0250	5.3377	16
17	0.1633	7.4376	41.4407	6.1249	45.5548	0.1345	0.0220	5.5718	17
18	0.1468	7.5844	43.9355	6.8140	51.6798	0.1318	0.0193	5.7929	18
19	0.1319	7.7163	46.3100	7.5805	58.4937	0.1296	0.0171	6.0016	19
20	0.1186	7.8349	48.5630	8.4334	66.0743	0.1276	0.0151	6.1983	20
21	0.1066	7.9415	50.6947	9.3821	74.5076	0.1259	0.0134	6.3836	21
22	0.0958	8.0373	52.7067	10.4376	83.8897	0.1244	0.0119	6.5578	22
23	0.0861	8.1234	54.6013	11.6118	94.3273	0.1231	0.0106	6.7215	23
24	0.0774	8.2008	56.3817	12.9182	105.9392	0.1219	0.0094	6.8752	24
25	0.0696	8.2704	58.0517	14.3714	118.8573	0.1209	0.0084	7.0192	25
26	0.0625	8.3329	59.6154	15.9882	133.2288	0.1200	0.0075	7.1542	26
27	0.0562	8.3891	61.0771	17.7869	149.2170	0.1192	0.0067	7.2805	27
28	0.0505	8.4397	62.4416	19.7879	167.0039	0.1185	0.0060	7.3986	28
29	0.0454	8.4851	63.7135	22.0141	186.7918	0.1179	0.0054	7.5089	29
30	0.0408	8.5259	64.8976	24.4907	208.8059	0.1173	0.0048	7.6118	30
31	0.0367	8.5626	65.9987	27.2459	233.2966	0.1168	0.0043	7.7078	31
32	0.0330	8.5956	67.0214	30.3110	260.5424	0.1163	0.0038	7.7971	32
33	0.0297	8.6253	67.9704	33.7210	290.8535	0.1159	0.0034	7.8804	33
34	0.0267	8.6519	68.8501	37.5146	324.5745	0.1156	0.0031	7.9578	34
35	0.0240	8.6759	69.6647	41.7350	362.0891	0.1153	0.0028	8.0297	35
36	0.0215	8.6974	70.4185	46.4302	403.8241	0.1150	0.0025	8.0965	36
37	0.0194	8.7168	71.1155	51.6536	450.2544	0.1147	0.0022	8.1584	37
38	0.0174	8.7342	71.7594	57.4646	501.9080	0.1145	0.0020	8.2159	38
39	0.0156	8.7498	72.3538	63.9294	559.3726	0.1143	0.0018	8.2691	39
40	0.0141	8.7639	72.9021	71.1215	623.3020	0.1141	0.0016	8.3185	40
41	0.0126	8.7765	73.4077	79.1226	694.4235	0.1139	0.0014	8.3641	41
42	0.0114	8.7879	73.8735	88.0239	773.5462	0.1138	0.0013	8.4063	42
43	0.0102	8.7981	74.3023	97.9266	861.5701	0.1137	0.0012	8.4453	43
44	0.0092	8.8073	74.6970	108.9434	959.4968	0.1135	0.0010	8.4813	44
45	0.0083	8.8155	75.0601	121.1995	1068.4401	0.1134	0.0009	8.5145	45
46	0.0074	8.8230	75.3938	134.8345	1189.6397	0.1133	0.0008	8.5452	46
47	0.0067	8.8296	75.7005	150.0033	1324.4741	0.1133	0.0008	8.5735	47
48	0.0060	8.8356	75.9821	166.8787	1474.4775	0.1132	0.0007	8.5995	48
49	0.0054	8.8410	76.2407	185.6526	1641.3562	0.1131	0.0006	8.6235	49
50	0.0048	8.8459	76.4779	206.5385	1827.0087	0.1130	0.0005	8.6456	50
51	0.0044	8.8502	76.6955	229.7741	2033.5472	0.1130	0.0005	8.6660	51
52	0.0039	8.8541	76.8950	255.6236	2263.3213	0.1129	0.0004	8.6847	52
53	0.0035	8.8576	77.0779	284.3813	2518.9449	0.1129	0.0004	8.7019	53
54	0.0032	8.8608	77.2454	316.3742	2803.3262	0.1129	0.0004	8.7177	54
55	0.0028	8.8636	77.3988	351.9663	3119.7004	0.1128	0.0003	8.7322	55
60	0.0017	8.8741	77.9914	599.7927	5322.6018	0.1127	0.0002	8.7887	60
65	0.0010	8.8802	78.3698	1022.1186	9076.6095	0.1126	0.0001	8.8252	65
70	0.0006	8.8838	78.6098	1741.8124	15473.8879	0.1126	0.0001	8.8487	70
75	0.0003	8.8859	78.7611	2968.2568	26375.6162	0.1125	0.0000	8.8636	75
80	0.0002	8.8871	78.8561	5058.2649	44953.4661	0.1125	0.0000	8.8731	80
85	0.0001	8.8879	78.9155	8619.8889	76612.3462	0.1125	0.0000	8.8790	85
90	0.0001	8.8883	78.9525	14689.3226	130562.8676	0.1125	0.0000	8.8828	90
95	0.0000	8.8885	78.9755	25032.3641	222501.0142	0.1125	0.0000	8.8851	95
100	0.0000	8.8887	78.9897	42658.1449	379174.6215	0.1125	0.0000	8.8865	100

$I = 11.50\%$

n	(P/F)	(P/A)	(P/G)	(F/P)	(F/A)	(A/P)	(A/F)	(A/G)	n
1	0.8969	0.8969	0.0000	1.1150	1.0000	1.1150	1.0000	0.0000	1
2	0.8044	1.7012	0.8044	1.2432	2.1150	0.5878	0.4728	0.4728	2
3	0.7214	2.4226	2.2472	1.3862	3.3582	0.4128	0.2978	0.9276	3
4	0.6470	3.0696	4.1881	1.5456	4.7444	0.3258	0.2108	1.3644	4
5	0.5803	3.6499	6.5092	1.7234	6.2900	0.2740	0.1590	1.7834	5
6	0.5204	4.1703	9.1113	1.9215	8.0134	0.2398	0.1248	2.1848	6
7	0.4667	4.6370	11.9117	2.1425	9.9349	0.2157	0.1007	2.5688	7
8	0.4186	5.0556	14.8419	2.3889	12.0774	0.1978	0.0828	2.9357	8
9	0.3754	5.4311	17.8454	2.6636	14.4663	0.1841	0.0691	3.2858	9
10	0.3367	5.7678	20.8757	2.9699	17.1300	0.1734	0.0584	3.6194	10
11	0.3020	6.0697	23.8955	3.3115	20.0999	0.1648	0.0498	3.9368	11
12	0.2708	6.3406	26.8747	3.6923	23.4114	0.1577	0.0427	4.2385	12
13	0.2429	6.5835	29.7895	4.1169	27.1037	0.1519	0.0369	4.5249	13
14	0.2178	6.8013	32.6215	4.5904	31.2207	0.1470	0.0320	4.7963	14
15	0.1954	6.9967	35.3568	5.1183	35.8110	0.1429	0.0279	5.0533	15
16	0.1752	7.1719	37.9852	5.7069	40.9293	0.1394	0.0244	5.2964	16
17	0.1572	7.3291	40.4997	6.3632	46.6362	0.1364	0.0214	5.5259	17
18	0.1409	7.4700	42.8957	7.0949	52.9993	0.1339	0.0189	5.7424	18
19	0.1264	7.5964	45.1711	7.9108	60.0942	0.1316	0.0166	5.9463	19
20	0.1134	7.7098	47.3252	8.8206	68.0051	0.1297	0.0147	6.1383	20
21	0.1017	7.8115	49.3587	9.8350	76.8257	0.1280	0.0130	6.3187	21
22	0.0912	7.9027	51.2737	10.9660	86.6606	0.1265	0.0115	6.4881	22
23	0.0818	7.9845	53.0730	12.2271	97.6266	0.1252	0.0102	6.6470	23
24	0.0734	8.0578	54.7601	13.6332	109.8536	0.1241	0.0091	6.7959	24
25	0.0658	8.1236	56.3389	15.2010	123.4868	0.1231	0.0081	6.9352	25
26	0.0590	8.1826	57.8139	16.9491	138.6878	0.1222	0.0072	7.0655	26
27	0.0529	8.2355	59.1897	18.8982	155.6369	0.1214	0.0064	7.1871	27
28	0.0475	8.2830	60.4711	21.0715	174.5351	0.1207	0.0057	7.3006	28
29	0.0426	8.3255	61.6628	23.4948	195.6067	0.1201	0.0051	7.4065	29
30	0.0382	8.3637	62.7698	26.1967	219.1014	0.1196	0.0046	7.5050	30
31	0.0342	8.3980	63.7969	29.2093	245.2981	0.1191	0.0041	7.5967	31
32	0.0307	8.4287	64.7487	32.5683	274.5074	0.1186	0.0036	7.6820	32
33	0.0275	8.4562	65.6300	36.3137	307.0757	0.1183	0.0033	7.7612	33
34	0.0247	8.4809	66.4450	40.4898	343.3895	0.1179	0.0029	7.8347	34
35	0.0222	8.5030	67.1981	45.1461	383.8792	0.1176	0.0026	7.9028	35
36	0.0199	8.5229	67.8934	50.3379	429.0254	0.1173	0.0023	7.9660	36
37	0.0178	8.5407	68.5348	56.1268	479.3633	0.1171	0.0021	8.0245	37
38	0.0160	8.5567	69.1260	62.5814	535.4900	0.1169	0.0019	8.0786	38
39	0.0143	8.5710	69.6706	69.7782	598.0714	0.1167	0.0017	8.1286	39
40	0.0129	8.5839	70.1719	77.8027	667.8496	0.1165	0.0015	8.1748	40
41	0.0115	8.5954	70.6330	86.7500	745.6523	0.1163	0.0013	8.2175	41
42	0.0103	8.6058	71.0568	96.7263	832.4023	0.1162	0.0012	8.2569	42
43	0.0093	8.6150	71.4463	107.8498	929.1286	0.1161	0.0011	8.2932	43
44	0.0083	8.6233	71.8039	120.2525	1036.9784	0.1160	0.0010	8.3267	44
45	0.0075	8.6308	72.1320	134.0816	1157.2309	0.1159	0.0009	8.3575	45
46	0.0067	8.6375	72.4330	149.5009	1291.3125	0.1158	0.0008	8.3859	46
47	0.0060	8.6435	72.7090	166.6935	1440.8134	0.1157	0.0007	8.4120	47
48	0.0054	8.6489	72.9618	185.8633	1607.5069	0.1156	0.0006	8.4360	48
49	0.0048	8.6537	73.1935	207.2376	1793.3702	0.1156	0.0006	8.4581	49
50	0.0043	8.6580	73.4055	231.0699	2000.6078	0.1155	0.0005	8.4783	50
51	0.0039	8.6619	73.5996	257.6429	2231.6777	0.1154	0.0004	8.4969	51
52	0.0035	8.6654	73.7771	287.2719	2489.3206	0.1154	0.0004	8.5140	52
53	0.0031	8.6685	73.9395	320.3081	2776.5925	0.1154	0.0004	8.5297	53
54	0.0028	8.6713	74.0879	357.1436	3096.9006	0.1153	0.0003	8.5440	54
55	0.0025	8.6738	74.2235	398.2151	3454.0442	0.1153	0.0003	8.5572	55
60	0.0015	8.6830	74.7439	686.2653	5958.8287	0.1152	0.0002	8.6081	60
65	0.0008	8.6883	75.0725	1182.6776	10275.4576	0.1151	0.0001	8.6406	65
70	0.0005	8.6914	75.2786	2038.1715	17714.5345	0.1151	0.0001	8.6613	70
75	0.0003	8.6932	75.4072	3512.4897	30534.6927	0.1150	0.0000	8.6743	75
80	0.0002	8.6942	75.4870	6053.2609	52628.3554	0.1150	0.0000	8.6824	80
85	0.0001	8.6948	75.5363	10431.9075	90703.5435	0.1150	0.0000	8.6875	85
90	0.0001	8.6952	75.5666	17977.8629	156320.5471	0.1150	0.0000	8.6906	90
95	0.0000	8.6954	75.5853	30982.2106	269401.8312	0.1150	0.0000	8.6926	95
100	0.0000	8.6955	75.5967	53393.2969	464280.8428	0.1150	0.0000	8.6938	100

$I = 11.75\%$

n	(P/F)	(P/A)	(P/G)	(F/P)	(F/A)	(A/P)	(A/F)	(A/G)	n
1	0.8949	0.8949	0.0000	1.1175	1.0000	1.1175	1.0000	0.0000	1
2	0.8008	1.6956	0.8008	1.2488	2.1175	0.5898	0.4723	0.4723	2
3	0.7166	2.4122	2.2339	1.3955	3.3663	0.4146	0.2971	0.9261	3
4	0.6412	3.0534	4.1576	1.5595	4.7618	0.3275	0.2100	1.3616	4
5	0.5738	3.6272	6.4528	1.7428	6.3214	0.2757	0.1582	1.7790	5
6	0.5135	4.1407	9.0201	1.9475	8.0641	0.2415	0.1240	2.1784	6
7	0.4595	4.6002	11.7770	2.1764	10.0117	0.2174	0.0999	2.5601	7
8	0.4112	5.0113	14.6552	2.4321	12.1880	0.1995	0.0820	2.9244	8
9	0.3679	5.3793	17.5987	2.7179	14.6201	0.1859	0.0684	3.2716	9
10	0.3292	5.7085	20.5619	3.0372	17.3380	0.1752	0.0577	3.6020	10
11	0.2946	6.0031	23.5082	3.3941	20.3752	0.1666	0.0491	3.9160	11
12	0.2637	6.2668	26.4084	3.7929	23.7693	0.1596	0.0421	4.2140	12
13	0.2359	6.5027	29.2395	4.2386	27.5622	0.1538	0.0363	4.4965	13
14	0.2111	6.7139	31.9841	4.7366	31.8007	0.1489	0.0314	4.7639	14
15	0.1889	6.9028	34.6291	5.2931	36.5373	0.1449	0.0274	5.0167	15
16	0.1691	7.0718	37.1650	5.9151	41.8305	0.1414	0.0239	5.2553	16
17	0.1513	7.2231	39.5855	6.6101	47.7455	0.1384	0.0209	5.4804	17
18	0.1354	7.3585	41.8869	7.3868	54.3556	0.1359	0.0184	5.6923	18
19	0.1211	7.4796	44.0675	8.2547	61.7424	0.1337	0.0162	5.8917	19
20	0.1084	7.5880	46.1272	9.2247	69.9972	0.1318	0.0143	6.0789	20
21	0.0970	7.6850	48.0673	10.3086	79.2218	0.1301	0.0126	6.2547	21
22	0.0868	7.7719	49.8902	11.5198	89.5304	0.1287	0.0112	6.4193	22
23	0.0777	7.8495	51.5992	12.8734	101.0502	0.1274	0.0099	6.5735	23
24	0.0695	7.9190	53.1980	14.3860	113.9236	0.1263	0.0088	6.7177	24
25	0.0622	7.9813	54.6908	16.0764	128.3096	0.1253	0.0078	6.8524	25
26	0.0557	8.0369	56.0824	17.9654	144.3860	0.1244	0.0069	6.9781	26
27	0.0498	8.0867	57.3775	20.0763	162.3514	0.1237	0.0062	7.0953	27
28	0.0446	8.1313	58.5809	22.4352	182.4276	0.1230	0.0055	7.2044	28
29	0.0399	8.1712	59.6977	25.0714	204.8629	0.1224	0.0049	7.3059	29
30	0.0357	8.2069	60.7328	28.0173	229.9343	0.1218	0.0043	7.4002	30
31	0.0319	8.2388	61.6910	31.3093	257.9516	0.1214	0.0039	7.4879	31
32	0.0286	8.2674	62.5770	34.9882	289.2609	0.1210	0.0035	7.5691	32
33	0.0256	8.2930	63.3954	39.0993	324.2490	0.1206	0.0031	7.6445	33
34	0.0229	8.3159	64.1507	43.6934	363.3483	0.1203	0.0028	7.7143	34
35	0.0205	8.3363	64.8470	48.8274	407.0417	0.1200	0.0025	7.7788	35
36	0.0183	8.3547	65.4885	54.5646	455.8691	0.1197	0.0022	7.8386	36
37	0.0164	8.3711	66.0789	60.9760	510.4337	0.1195	0.0020	7.8937	37
38	0.0147	8.3857	66.6219	68.1406	571.4097	0.1193	0.0018	7.9447	38
39	0.0131	8.3989	67.1209	76.1472	639.5503	0.1191	0.0016	7.9917	39
40	0.0118	8.4106	67.5792	85.0945	715.6975	0.1189	0.0014	8.0350	40
41	0.0105	8.4211	67.9999	95.0931	800.7920	0.1187	0.0012	8.0749	41
42	0.0094	8.4306	68.3857	106.2665	895.8850	0.1186	0.0011	8.1117	42
43	0.0084	8.4390	68.7394	118.7528	1002.1515	0.1185	0.0010	8.1455	43
44	0.0075	8.4465	69.0634	132.7063	1120.9043	0.1184	0.0009	8.1766	44
45	0.0067	8.4533	69.3601	148.2992	1253.6106	0.1183	0.0008	8.2051	45
46	0.0060	8.4593	69.6316	165.7244	1401.9098	0.1182	0.0007	8.2314	46
47	0.0054	8.4647	69.8800	185.1970	1567.6342	0.1181	0.0006	8.2555	47
48	0.0048	8.4695	70.1071	206.9577	1752.8312	0.1181	0.0006	8.2776	48
49	0.0043	8.4738	70.3146	231.2752	1959.7889	0.1180	0.0005	8.2978	49
50	0.0039	8.4777	70.5042	258.4500	2191.0641	0.1180	0.0005	8.3164	50
51	0.0035	8.4812	70.6774	288.8179	2449.5141	0.1179	0.0004	8.3334	51
52	0.0031	8.4843	70.8354	322.7540	2738.3320	0.1179	0.0004	8.3490	52
53	0.0028	8.4870	70.9795	360.6776	3061.0861	0.1178	0.0003	8.3633	53
54	0.0025	8.4895	71.1110	403.0572	3421.7637	0.1178	0.0003	8.3763	54
55	0.0022	8.4917	71.2309	450.4165	3824.8209	0.1178	0.0003	8.3883	55
60	0.0013	8.4998	71.6882	784.9679	6672.0674	0.1176	0.0001	8.4341	60
65	0.0007	8.5044	71.9736	1368.0109	11634.1356	0.1176	0.0001	8.4631	65
70	0.0004	8.5071	72.1507	2384.1151	20281.8310	0.1175	0.0000	8.4813	70
75	0.0002	8.5086	72.2599	4154.9412	35352.6913	0.1175	0.0000	8.4926	75
80	0.0001	8.5095	72.3269	7241.0666	61617.5884	0.1175	0.0000	8.4996	80
85	0.0001	8.5100	72.3679	12619.4435	107391.0085	0.1175	0.0000	8.5039	85
90	0.0000	8.5103	72.3928	21992.6652	187163.1079	0.1175	0.0000	8.5065	90
95	0.0000	8.5104	72.4080	38327.9438	326186.7560	0.1175	0.0000	8.5082	95
100	0.0000	8.5105	72.4171	66796.4190	568471.6512	0.1175	0.0000	8.5091	100

$I = 12.00\%$

n	(P/F)	(P/A)	(P/G)	(F/P)	(F/A)	(A/P)	(A/F)	(A/G)	n
1	0.8929	0.8929	0.0000	1.1200	1.0000	1.1200	1.0000	0.0000	1
2	0.7972	1.6901	0.7972	1.2544	2.1200	0.5917	0.4717	0.4717	2
3	0.7118	2.4018	2.2208	1.4049	3.3744	0.4163	0.2963	0.9246	3
4	0.6355	3.0373	4.1273	1.5735	4.7793	0.3292	0.2092	1.3589	4
5	0.5674	3.6048	6.3970	1.7623	6.3528	0.2774	0.1574	1.7746	5
6	0.5066	4.1114	8.9302	1.9738	8.1152	0.2432	0.1232	2.1720	6
7	0.4523	4.5638	11.6443	2.2107	10.0890	0.2191	0.0991	2.5515	7
8	0.4039	4.9676	14.4714	2.4760	12.2997	0.2013	0.0813	2.9131	8
9	0.3606	5.3282	17.3563	2.7731	14.7757	0.1877	0.0677	3.2574	9
10	0.3220	5.6502	20.2541	3.1058	17.5487	0.1770	0.0570	3.5847	10
11	0.2875	5.9377	23.1288	3.4785	20.6546	0.1684	0.0484	3.8953	11
12	0.2567	6.1944	25.9523	3.8960	24.1331	0.1614	0.0414	4.1897	12
13	0.2292	6.4235	28.7024	4.3635	28.0291	0.1557	0.0357	4.4683	13
14	0.2046	6.6282	31.3624	4.8871	32.3926	0.1509	0.0309	4.7317	14
15	0.1827	6.8109	33.9202	5.4736	37.2797	0.1468	0.0268	4.9803	15
16	0.1631	6.9740	36.3670	6.1304	42.7533	0.1434	0.0234	5.2147	16
17	0.1456	7.1196	38.6973	6.8660	48.8837	0.1405	0.0205	5.4353	17
18	0.1300	7.2497	40.9080	7.6900	55.7497	0.1379	0.0179	5.6427	18
19	0.1161	7.3658	42.9979	8.6128	63.4397	0.1358	0.0158	5.8375	19
20	0.1037	7.4694	44.9676	9.6463	72.0524	0.1339	0.0139	6.0202	20
21	0.0926	7.5620	46.8188	10.8038	81.6987	0.1322	0.0122	6.1913	21
22	0.0826	7.6446	48.5543	12.1003	92.5026	0.1308	0.0108	6.3514	22
23	0.0738	7.7184	50.1776	13.5523	104.6029	0.1296	0.0096	6.5010	23
24	0.0659	7.7843	51.6929	15.1786	118.1552	0.1285	0.0085	6.6406	24
25	0.0588	7.8431	53.1046	17.0001	133.3339	0.1275	0.0075	6.7708	25
26	0.0525	7.8957	54.4177	19.0401	150.3339	0.1267	0.0067	6.8921	26
27	0.0469	7.9426	55.6369	21.3249	169.3740	0.1259	0.0059	7.0049	27
28	0.0419	7.9844	56.7674	23.8839	190.6989	0.1252	0.0052	7.1098	28
29	0.0374	8.0218	57.8141	26.7499	214.5828	0.1247	0.0047	7.2071	29
30	0.0334	8.0552	58.7821	29.9599	241.3327	0.1241	0.0041	7.2974	30
31	0.0298	8.0850	59.6761	33.5551	271.2926	0.1237	0.0037	7.3811	31
32	0.0266	8.1116	60.5010	37.5817	304.8477	0.1233	0.0033	7.4586	32
33	0.0238	8.1354	61.2612	42.0915	342.4294	0.1229	0.0029	7.5302	33
34	0.0212	8.1566	61.9612	47.1425	384.5210	0.1226	0.0026	7.5965	34
35	0.0189	8.1755	62.6052	52.7996	431.6635	0.1223	0.0023	7.6577	35
36	0.0169	8.1924	63.1970	59.1356	484.4631	0.1221	0.0021	7.7141	36
37	0.0151	8.2075	63.7406	66.2318	543.5987	0.1218	0.0018	7.7661	37
38	0.0135	8.2210	64.2394	74.1797	609.8305	0.1216	0.0016	7.8141	38
39	0.0120	8.2330	64.6967	83.0812	684.0102	0.1215	0.0015	7.8582	39
40	0.0107	8.2438	65.1159	93.0510	767.0914	0.1213	0.0013	7.8988	40
41	0.0096	8.2534	65.4997	104.2171	860.1424	0.1212	0.0012	7.9361	41
42	0.0086	8.2619	65.8509	116.7231	964.3595	0.1210	0.0010	7.9704	42
43	0.0076	8.2696	66.1722	130.7299	1081.0826	0.1209	0.0009	8.0019	43
44	0.0068	8.2764	66.4659	146.4175	1211.8125	0.1208	0.0008	8.0308	44
45	0.0061	8.2825	66.7342	163.9876	1358.2300	0.1207	0.0007	8.0572	45
46	0.0054	8.2880	66.9792	183.6661	1522.2176	0.1207	0.0007	8.0815	46
47	0.0049	8.2928	67.2028	205.7061	1705.8838	0.1206	0.0006	8.1037	47
48	0.0043	8.2972	67.4068	230.3908	1911.5898	0.1205	0.0005	8.1241	48
49	0.0039	8.3010	67.5929	258.0377	2141.9806	0.1205	0.0005	8.1427	49
50	0.0035	8.3045	67.7624	289.0022	2400.0182	0.1204	0.0004	8.1597	50
51	0.0031	8.3076	67.9169	323.6825	2689.0204	0.1204	0.0004	8.1753	51
52	0.0028	8.3103	68.0576	362.5243	3012.7029	0.1203	0.0003	8.1895	52
53	0.0025	8.3128	68.1856	406.0273	3375.2272	0.1203	0.0003	8.2025	53
54	0.0022	8.3150	68.3022	454.7505	3781.2545	0.1203	0.0003	8.2143	54
55	0.0020	8.3170	68.4082	509.3206	4236.0050	0.1202	0.0002	8.2251	55
60	0.0011	8.3240	68.8100	897.5969	7471.6411	0.1201	0.0001	8.2664	60
65	0.0006	8.3281	69.0581	1581.8725	13173.9374	0.1201	0.0001	8.2922	65
70	0.0004	8.3303	69.2103	2787.7998	23223.3319	0.1200	0.0000	8.3082	70
75	0.0002	8.3316	69.3031	4913.0558	40933.7987	0.1200	0.0000	8.3181	75
80	0.0001	8.3324	69.3594	8658.4831	72145.6925	0.1200	0.0000	8.3241	80
85	0.0001	8.3328	69.3935	15259.2057	127151.7140	0.1200	0.0000	8.3278	85
90	0.0000	8.3330	69.4140	26891.9342	224091.1185	0.1200	0.0000	8.3300	90
95	0.0000	8.3332	69.4263	47392.7766	394931.4719	0.1200	0.0000	8.3313	95
100	0.0000	8.3332	69.4336	83522.2657	696010.5477	0.1200	0.0000	8.3321	100

$I = 12.25\%$

n	(P/F)	(P/A)	(P/G)	(F/P)	(F/A)	(A/P)	(A/F)	(A/G)	n
1	0.8909	0.8909	0.0000	1.1225	1.0000	1.1225	1.0000	0.0000	1
2	0.7936	1.6845	0.7936	1.2600	2.1225	0.5936	0.4711	0.4711	2
3	0.7070	2.3916	2.2077	1.4144	3.3825	0.4181	0.2956	0.9231	3
4	0.6299	3.0214	4.0973	1.5876	4.7969	0.3310	0.2085	1.3561	4
5	0.5611	3.5826	6.3419	1.7821	6.3845	0.2791	0.1566	1.7702	5
6	0.4999	4.0825	8.8414	2.0004	8.1666	0.2450	0.1225	2.1657	6
7	0.4453	4.5278	11.5134	2.2455	10.1670	0.2209	0.0984	2.5428	7
8	0.3967	4.9245	14.2906	2.5205	12.4124	0.2031	0.0806	2.9019	8
9	0.3534	5.2780	17.1182	2.8293	14.9330	0.1895	0.0670	3.2433	9
10	0.3149	5.5929	19.9521	3.1759	17.7623	0.1788	0.0563	3.5674	10
11	0.2805	5.8734	22.7572	3.5649	20.9381	0.1703	0.0478	3.8746	11
12	0.2499	6.1233	25.5061	4.0016	24.5030	0.1633	0.0408	4.1654	12
13	0.2226	6.3459	28.1776	4.4918	28.5047	0.1576	0.0351	4.4403	13
14	0.1983	6.5442	30.7559	5.0421	32.9965	0.1528	0.0303	4.6997	14
15	0.1767	6.7209	33.2295	5.6597	38.0386	0.1488	0.0263	4.9442	15
16	0.1574	6.8783	35.5906	6.3530	43.6983	0.1454	0.0229	5.1743	16
17	0.1402	7.0186	37.8342	7.1313	50.0513	0.1425	0.0200	5.3906	17
18	0.1249	7.1435	39.9579	8.0049	57.1826	0.1400	0.0175	5.5936	18
19	0.1113	7.2548	41.9612	8.9855	65.1875	0.1378	0.0153	5.7839	19
20	0.0991	7.3539	43.8449	10.0862	74.1730	0.1360	0.0135	5.9621	20
21	0.0883	7.4422	45.6114	11.3217	84.2591	0.1344	0.0119	6.1287	21
22	0.0787	7.5209	47.2639	12.7087	95.5809	0.1330	0.0105	6.2843	22
23	0.0701	7.5910	48.8060	14.2655	108.2895	0.1317	0.0092	6.4294	23
24	0.0624	7.6535	50.2424	16.0130	122.5550	0.1307	0.0082	6.5646	24
25	0.0556	7.7091	51.5776	17.9746	138.5680	0.1297	0.0072	6.6905	25
26	0.0496	7.7587	52.8167	20.1765	156.5426	0.1289	0.0064	6.8074	26
27	0.0442	7.8028	53.9647	22.6481	176.7190	0.1282	0.0057	6.9160	27
28	0.0393	7.8422	55.0267	25.4225	199.3671	0.1275	0.0050	7.0168	28
29	0.0350	7.8772	56.0079	28.5367	224.7896	0.1269	0.0044	7.1101	29
30	0.0312	7.9084	56.9132	32.0325	253.3263	0.1264	0.0039	7.1965	30
31	0.0278	7.9362	57.7476	35.9565	285.3588	0.1260	0.0035	7.2764	31
32	0.0248	7.9610	58.5157	40.3611	321.3153	0.1256	0.0031	7.3503	32
33	0.0221	7.9831	59.2220	45.3054	361.6764	0.1253	0.0028	7.4184	33
34	0.0197	8.0027	59.8709	50.8553	406.9817	0.1250	0.0025	7.4813	34
35	0.0175	8.0203	60.4665	57.0850	457.8370	0.1247	0.0022	7.5392	35
36	0.0156	8.0359	61.0127	64.0779	514.9220	0.1244	0.0019	7.5925	36
37	0.0139	8.0498	61.5132	71.9275	579.0000	0.1242	0.0017	7.6416	37
38	0.0124	8.0622	61.9715	80.7386	650.9275	0.1240	0.0015	7.6867	38
39	0.0110	8.0732	62.3907	90.6291	731.6661	0.1239	0.0014	7.7281	39
40	0.0098	8.0830	62.7741	101.7312	822.2952	0.1237	0.0012	7.7662	40
41	0.0088	8.0918	63.1244	114.1932	924.0263	0.1236	0.0011	7.8011	41
42	0.0078	8.0996	63.4443	128.1819	1038.2195	0.1235	0.0010	7.8330	42
43	0.0070	8.1065	63.7362	143.8842	1166.4014	0.1234	0.0009	7.8623	43
44	0.0062	8.1127	64.0024	161.5100	1310.2856	0.1233	0.0008	7.8891	44
45	0.0055	8.1182	64.2451	181.2950	1471.7956	0.1232	0.0007	7.9137	45
46	0.0049	8.1232	64.4662	203.5036	1653.0906	0.1231	0.0006	7.9361	46
47	0.0044	8.1275	64.6676	228.4328	1856.5942	0.1230	0.0005	7.9566	47
48	0.0039	8.1314	64.8509	256.4158	2085.0269	0.1230	0.0005	7.9753	48
49	0.0035	8.1349	65.0177	287.8267	2341.4428	0.1229	0.0004	7.9924	49
50	0.0031	8.1380	65.1693	323.0855	2629.2695	0.1229	0.0004	8.0080	50
51	0.0028	8.1408	65.3072	362.6635	2952.3550	0.1228	0.0003	8.0223	51
52	0.0025	8.1432	65.4325	407.0898	3315.0185	0.1228	0.0003	8.0352	52
53	0.0022	8.1454	65.5463	456.9583	3722.1083	0.1228	0.0003	8.0470	53
54	0.0019	8.1474	65.6496	512.9356	4179.0665	0.1227	0.0002	8.0578	54
55	0.0017	8.1491	65.7434	575.7703	4692.0022	0.1227	0.0002	8.0676	55
60	0.0010	8.1553	66.0966	1026.0794	8367.9953	0.1226	0.0001	8.1047	60
65	0.0005	8.1588	66.3123	1828.5748	14918.9779	0.1226	0.0001	8.1277	65
70	0.0003	8.1608	66.4431	3258.7007	26593.4753	0.1225	0.0000	8.1418	70
75	0.0002	8.1619	66.5220	5807.3263	47398.5817	0.1225	0.0000	8.1503	75
80	0.0001	8.1625	66.5694	10349.2285	84475.3343	0.1225	0.0000	8.1555	80
85	0.0001	8.1628	66.5977	18443.3464	150549.7663	0.1225	0.0000	8.1587	85
90	0.0000	8.1630	66.6145	32867.8632	268300.9241	0.1225	0.0000	8.1605	90
95	0.0000	8.1631	66.6245	58573.7756	478145.1070	0.1225	0.0000	8.1616	95
100	0.0000	8.1632	66.6304	104384.2482	852108.1483	0.1225	0.0000	8.1623	100

$I = 12.50\%$

n	(P/F)	(P/A)	(P/G)	(F/P)	(F/A)	(A/P)	(A/F)	(A/G)	n
1	0.8889	0.8889	0.0000	1.1250	1.0000	1.1250	1.0000	0.0000	1
2	0.7901	1.6790	0.7901	1.2656	2.1250	0.5956	0.4706	0.4706	2
3	0.7023	2.3813	2.1948	1.4238	3.3906	0.4199	0.2949	0.9217	3
4	0.6243	3.0056	4.0677	1.6018	4.8145	0.3327	0.2077	1.3533	4
5	0.5549	3.5606	6.2874	1.8020	6.4163	0.2809	0.1559	1.7658	5
6	0.4933	4.0538	8.7537	2.0273	8.2183	0.2467	0.1217	2.1594	6
7	0.4385	4.4923	11.3845	2.2807	10.2456	0.2226	0.0976	2.5342	7
8	0.3897	4.8820	14.1127	2.5658	12.5263	0.2048	0.0798	2.8907	8
9	0.3464	5.2285	16.8842	2.8865	15.0921	0.1913	0.0663	3.2293	9
10	0.3079	5.5364	19.6558	3.2473	17.9786	0.1806	0.0556	3.5503	10
11	0.2737	5.8102	22.3931	3.6532	21.2259	0.1721	0.0471	3.8541	11
12	0.2433	6.0535	25.0695	4.1099	24.8791	0.1652	0.0402	4.1413	12
13	0.2163	6.2698	27.6649	4.6236	28.9890	0.1595	0.0345	4.4124	13
14	0.1922	6.4620	30.1641	5.2016	33.6126	0.1548	0.0298	4.6679	14
15	0.1709	6.6329	32.5566	5.8518	38.8142	0.1508	0.0258	4.9083	15
16	0.1519	6.7848	34.8351	6.5833	44.6660	0.1474	0.0224	5.1343	16
17	0.1350	6.9198	36.9954	7.4062	51.2493	0.1445	0.0195	5.3463	17
18	0.1200	7.0398	39.0358	8.3319	58.6554	0.1420	0.0170	5.5450	18
19	0.1067	7.1465	40.9561	9.3734	66.9873	0.1399	0.0149	5.7309	19
20	0.0948	7.2414	42.7579	10.5451	76.3608	0.1381	0.0131	5.9047	20
21	0.0843	7.3256	44.4438	11.8632	86.9058	0.1365	0.0115	6.0669	21
22	0.0749	7.4006	46.0173	13.3461	98.7691	0.1351	0.0101	6.2181	22
23	0.0666	7.4672	47.4825	15.0144	112.1152	0.1339	0.0089	6.3588	23
24	0.0592	7.5264	48.8442	16.8912	127.1296	0.1329	0.0079	6.4897	24
25	0.0526	7.5790	50.1072	19.0026	144.0208	0.1319	0.0069	6.6113	25
26	0.0468	7.6258	51.2766	21.3779	163.0234	0.1311	0.0061	6.7241	26
27	0.0416	7.6674	52.3577	24.0502	184.4013	0.1304	0.0054	6.8286	27
28	0.0370	7.7043	53.3556	27.0564	208.4515	0.1298	0.0048	6.9254	28
29	0.0329	7.7372	54.2755	30.4385	235.5079	0.1292	0.0042	7.0149	29
30	0.0292	7.7664	55.1224	34.2433	265.9464	0.1288	0.0038	7.0976	30
31	0.0260	7.7923	55.9011	38.5237	300.1897	0.1283	0.0033	7.1739	31
32	0.0231	7.8154	56.6164	43.3392	338.7135	0.1280	0.0030	7.2442	32
33	0.0205	7.8359	57.2727	48.7566	382.0526	0.1276	0.0026	7.3090	33
34	0.0182	7.8542	57.8743	54.8512	430.8092	0.1273	0.0023	7.3686	34
35	0.0162	7.8704	58.4253	61.7075	485.6604	0.1271	0.0021	7.4235	35
36	0.0144	7.8848	58.9295	69.4210	547.3679	0.1268	0.0018	7.4738	36
37	0.0128	7.8976	59.3904	78.0986	616.7889	0.1266	0.0016	7.5201	37
38	0.0114	7.9089	59.8116	87.8609	694.8875	0.1264	0.0014	7.5625	38
39	0.0101	7.9191	60.1960	98.8436	782.7485	0.1263	0.0013	7.6014	39
40	0.0090	7.9281	60.5467	111.1990	881.5920	0.1261	0.0011	7.6370	40
41	0.0080	7.9361	60.8665	125.0989	992.7910	0.1260	0.0010	7.6696	41
42	0.0071	7.9432	61.1578	140.7362	1117.8899	0.1259	0.0009	7.6994	42
43	0.0063	7.9495	61.4231	158.3283	1258.6262	0.1258	0.0008	7.7267	43
44	0.0056	7.9551	61.6645	178.1193	1416.9544	0.1257	0.0007	7.7516	44
45	0.0050	7.9601	61.8841	200.3842	1595.0737	0.1256	0.0006	7.7743	45
46	0.0044	7.9645	62.0837	225.4322	1795.4579	0.1256	0.0006	7.7950	46
47	0.0039	7.9685	62.2651	253.6113	2020.8902	0.1255	0.0005	7.8139	47
48	0.0035	7.9720	62.4298	285.3127	2274.5015	0.1254	0.0004	7.8312	48
49	0.0031	7.9751	62.5793	320.9768	2559.8141	0.1254	0.0004	7.8469	49
50	0.0028	7.9778	62.7150	361.0989	2880.7909	0.1253	0.0003	7.8611	50
51	0.0025	7.9803	62.8381	406.2362	3241.8898	0.1253	0.0003	7.8741	51
52	0.0022	7.9825	62.9497	457.0157	3648.1260	0.1253	0.0003	7.8860	52
53	0.0019	7.9844	63.0508	514.1427	4105.1417	0.1252	0.0002	7.8967	53
54	0.0017	7.9862	63.1425	578.4106	4619.2845	0.1252	0.0002	7.9065	54
55	0.0015	7.9877	63.2255	650.7119	5197.6950	0.1252	0.0002	7.9153	55
60	0.0009	7.9932	63.5361	1172.6039	9372.8315	0.1251	0.0001	7.9488	60
65	0.0005	7.9962	63.7236	2113.0704	16896.5629	0.1251	0.0001	7.9692	65
70	0.0003	7.9979	63.8361	3807.8214	30454.5713	0.1250	0.0000	7.9816	70
75	0.0001	7.9988	63.9032	6861.8178	54886.5426	0.1250	0.0000	7.9891	75
80	0.0001	7.9994	63.9431	12365.2185	98913.7482	0.1250	0.0000	7.9935	80
85	0.0000	7.9996	63.9666	22282.5253	178252.2023	0.1250	0.0000	7.9962	85
90	0.0000	7.9998	63.9805	40153.8341	321222.6728	0.1250	0.0000	7.9978	90
95	0.0000	7.9999	63.9886	72358.5129	578860.1029	0.1250	0.0000	7.9987	95
100	0.0000	7.9999	63.9934	130392.3897	1043131.1177	0.1250	0.0000	7.9992	100

$I = 12.75\%$

n	(P/F)	(P/A)	(P/G)	(F/P)	(F/A)	(A/P)	(A/F)	(A/G)	n
1	0.8869	0.8869	0.0000	1.1275	1.0000	1.1275	1.0000	0.0000	1
2	0.7866	1.6735	0.7866	1.2713	2.1275	0.5975	0.4700	0.4700	2
3	0.6977	2.3712	2.1820	1.4333	3.3988	0.4217	0.2942	0.9202	3
4	0.6188	2.9900	4.0383	1.6161	4.8321	0.3344	0.2069	1.3506	4
5	0.5488	3.5388	6.2335	1.8221	6.4482	0.2826	0.1551	1.7615	5
6	0.4867	4.0255	8.6672	2.0545	8.2703	0.2484	0.1209	2.1531	6
7	0.4317	4.4572	11.2574	2.3164	10.3248	0.2244	0.0969	2.5257	7
8	0.3829	4.8401	13.9376	2.6118	12.6412	0.2066	0.0791	2.8796	8
9	0.3396	5.1797	16.6543	2.9448	15.2530	0.1931	0.0656	3.2153	9
10	0.3012	5.4809	19.3650	3.3202	18.1977	0.1825	0.0550	3.5332	10
11	0.2671	5.7480	22.0363	3.7435	21.5179	0.1740	0.0465	3.8337	11
12	0.2369	5.9849	24.6424	4.2208	25.2615	0.1671	0.0396	4.1174	12
13	0.2101	6.1951	27.1639	4.7590	29.4823	0.1614	0.0339	4.3848	13
14	0.1864	6.3814	29.5867	5.3658	34.2413	0.1567	0.0292	4.6364	14
15	0.1653	6.5467	31.9008	6.0499	39.6071	0.1527	0.0252	4.8728	15
16	0.1466	6.6933	34.0998	6.8213	45.6570	0.1494	0.0219	5.0946	16
17	0.1300	6.8234	36.1802	7.6910	52.4782	0.1466	0.0191	5.3024	17
18	0.1153	6.9387	38.1406	8.6716	60.1692	0.1441	0.0166	5.4968	18
19	0.1023	7.0410	39.9816	9.7772	68.8408	0.1420	0.0145	5.6784	19
20	0.0907	7.1317	41.7051	11.0238	78.6180	0.1402	0.0127	5.8479	20
21	0.0805	7.2121	43.3142	12.4293	89.6418	0.1387	0.0112	6.0058	21
22	0.0714	7.2835	44.8127	14.0141	102.0711	0.1373	0.0098	6.1527	22
23	0.0633	7.3468	46.2051	15.8009	116.0852	0.1361	0.0086	6.2892	23
24	0.0561	7.4029	47.4961	17.8155	131.8860	0.1351	0.0076	6.4159	24
25	0.0498	7.4527	48.6909	20.0869	149.7015	0.1342	0.0067	6.5333	25
26	0.0442	7.4968	49.7947	22.6480	169.7884	0.1334	0.0059	6.6421	26
27	0.0392	7.5360	50.8129	25.5356	192.4364	0.1327	0.0052	6.7427	27
28	0.0347	7.5707	51.7507	28.7914	217.9721	0.1321	0.0046	6.8356	28
29	0.0308	7.6015	52.6132	32.4624	246.7635	0.1316	0.0041	6.9214	29
30	0.0273	7.6289	53.4056	36.6013	279.2259	0.1311	0.0036	7.0005	30
31	0.0242	7.6531	54.1325	41.2680	315.8272	0.1307	0.0032	7.0733	31
32	0.0215	7.6746	54.7988	46.5296	357.0952	0.1303	0.0028	7.1403	32
33	0.0191	7.6936	55.4087	52.4622	403.6248	0.1300	0.0025	7.2019	33
34	0.0169	7.7105	55.9666	59.1511	456.0869	0.1297	0.0022	7.2585	34
35	0.0150	7.7255	56.4764	66.6928	515.2380	0.1294	0.0019	7.3104	35
36	0.0133	7.7388	56.9419	75.1962	581.9309	0.1292	0.0017	7.3579	36
37	0.0118	7.7506	57.3665	84.7837	657.1271	0.1290	0.0015	7.4015	37
38	0.0105	7.7611	57.7535	95.5936	741.9108	0.1288	0.0013	7.4414	38
39	0.0093	7.7704	58.1061	107.7818	837.5044	0.1287	0.0012	7.4779	39
40	0.0082	7.7786	58.4270	121.5240	945.2862	0.1286	0.0011	7.5113	40
41	0.0073	7.7859	58.7189	137.0183	1066.8102	0.1284	0.0009	7.5417	41
42	0.0065	7.7924	58.9843	154.4881	1203.8285	0.1283	0.0008	7.5695	42
43	0.0057	7.7981	59.2255	174.1854	1358.3166	0.1282	0.0007	7.5948	43
44	0.0051	7.8032	59.4444	196.3940	1532.5020	0.1282	0.0007	7.6180	44
45	0.0045	7.8077	59.6431	221.4342	1728.8960	0.1281	0.0006	7.6390	45
46	0.0040	7.8117	59.8234	249.6671	1950.3302	0.1280	0.0005	7.6582	46
47	0.0036	7.8153	59.9868	281.4997	2199.9973	0.1280	0.0005	7.6756	47
48	0.0032	7.8184	60.1348	317.3909	2481.4970	0.1279	0.0004	7.6914	48
49	0.0028	7.8212	60.2690	357.8582	2798.8879	0.1279	0.0004	7.7058	49
50	0.0025	7.8237	60.3904	403.4851	3156.7461	0.1278	0.0003	7.7189	50
51	0.0022	7.8259	60.5003	454.9295	3560.2312	0.1278	0.0003	7.7308	51
52	0.0019	7.8278	60.5998	512.9330	4015.1607	0.1277	0.0002	7.7416	52
53	0.0017	7.8296	60.6897	578.3319	4528.0937	0.1277	0.0002	7.7513	53
54	0.0015	7.8311	60.7709	652.0693	5106.4256	0.1277	0.0002	7.7602	54
55	0.0014	7.8325	60.8444	735.2081	5758.4949	0.1277	0.0002	7.7682	55
60	0.0007	7.8373	61.1176	1339.6552	10499.2564	0.1276	0.0001	7.7983	60
65	0.0004	7.8399	61.2808	2441.0450	19137.6079	0.1276	0.0001	7.8165	65
70	0.0002	7.8414	61.3775	4447.9361	34877.9301	0.1275	0.0000	7.8274	70
75	0.0001	7.8422	61.4346	8104.7811	63559.0677	0.1275	0.0000	7.8339	75
80	0.0001	7.8426	61.4681	14768.0803	115820.2373	0.1275	0.0000	7.8377	80
85	0.0000	7.8428	61.4877	26909.5724	211047.6267	0.1275	0.0000	7.8400	85
90	0.0000	7.8430	61.4992	49033.1224	384565.6657	0.1275	0.0000	7.8413	90
95	0.0000	7.8430	61.5058	89345.4215	700740.5612	0.1275	0.0000	7.8421	95
100	0.0000	7.8431	61.5096	162800.2453	1276856.8261	0.1275	0.0000	7.8425	100

$I = 13.00\%$

n	(P/F)	(P/A)	(P/G)	(F/P)	(F/A)	(A/P)	(A/F)	(A/G)	n
1	0.8850	0.8850	0.0000	1.1300	1.0000	1.1300	1.0000	0.0000	1
2	0.7831	1.6681	0.7831	1.2769	2.1300	0.5995	0.4695	0.4695	2
3	0.6931	2.3612	2.1692	1.4429	3.4069	0.4235	0.2935	0.9187	3
4	0.6133	2.9745	4.0092	1.6305	4.8498	0.3362	0.2062	1.3479	4
5	0.5428	3.5172	6.1802	1.8424	6.4803	0.2843	0.1543	1.7571	5
6	0.4803	3.9975	8.5818	2.0820	8.3227	0.2502	0.1202	2.1468	6
7	0.4251	4.4226	11.1322	2.3526	10.4047	0.2261	0.0961	2.5171	7
8	0.3762	4.7988	13.7653	2.6584	12.7573	0.2084	0.0784	2.8685	8
9	0.3329	5.1317	16.4284	3.0040	15.4157	0.1949	0.0649	3.2014	9
10	0.2946	5.4262	19.0797	3.3946	18.4197	0.1843	0.0543	3.5162	10
11	0.2607	5.6869	21.6867	3.8359	21.8143	0.1758	0.0458	3.8134	11
12	0.2307	5.9176	24.2244	4.3345	25.6502	0.1690	0.0390	4.0936	12
13	0.2042	6.1218	26.6744	4.8980	29.9847	0.1634	0.0334	4.3573	13
14	0.1807	6.3025	29.0232	5.5348	34.8827	0.1587	0.0287	4.6050	14
15	0.1599	6.4624	31.2617	6.2543	40.4175	0.1547	0.0247	4.8375	15
16	0.1415	6.6039	33.3841	7.0673	46.6717	0.1514	0.0214	5.0552	16
17	0.1252	6.7291	35.3876	7.9861	53.7391	0.1486	0.0186	5.2589	17
18	0.1108	6.8399	37.2714	9.0243	61.7251	0.1462	0.0162	5.4491	18
19	0.0981	6.9380	39.0366	10.1974	70.7494	0.1441	0.0141	5.6265	19
20	0.0868	7.0248	40.6854	11.5231	80.9468	0.1424	0.0124	5.7917	20
21	0.0768	7.1016	42.2214	13.0211	92.4699	0.1408	0.0108	5.9454	21
22	0.0680	7.1695	43.6486	14.7138	105.4910	0.1395	0.0095	6.0881	22
23	0.0601	7.2297	44.9718	16.6266	120.2048	0.1383	0.0083	6.2205	23
24	0.0532	7.2829	46.1960	18.7881	136.8315	0.1373	0.0073	6.3431	24
25	0.0471	7.3300	47.3264	21.2305	155.6196	0.1364	0.0064	6.4566	25
26	0.0417	7.3717	48.3685	23.9905	176.8501	0.1357	0.0057	6.5614	26
27	0.0369	7.4086	49.3276	27.1093	200.8406	0.1350	0.0050	6.6582	27
28	0.0326	7.4412	50.2090	30.6335	227.9499	0.1344	0.0044	6.7474	28
29	0.0289	7.4701	51.0179	34.6158	258.5834	0.1339	0.0039	6.8296	29
30	0.0256	7.4957	51.7592	39.1159	293.1992	0.1334	0.0034	6.9052	30
31	0.0226	7.5183	52.4380	44.2010	332.3151	0.1330	0.0030	6.9747	31
32	0.0200	7.5383	53.0586	49.9471	376.5161	0.1327	0.0027	7.0385	32
33	0.0177	7.5560	53.6256	56.4402	426.4632	0.1323	0.0023	7.0971	33
34	0.0157	7.5717	54.1430	63.7774	482.9034	0.1321	0.0021	7.1507	34
35	0.0139	7.5856	54.6148	72.0685	546.6808	0.1318	0.0018	7.1998	35
36	0.0123	7.5979	55.0446	81.4374	618.7493	0.1316	0.0016	7.2448	36
37	0.0109	7.6087	55.4358	92.0243	700.1867	0.1314	0.0014	7.2858	37
38	0.0096	7.6183	55.7916	103.9874	792.2110	0.1313	0.0013	7.3233	38
39	0.0085	7.6268	56.1150	117.5058	896.1984	0.1311	0.0011	7.3576	39
40	0.0075	7.6344	56.4087	132.7816	1013.7042	0.1310	0.0010	7.3888	40
41	0.0067	7.6410	56.6753	150.0432	1146.4858	0.1309	0.0009	7.4172	41
42	0.0059	7.6469	56.9171	169.5488	1296.5289	0.1308	0.0008	7.4431	42
43	0.0052	7.6522	57.1363	191.5901	1466.0777	0.1307	0.0007	7.4667	43
44	0.0046	7.6568	57.3349	216.4968	1657.6678	0.1306	0.0006	7.4881	44
45	0.0041	7.6609	57.5148	244.6414	1874.1646	0.1305	0.0005	7.5076	45
46	0.0036	7.6645	57.6776	276.4448	2118.8060	0.1305	0.0005	7.5253	46
47	0.0032	7.6677	57.8248	312.3826	2395.2508	0.1304	0.0004	7.5414	47
48	0.0028	7.6705	57.9580	352.9923	2707.6334	0.1304	0.0004	7.5559	48
49	0.0025	7.6730	58.0783	398.8813	3060.6258	0.1303	0.0003	7.5692	49
50	0.0022	7.6752	58.1870	450.7359	3459.5071	0.1303	0.0003	7.5811	50
51	0.0020	7.6772	58.2852	509.3316	3910.2430	0.1303	0.0003	7.5920	51
52	0.0017	7.6789	58.3738	575.5447	4419.5746	0.1302	0.0002	7.6018	52
53	0.0015	7.6805	58.4537	650.3655	4995.1193	0.1302	0.0002	7.6107	53
54	0.0014	7.6818	58.5259	734.9130	5645.4849	0.1302	0.0002	7.6187	54
55	0.0012	7.6830	58.5909	830.4517	6380.3979	0.1302	0.0002	7.6260	55
60	0.0007	7.6873	58.8313	1530.0535	11761.9498	0.1301	0.0001	7.6531	60
65	0.0004	7.6896	58.9732	2819.0243	21677.1103	0.1300	0.0000	7.6692	65
70	0.0002	7.6908	59.0565	5193.8696	39945.1510	0.1300	0.0000	7.6788	70
75	0.0001	7.6915	59.1051	9569.3681	73602.8316	0.1300	0.0000	7.6845	75
80	0.0001	7.6919	59.1333	17630.9405	135614.9266	0.1300	0.0000	7.6878	80
85	0.0000	7.6921	59.1496	32483.8649	249868.1918	0.1300	0.0000	7.6897	85
90	0.0000	7.6922	59.1590	59849.4155	460372.4271	0.1300	0.0000	7.6908	90
95	0.0000	7.6922	59.1644	110268.6686	848212.8355	0.1300	0.0000	7.6914	95
100	0.0000	7.6923	59.1675	203162.8742	1562783.6479	0.1300	0.0000	7.6918	100

$I = 13.25\%$

n	(P/F)	(P/A)	(P/G)	(F/P)	(F/A)	(A/P)	(A/F)	(A/G)	n
1	0.8830	0.8830	0.0000	1.1325	1.0000	1.1325	1.0000	0.0000	1
2	0.7797	1.6627	0.7797	1.2826	2.1325	0.6014	0.4689	0.4689	2
3	0.6885	2.3512	2.1566	1.4525	3.4151	0.4253	0.2928	0.9173	3
4	0.6079	2.9591	3.9804	1.6450	4.8676	0.3379	0.2054	1.3451	4
5	0.5368	3.4959	6.1276	1.8629	6.5125	0.2861	0.1536	1.7528	5
6	0.4740	3.9699	8.4975	2.1097	8.3754	0.2519	0.1194	2.1405	6
7	0.4185	4.3884	11.0088	2.3893	10.4851	0.2279	0.0954	2.5086	7
8	0.3696	4.7580	13.5957	2.7059	12.8744	0.2102	0.0777	2.8575	8
9	0.3263	5.0843	16.2064	3.0644	15.5803	0.1967	0.0642	3.1875	9
10	0.2881	5.3725	18.7997	3.4704	18.6447	0.1861	0.0536	3.4993	10
11	0.2544	5.6269	21.3441	3.9303	22.1151	0.1777	0.0452	3.7932	11
12	0.2247	5.8516	23.8154	4.4510	26.0454	0.1709	0.0384	4.0699	12
13	0.1984	6.0499	26.1960	5.0408	30.4964	0.1653	0.0328	4.3300	13
14	0.1752	6.2251	28.4733	5.7087	35.5371	0.1606	0.0281	4.5739	14
15	0.1547	6.3798	30.6387	6.4651	41.2458	0.1567	0.0242	4.8025	15
16	0.1366	6.5164	32.6874	7.3217	47.7109	0.1535	0.0210	5.0162	16
17	0.1206	6.6370	34.6171	8.2918	55.0326	0.1507	0.0182	5.2158	17
18	0.1065	6.7435	36.4274	9.3905	63.3244	0.1483	0.0158	5.4019	18
19	0.0940	6.8375	38.1200	10.6347	72.7149	0.1463	0.0138	5.5751	19
20	0.0830	6.9205	39.6975	12.0438	83.3496	0.1445	0.0120	5.7362	20
21	0.0733	6.9938	41.1693	13.6396	95.3934	0.1430	0.0105	5.8857	21
22	0.0647	7.0586	42.5234	15.4469	109.0330	0.1417	0.0092	6.0243	22
23	0.0572	7.1157	43.7810	17.4936	124.4799	0.1405	0.0080	6.1527	23
24	0.0505	7.1662	44.9419	19.8115	141.9735	0.1395	0.0070	6.2714	24
25	0.0446	7.2108	46.0116	22.4365	161.7850	0.1387	0.0062	6.3809	25
26	0.0394	7.2501	46.9955	25.4093	184.2215	0.1379	0.0054	6.4820	26
27	0.0348	7.2849	47.8990	28.7761	209.6308	0.1373	0.0048	6.5751	27
28	0.0307	7.3156	48.7275	32.5889	238.4069	0.1367	0.0042	6.6608	28
29	0.0271	7.3427	49.4862	36.9069	270.9958	0.1362	0.0037	6.7395	29
30	0.0239	7.3666	50.1800	41.7971	307.9028	0.1357	0.0032	6.8118	30
31	0.0211	7.3877	50.8138	47.3352	349.6999	0.1354	0.0029	6.8781	31
32	0.0187	7.4064	51.3921	53.6072	397.0351	0.1350	0.0025	6.9389	32
33	0.0165	7.4229	51.9192	60.7101	450.6423	0.1347	0.0022	6.9945	33
34	0.0145	7.4374	52.3991	68.7542	511.3524	0.1345	0.0020	7.0454	34
35	0.0128	7.4502	52.8358	77.8641	580.1066	0.1342	0.0017	7.0918	35
36	0.0113	7.4616	53.2327	88.1811	657.9707	0.1340	0.0015	7.1342	36
37	0.0100	7.4716	53.5932	99.8651	746.1518	0.1338	0.0013	7.1729	37
38	0.0088	7.4804	53.9203	113.0972	846.0169	0.1337	0.0012	7.2082	38
39	0.0078	7.4882	54.2170	128.0826	959.1141	0.1335	0.0010	7.2403	39
40	0.0069	7.4951	54.4859	145.0536	1087.1968	0.1334	0.0009	7.2695	40
41	0.0061	7.5012	54.7294	164.2732	1232.2503	0.1333	0.0008	7.2961	41
42	0.0054	7.5066	54.9498	186.0394	1396.5235	0.1332	0.0007	7.3202	42
43	0.0047	7.5113	55.1491	210.6896	1582.5629	0.1331	0.0006	7.3421	43
44	0.0042	7.5155	55.3293	238.6060	1793.2525	0.1331	0.0006	7.3620	44
45	0.0037	7.5192	55.4922	270.2212	2031.8584	0.1330	0.0005	7.3800	45
46	0.0033	7.5225	55.6392	306.0256	2302.0796	0.1329	0.0004	7.3964	46
47	0.0029	7.5254	55.7719	346.5739	2608.1052	0.1329	0.0004	7.4112	47
48	0.0025	7.5279	55.8917	392.4950	2954.6791	0.1328	0.0003	7.4246	48
49	0.0022	7.5302	55.9997	444.5006	3347.1741	0.1328	0.0003	7.4367	49
50	0.0020	7.5322	56.0970	503.3969	3791.6747	0.1328	0.0003	7.4476	50
51	0.0018	7.5339	56.1847	570.0970	4295.0716	0.1327	0.0002	7.4576	51
52	0.0015	7.5355	56.2637	645.6348	4865.1686	0.1327	0.0002	7.4665	52
53	0.0014	7.5368	56.3348	731.1815	5510.8034	0.1327	0.0002	7.4746	53
54	0.0012	7.5381	56.3988	828.0630	6241.9849	0.1327	0.0002	7.4819	54
55	0.0011	7.5391	56.4564	937.7813	7070.0479	0.1326	0.0001	7.4885	55
60	0.0006	7.5428	56.6680	1746.9989	13177.3505	0.1326	0.0001	7.5128	60
65	0.0003	7.5449	56.7915	3254.4956	24554.6841	0.1325	0.0000	7.5272	65
70	0.0002	7.5459	56.8632	6062.8210	45749.5926	0.1325	0.0000	7.5356	70
75	0.0001	7.5465	56.9046	11294.4686	85233.7251	0.1325	0.0000	7.5405	75
80	0.0000	7.5468	56.9284	21040.5387	158788.9713	0.1325	0.0000	7.5434	80
85	0.0000	7.5470	56.9420	39196.5559	295815.5161	0.1325	0.0000	7.5450	85
90	0.0000	7.5471	56.9497	73019.5180	551083.1544	0.1325	0.0000	7.5459	90
95	0.0000	7.5471	56.9541	136028.5332	1026622.8921	0.1325	0.0000	7.5465	95
100	0.0000	7.5471	56.9566	253408.4360	1912508.9508	0.1325	0.0000	7.5468	100

$I = 13.50\%$

n	(P/F)	(P/A)	(P/G)	(F/P)	(F/A)	(A/P)	(A/F)	(A/G)	n
1	0.8811	0.8811	0.0000	1.1350	1.0000	1.1350	1.0000	0.0000	1
2	0.7763	1.6573	0.7763	1.2882	2.1350	0.6034	0.4684	0.4684	2
3	0.6839	2.3413	2.1441	1.4621	3.4232	0.4271	0.2921	0.9158	3
4	0.6026	2.9438	3.9519	1.6595	4.8854	0.3397	0.2047	1.3424	4
5	0.5309	3.4747	6.0755	1.8836	6.5449	0.2878	0.1528	1.7485	5
6	0.4678	3.9425	8.4143	2.1378	8.4284	0.2536	0.1186	2.1343	6
7	0.4121	4.3546	10.8871	2.4264	10.5663	0.2296	0.0946	2.5001	7
8	0.3631	4.7177	13.4288	2.7540	12.9927	0.2120	0.0770	2.8465	8
9	0.3199	5.0377	15.9881	3.1258	15.7468	0.1985	0.0635	3.1737	9
10	0.2819	5.3195	18.5249	3.5478	18.8726	0.1880	0.0530	3.4824	10
11	0.2483	5.5679	21.0083	4.0267	22.4204	0.1796	0.0446	3.7731	11
12	0.2188	5.7867	23.4151	4.5704	26.4471	0.1728	0.0378	4.0464	12
13	0.1928	5.9794	25.7285	5.1874	31.0175	0.1672	0.0322	4.3028	13
14	0.1698	6.1493	27.9365	5.8877	36.2048	0.1626	0.0276	4.5430	14
15	0.1496	6.2989	30.0315	6.6825	42.0925	0.1588	0.0238	4.7677	15
16	0.1318	6.4308	32.0092	7.5846	48.7750	0.1555	0.0205	4.9775	16
17	0.1162	6.5469	33.8678	8.6085	56.3596	0.1527	0.0177	5.1731	17
18	0.1023	6.6493	35.6077	9.7707	64.9681	0.1504	0.0154	5.3551	18
19	0.0902	6.7395	37.2308	11.0897	74.7388	0.1484	0.0134	5.5243	19
20	0.0794	6.8189	38.7403	12.5869	85.8286	0.1467	0.0117	5.6813	20
21	0.0700	6.8889	40.1403	14.2861	98.4154	0.1452	0.0102	5.8268	21
22	0.0617	6.9506	41.4354	16.2147	112.7015	0.1439	0.0089	5.9614	22
23	0.0543	7.0049	42.6308	18.4037	128.9162	0.1428	0.0078	6.0858	23
24	0.0479	7.0528	43.7319	20.8882	147.3199	0.1418	0.0068	6.2007	24
25	0.0422	7.0950	44.7442	23.7081	168.2081	0.1409	0.0059	6.3065	25
26	0.0372	7.1321	45.6733	26.9087	191.9162	0.1402	0.0052	6.4039	26
27	0.0327	7.1649	46.5246	30.5414	218.8248	0.1396	0.0046	6.4934	27
28	0.0288	7.1937	47.3035	34.6644	249.3662	0.1390	0.0040	6.5757	28
29	0.0254	7.2191	48.0152	39.3441	284.0306	0.1385	0.0035	6.6511	29
30	0.0224	7.2415	48.6646	44.6556	323.3748	0.1381	0.0031	6.7202	30
31	0.0197	7.2613	49.2565	50.6841	368.0303	0.1377	0.0027	6.7835	31
32	0.0174	7.2786	49.7954	57.5264	418.7144	0.1374	0.0024	6.8413	32
33	0.0153	7.2940	50.2855	65.2925	476.2409	0.1371	0.0021	6.8941	33
34	0.0135	7.3075	50.7308	74.1070	541.5334	0.1368	0.0018	6.9423	34
35	0.0119	7.3193	51.1350	84.1115	615.6404	0.1366	0.0016	6.9863	35
36	0.0105	7.3298	51.5016	95.4665	699.7519	0.1364	0.0014	7.0263	36
37	0.0092	7.3390	51.8339	108.3545	795.2184	0.1363	0.0013	7.0628	37
38	0.0081	7.3472	52.1347	122.9823	903.5729	0.1361	0.0011	7.0959	38
39	0.0072	7.3543	52.4070	139.5850	1026.5552	0.1360	0.0010	7.1260	39
40	0.0063	7.3607	52.6531	158.4289	1166.1401	0.1359	0.0009	7.1533	40
41	0.0056	7.3662	52.8756	179.8168	1324.5691	0.1358	0.0008	7.1781	41
42	0.0049	7.3711	53.0765	204.0921	1504.3859	0.1357	0.0007	7.2006	42
43	0.0043	7.3754	53.2578	231.6445	1708.4780	0.1356	0.0006	7.2210	43
44	0.0038	7.3792	53.4213	262.9165	1940.1225	0.1355	0.0005	7.2394	44
45	0.0034	7.3826	53.5688	298.4103	2203.0391	0.1355	0.0005	7.2561	45
46	0.0030	7.3855	53.7016	338.6957	2501.4493	0.1354	0.0004	7.2712	46
47	0.0026	7.3881	53.8213	384.4196	2840.1450	0.1354	0.0004	7.2848	47
48	0.0023	7.3904	53.9290	436.3162	3224.5646	0.1353	0.0003	7.2971	48
49	0.0020	7.3924	54.0260	495.2189	3660.8808	0.1353	0.0003	7.3083	49
50	0.0018	7.3942	54.1131	562.0735	4156.0997	0.1352	0.0002	7.3183	50
51	0.0016	7.3958	54.1915	637.9534	4718.1731	0.1352	0.0002	7.3273	51
52	0.0014	7.3972	54.2619	724.0771	5356.1265	0.1352	0.0002	7.3355	52
53	0.0012	7.3984	54.3252	821.8275	6080.2036	0.1352	0.0002	7.3428	53
54	0.0011	7.3995	54.3820	932.7742	6902.0311	0.1351	0.0001	7.3495	54
55	0.0009	7.4004	54.4330	1058.6987	7834.8053	0.1351	0.0001	7.3554	55
60	0.0005	7.4037	54.6193	1994.1218	14763.8655	0.1351	0.0001	7.3773	60
65	0.0003	7.4054	54.7269	3756.0468	27815.1618	0.1350	0.0000	7.3901	65
70	0.0001	7.4064	54.7886	7074.7371	52398.0527	0.1350	0.0000	7.3975	70
75	0.0001	7.4069	54.8239	13325.6872	98701.3867	0.1350	0.0000	7.4018	75
80	0.0000	7.4071	54.8439	25099.7226	185916.4640	0.1350	0.0000	7.4042	80
85	0.0000	7.4073	54.8552	47276.8171	350191.2377	0.1350	0.0000	7.4056	85
90	0.0000	7.4073	54.8616	89048.6906	659612.5226	0.1350	0.0000	7.4064	90
95	0.0000	7.4074	54.8652	167728.4931	1242425.8749	0.1350	0.0000	7.4068	95
100	0.0000	7.4074	54.8672	315926.5704	2340189.4102	0.1350	0.0000	7.4071	100

$I = 13.75\%$

n	(P/F)	(P/A)	(P/G)	(F/P)	(F/A)	(A/P)	(A/F)	(A/G)	n
1	0.8791	0.8791	0.0000	1.1375	1.0000	1.1375	1.0000	0.0000	1
2	0.7729	1.6520	0.7729	1.2939	2.1375	0.6053	0.4678	0.4678	2
3	0.6794	2.3314	2.1317	1.4718	3.4314	0.4289	0.2914	0.9143	3
4	0.5973	2.9287	3.9236	1.6742	4.9032	0.3414	0.2039	1.3397	4
5	0.5251	3.4538	6.0240	1.9044	6.5774	0.2895	0.1520	1.7442	5
6	0.4616	3.9154	8.3322	2.1662	8.4818	0.2554	0.1179	2.1280	6
7	0.4058	4.3213	10.7671	2.4641	10.6481	0.2314	0.0939	2.4917	7
8	0.3568	4.6780	13.2645	2.8029	13.1122	0.2138	0.0763	2.8355	8
9	0.3136	4.9917	15.7737	3.1883	15.9151	0.2003	0.0628	3.1600	9
10	0.2757	5.2674	18.2553	3.6267	19.1034	0.1898	0.0523	3.4657	10
11	0.2424	5.5098	20.6793	4.1254	22.7301	0.1815	0.0440	3.7532	11
12	0.2131	5.7229	23.0234	4.6926	26.8555	0.1747	0.0372	4.0230	12
13	0.1873	5.9103	25.2714	5.3379	31.5482	0.1692	0.0317	4.2759	13
14	0.1647	6.0749	27.4125	6.0718	36.8860	0.1646	0.0271	4.5124	14
15	0.1448	6.2197	29.4395	6.9067	42.9579	0.1608	0.0233	4.7332	15
16	0.1273	6.3470	31.3488	7.8564	49.8646	0.1576	0.0201	4.9391	16
17	0.1119	6.4589	33.1392	8.9366	57.7210	0.1548	0.0173	5.1308	17
18	0.0984	6.5573	34.8115	10.1654	66.6576	0.1525	0.0150	5.3088	18
19	0.0865	6.6438	36.3682	11.5632	76.8230	0.1505	0.0130	5.4740	19
20	0.0760	6.7198	37.8127	13.1531	88.3862	0.1488	0.0113	5.6271	20
21	0.0668	6.7866	39.1494	14.9617	101.5393	0.1473	0.0098	5.7686	21
22	0.0588	6.8454	40.3834	17.0189	116.5009	0.1461	0.0086	5.8993	22
23	0.0517	6.8970	41.5198	19.3590	133.5198	0.1450	0.0075	6.0199	23
24	0.0454	6.9425	42.5643	22.0208	152.8788	0.1440	0.0065	6.1310	24
25	0.0399	6.9824	43.5224	25.0487	174.8996	0.1432	0.0057	6.2332	25
26	0.0351	7.0175	44.3998	28.4929	199.9483	0.1425	0.0050	6.3270	26
27	0.0309	7.0483	45.2020	32.4107	228.4412	0.1419	0.0044	6.4131	27
28	0.0271	7.0755	45.9344	36.8671	260.8519	0.1413	0.0038	6.4921	28
29	0.0238	7.0993	46.6020	41.9364	297.7190	0.1409	0.0034	6.5643	29
30	0.0210	7.1203	47.2100	47.7026	339.6554	0.1404	0.0029	6.6304	30
31	0.0184	7.1387	47.7628	54.2617	387.3580	0.1401	0.0026	6.6907	31
32	0.0162	7.1549	48.2651	61.7227	441.6197	0.1398	0.0023	6.7457	32
33	0.0142	7.1691	48.7209	70.2096	503.3424	0.1395	0.0020	6.7959	33
34	0.0125	7.1817	49.1341	79.8634	573.5520	0.1392	0.0017	6.8416	34
35	0.0110	7.1927	49.5083	90.8446	653.4154	0.1390	0.0015	6.8832	35
36	0.0097	7.2023	49.8470	103.3357	744.2600	0.1388	0.0013	6.9209	36
37	0.0085	7.2109	50.1533	117.5444	847.5957	0.1387	0.0012	6.9553	37
38	0.0075	7.2183	50.4300	133.7068	965.1402	0.1385	0.0010	6.9864	38
39	0.0066	7.2249	50.6799	152.0915	1098.8469	0.1384	0.0009	7.0146	39
40	0.0058	7.2307	50.9053	173.0040	1250.9384	0.1383	0.0008	7.0402	40
41	0.0051	7.2358	51.1086	196.7921	1423.9424	0.1382	0.0007	7.0633	41
42	0.0045	7.2402	51.2917	223.8510	1620.7345	0.1381	0.0006	7.0843	42
43	0.0039	7.2442	51.4567	254.6305	1844.5855	0.1380	0.0005	7.1032	43
44	0.0035	7.2476	51.6051	289.6422	2099.2160	0.1380	0.0005	7.1203	44
45	0.0030	7.2507	51.7387	329.4680	2388.8582	0.1379	0.0004	7.1357	45
46	0.0027	7.2533	51.8588	374.7699	2718.3262	0.1379	0.0004	7.1497	46
47	0.0023	7.2557	51.9667	426.3007	3093.0960	0.1378	0.0003	7.1622	47
48	0.0021	7.2577	52.0636	484.9171	3519.3967	0.1378	0.0003	7.1735	48
49	0.0018	7.2595	52.1506	551.5931	4004.3138	0.1377	0.0002	7.1837	49
50	0.0016	7.2611	52.2287	627.4372	4555.9069	0.1377	0.0002	7.1929	50
51	0.0014	7.2625	52.2988	713.7098	5183.3441	0.1377	0.0002	7.2012	51
52	0.0012	7.2638	52.3616	811.8449	5897.0540	0.1377	0.0002	7.2086	52
53	0.0011	7.2649	52.4179	923.4736	6708.8989	0.1376	0.0001	7.2153	53
54	0.0010	7.2658	52.4683	1050.4512	7632.3725	0.1376	0.0001	7.2213	54
55	0.0008	7.2666	52.5135	1194.8883	8682.8237	0.1376	0.0001	7.2267	55
60	0.0004	7.2695	52.6776	2275.5392	16542.1032	0.1376	0.0001	7.2463	60
65	0.0002	7.2710	52.7713	4333.5254	31509.2756	0.1375	0.0000	7.2577	65
70	0.0001	7.2718	52.8245	8252.7440	60012.6836	0.1375	0.0000	7.2642	70
75	0.0001	7.2723	52.8545	15716.4842	114294.4304	0.1375	0.0000	7.2680	75
80	0.0000	7.2725	52.8714	29930.3935	217668.3162	0.1375	0.0000	7.2701	80
85	0.0000	7.2726	52.8808	56999.2909	414533.0248	0.1375	0.0000	7.2712	85
90	0.0000	7.2727	52.8860	108549.1631	789441.1865	0.1375	0.0000	7.2719	90
95	0.0000	7.2727	52.8890	206720.4807	1503414.4053	0.1375	0.0000	7.2723	95
100	0.0000	7.2727	52.8906	393677.4445	2863101.4148	0.1375	0.0000	7.2725	100

$I = 14.00\%$

n	(P/F)	(P/A)	(P/G)	(F/P)	(F/A)	(A/P)	(A/F)	(A/G)	n
1	0.8772	0.8772	0.0000	1.1400	1.0000	1.1400	1.0000	0.0000	1
2	0.7695	1.6467	0.7695	1.2996	2.1400	0.6073	0.4673	0.4673	2
3	0.6750	2.3216	2.1194	1.4815	3.4396	0.4307	0.2907	0.9129	3
4	0.5921	2.9137	3.8957	1.6890	4.9211	0.3432	0.2032	1.3370	4
5	0.5194	3.4331	5.9731	1.9254	6.6101	0.2913	0.1513	1.7399	5
6	0.4556	3.8887	8.2511	2.1950	8.5355	0.2572	0.1172	2.1218	6
7	0.3996	4.2883	10.6489	2.5023	10.7305	0.2332	0.0932	2.4832	7
8	0.3506	4.6389	13.1028	2.8526	13.2328	0.2156	0.0756	2.8246	8
9	0.3075	4.9464	15.5629	3.2519	16.0853	0.2022	0.0622	3.1463	9
10	0.2697	5.2161	17.9906	3.7072	19.3373	0.1917	0.0517	3.4490	10
11	0.2366	5.4527	20.3567	4.2262	23.0445	0.1834	0.0434	3.7333	11
12	0.2076	5.6603	22.6399	4.8179	27.2707	0.1767	0.0367	3.9998	12
13	0.1821	5.8424	24.8247	5.4924	32.0887	0.1712	0.0312	4.2491	13
14	0.1597	6.0021	26.9009	6.2613	37.5811	0.1666	0.0266	4.4819	14
15	0.1401	6.1422	28.8623	7.1379	43.8424	0.1628	0.0228	4.6990	15
16	0.1229	6.2651	30.7057	8.1372	50.9804	0.1596	0.0196	4.9011	16
17	0.1078	6.3729	32.4305	9.2765	59.1176	0.1569	0.0169	5.0888	17
18	0.0946	6.4674	34.0380	10.5752	68.3941	0.1546	0.0146	5.2630	18
19	0.0829	6.5504	35.5311	12.0557	78.9692	0.1527	0.0127	5.4243	19
20	0.0728	6.6231	36.9135	13.7435	91.0249	0.1510	0.0110	5.5734	20
21	0.0638	6.6870	38.1901	15.6676	104.7684	0.1495	0.0095	5.7111	21
22	0.0560	6.7429	39.3658	17.8610	120.4360	0.1483	0.0083	5.8381	22
23	0.0491	6.7921	40.4463	20.3616	138.2970	0.1472	0.0072	5.9549	23
24	0.0431	6.8351	41.4371	23.2122	158.6586	0.1463	0.0063	6.0624	24
25	0.0378	6.8729	42.3441	26.4619	181.8708	0.1455	0.0055	6.1610	25
26	0.0331	6.9061	43.1728	30.1666	208.3327	0.1448	0.0048	6.2514	26
27	0.0291	6.9352	43.9289	34.3899	238.4993	0.1442	0.0042	6.3342	27
28	0.0255	6.9607	44.6176	39.2045	272.8892	0.1437	0.0037	6.4100	28
29	0.0224	6.9830	45.2441	44.6931	312.0937	0.1432	0.0032	6.4791	29
30	0.0196	7.0027	45.8132	50.9502	356.7868	0.1428	0.0028	6.5423	30
31	0.0172	7.0199	46.3297	58.0832	407.7370	0.1425	0.0025	6.5998	31
32	0.0151	7.0350	46.7979	66.2148	465.8202	0.1421	0.0021	6.6522	32
33	0.0132	7.0482	47.2218	75.4849	532.0350	0.1419	0.0019	6.6998	33
34	0.0116	7.0599	47.6053	86.0528	607.5199	0.1416	0.0016	6.7431	34
35	0.0102	7.0700	47.9519	98.1002	693.5727	0.1414	0.0014	6.7824	35
36	0.0089	7.0790	48.2649	111.8342	791.6729	0.1413	0.0013	6.8180	36
37	0.0078	7.0868	48.5472	127.4910	903.5071	0.1411	0.0011	6.8503	37
38	0.0069	7.0937	48.8018	145.3397	1030.9981	0.1410	0.0010	6.8796	38
39	0.0060	7.0997	49.0312	165.6873	1176.3378	0.1409	0.0009	6.9060	39
40	0.0053	7.1050	49.2376	188.8835	1342.0251	0.1407	0.0007	6.9300	40
41	0.0046	7.1097	49.4234	215.3272	1530.9086	0.1407	0.0007	6.9516	41
42	0.0041	7.1138	49.5904	245.4730	1746.2358	0.1406	0.0006	6.9711	42
43	0.0036	7.1173	49.7405	279.8392	1991.7088	0.1405	0.0005	6.9886	43
44	0.0031	7.1205	49.8753	319.0167	2271.5481	0.1404	0.0004	7.0045	44
45	0.0027	7.1232	49.9963	363.6791	2590.5648	0.1404	0.0004	7.0188	45
46	0.0024	7.1256	50.1048	414.5941	2954.2439	0.1403	0.0003	7.0316	46
47	0.0021	7.1277	50.2022	472.6373	3368.8380	0.1403	0.0003	7.0432	47
48	0.0019	7.1296	50.2894	538.8065	3841.4753	0.1403	0.0003	7.0536	48
49	0.0016	7.1312	50.3675	614.2395	4380.2819	0.1402	0.0002	7.0630	49
50	0.0014	7.1327	50.4375	700.2330	4994.5213	0.1402	0.0002	7.0714	50
51	0.0013	7.1339	50.5001	798.2656	5694.7543	0.1402	0.0002	7.0789	51
52	0.0011	7.1350	50.5562	910.0228	6493.0199	0.1402	0.0002	7.0857	52
53	0.0010	7.1360	50.6063	1037.4260	7403.0427	0.1401	0.0001	7.0917	53
54	0.0008	7.1368	50.6511	1182.6656	8440.4687	0.1401	0.0001	7.0972	54
55	0.0007	7.1376	50.6912	1348.2388	9623.1343	0.1401	0.0001	7.1020	55
60	0.0004	7.1401	50.8357	2595.9187	18535.1333	0.1401	0.0001	7.1197	60
65	0.0002	7.1414	50.9173	4998.2196	35694.4260	0.1400	0.0000	7.1298	65
70	0.0001	7.1421	50.9632	9623.6450	68733.1785	0.1400	0.0000	7.1356	70
75	0.0001	7.1425	50.9887	18529.5064	132346.4742	0.1400	0.0000	7.1388	75
80	0.0000	7.1427	51.0030	35676.9818	254828.4415	0.1400	0.0000	7.1406	80
85	0.0000	7.1428	51.0108	68692.9810	490657.0073	0.1400	0.0000	7.1416	85
90	0.0000	7.1428	51.0152	132262.4674	944724.7670	0.1400	0.0000	7.1422	90
95	0.0000	7.1428	51.0175	254660.0834	1818993.4528	0.1400	0.0000	7.1425	95
100	0.0000	7.1428	51.0188	490326.2381	3502323.1295	0.1400	0.0000	7.1427	100

$I = 14.25\%$

n	(P/F)	(P/A)	(P/G)	(F/P)	(F/A)	(A/P)	(A/F)	(A/G)	n
1	0.8753	0.8753	0.0000	1.1425	1.0000	1.1425	1.0000	0.0000	1
2	0.7661	1.6414	0.7661	1.3053	2.1425	0.6092	0.4667	0.4667	2
3	0.6706	2.3119	2.1072	1.4913	3.4478	0.4325	0.2900	0.9114	3
4	0.5869	2.8988	3.8679	1.7038	4.9391	0.3450	0.2025	1.3343	4
5	0.5137	3.4126	5.9228	1.9466	6.6429	0.2930	0.1505	1.7356	5
6	0.4496	3.8622	8.1710	2.2240	8.5896	0.2589	0.1164	2.1156	6
7	0.3936	4.2557	10.5323	2.5409	10.8136	0.2350	0.0925	2.4748	7
8	0.3445	4.6002	12.9436	2.9030	13.3545	0.2174	0.0749	2.8137	8
9	0.3015	4.9017	15.3556	3.3167	16.2575	0.2040	0.0615	3.1327	9
10	0.2639	5.1656	17.7307	3.7893	19.5742	0.1936	0.0511	3.4324	10
11	0.2310	5.3966	20.0406	4.3293	23.3636	0.1853	0.0428	3.7136	11
12	0.2022	5.5988	22.2645	4.9462	27.6929	0.1786	0.0361	3.9767	12
13	0.1770	5.7757	24.3880	5.6511	32.6391	0.1731	0.0306	4.2225	13
14	0.1549	5.9306	26.4015	6.4563	38.2902	0.1686	0.0261	4.4517	14
15	0.1356	6.0662	28.2995	7.3764	44.7465	0.1648	0.0223	4.6651	15
16	0.1187	6.1848	30.0793	8.4275	52.1229	0.1617	0.0192	4.8634	16
17	0.1039	6.2887	31.7411	9.6284	60.5504	0.1590	0.0165	5.0473	17
18	0.0909	6.3796	33.2865	11.0005	70.1788	0.1567	0.0142	5.2176	18
19	0.0796	6.4592	34.7187	12.5681	81.1793	0.1548	0.0123	5.3751	19
20	0.0696	6.5288	36.0419	14.3590	93.7474	0.1532	0.0107	5.5204	20
21	0.0610	6.5898	37.2610	16.4052	108.1064	0.1518	0.0093	5.6544	21
22	0.0534	6.6431	38.3814	18.7429	124.5115	0.1505	0.0080	5.7776	22
23	0.0467	6.6898	39.4088	21.4138	143.2544	0.1495	0.0070	5.8909	23
24	0.0409	6.7307	40.3489	24.4652	164.6682	0.1486	0.0061	5.9948	24
25	0.0358	6.7665	41.2076	27.9515	189.1334	0.1478	0.0053	6.0900	25
26	0.0313	6.7978	41.9904	31.9346	217.0849	0.1471	0.0046	6.1771	26
27	0.0274	6.8252	42.7030	36.4853	249.0195	0.1465	0.0040	6.2567	27
28	0.0240	6.8492	43.3507	41.6844	285.5048	0.1460	0.0035	6.3293	28
29	0.0210	6.8702	43.9387	47.6245	327.1892	0.1456	0.0031	6.3956	29
30	0.0184	6.8886	44.4717	54.4109	374.8136	0.1452	0.0027	6.4559	30
31	0.0161	6.9047	44.9542	62.1645	429.2246	0.1448	0.0023	6.5107	31
32	0.0141	6.9187	45.3907	71.0229	491.3891	0.1445	0.0020	6.5606	32
33	0.0123	6.9311	45.7851	81.1437	562.4120	0.1443	0.0018	6.6058	33
34	0.0108	6.9418	46.1410	92.7067	643.5558	0.1441	0.0016	6.6468	34
35	0.0094	6.9513	46.4621	105.9174	736.2624	0.1439	0.0014	6.6839	35
36	0.0083	6.9596	46.7513	121.0106	842.1798	0.1437	0.0012	6.7176	36
37	0.0072	6.9668	47.0117	138.2546	963.1905	0.1435	0.0010	6.7480	37
38	0.0063	6.9731	47.2459	157.9559	1101.4451	0.1434	0.0009	6.7754	38
39	0.0055	6.9787	47.4565	180.4646	1259.4010	0.1433	0.0008	6.8002	39
40	0.0049	6.9835	47.6456	206.1809	1439.8657	0.1432	0.0007	6.8226	40
41	0.0042	6.9878	47.8154	235.5616	1646.0466	0.1431	0.0006	6.8427	41
42	0.0037	6.9915	47.9678	269.1292	1881.6082	0.1430	0.0005	6.8609	42
43	0.0033	6.9947	48.1044	307.4801	2150.7374	0.1430	0.0005	6.8772	43
44	0.0028	6.9976	48.2268	351.2960	2458.2174	0.1429	0.0004	6.8919	44
45	0.0025	7.0001	48.3364	401.3557	2809.5134	0.1429	0.0004	6.9051	45
46	0.0022	7.0022	48.4346	458.5488	3210.8691	0.1428	0.0003	6.9170	46
47	0.0019	7.0041	48.5224	523.8921	3669.4179	0.1428	0.0003	6.9277	47
48	0.0017	7.0058	48.6009	598.5467	4193.3100	0.1427	0.0002	6.9372	48
49	0.0015	7.0073	48.6711	683.8396	4791.8566	0.1427	0.0002	6.9458	49
50	0.0013	7.0086	48.7338	781.2867	5475.6962	0.1427	0.0002	6.9535	50
51	0.0011	7.0097	48.7898	892.6201	6256.9829	0.1427	0.0002	6.9603	51
52	0.0010	7.0107	48.8398	1019.8184	7149.6030	0.1426	0.0001	6.9665	52
53	0.0009	7.0115	48.8844	1165.1426	8169.4214	0.1426	0.0001	6.9720	53
54	0.0008	7.0123	48.9243	1331.1754	9334.5640	0.1426	0.0001	6.9769	54
55	0.0007	7.0129	48.9598	1520.8679	10665.7393	0.1426	0.0001	6.9814	55
60	0.0003	7.0152	49.0871	2960.5508	20768.7779	0.1425	0.0000	6.9973	60
65	0.0002	7.0163	49.1582	5763.0657	40435.5485	0.1425	0.0000	7.0063	65
70	0.0001	7.0169	49.1977	11218.4953	78719.2656	0.1425	0.0000	7.0113	70
75	0.0000	7.0172	49.2196	21838.1405	153243.0912	0.1425	0.0000	7.0141	75
80	0.0000	7.0174	49.2316	42510.5476	298312.6147	0.1425	0.0000	7.0157	80
85	0.0000	7.0175	49.2381	82751.8560	580707.7613	0.1425	0.0000	7.0165	85
90	0.0000	7.0175	49.2417	161086.3669	1130423.6277	0.1425	0.0000	7.0170	90
95	0.0000	7.0175	49.2436	313573.8444	2200511.1889	0.1425	0.0000	7.0172	95
100	0.0000	7.0175	49.2447	610408.9239	4283564.3783	0.1425	0.0000	7.0174	100

$I = 14.50\%$

n	(P/F)	(P/A)	(P/G)	(F/P)	(F/A)	(A/P)	(A/F)	(A/G)	n
1	0.8734	0.8734	0.0000	1.1450	1.0000	1.1450	1.0000	0.0000	1
2	0.7628	1.6361	0.7628	1.3110	2.1450	0.6112	0.4662	0.4662	2
3	0.6662	2.3023	2.0951	1.5011	3.4560	0.4343	0.2893	0.9100	3
4	0.5818	2.8841	3.8405	1.7188	4.9571	0.3467	0.2017	1.3316	4
5	0.5081	3.3922	5.8730	1.9680	6.6759	0.2948	0.1498	1.7313	5
6	0.4438	3.8360	8.0919	2.2534	8.6439	0.2607	0.1157	2.1095	6
7	0.3876	4.2236	10.4174	2.5801	10.8973	0.2368	0.0918	2.4665	7
8	0.3385	4.5621	12.7869	2.9542	13.4774	0.2192	0.0742	2.8029	8
9	0.2956	4.8577	15.1519	3.3826	16.4317	0.2059	0.0609	3.1192	9
10	0.2582	5.1159	17.4757	3.8731	19.8142	0.1955	0.0505	3.4159	10
11	0.2255	5.3414	19.7306	4.4347	23.6873	0.1872	0.0422	3.6939	11
12	0.1969	5.5383	21.8970	5.0777	28.1220	0.1806	0.0356	3.9537	12
13	0.1720	5.7103	23.9610	5.8140	33.1997	0.1751	0.0301	4.1961	13
14	0.1502	5.8606	25.9138	6.6570	39.0136	0.1706	0.0256	4.4217	14
15	0.1312	5.9918	27.7506	7.6222	45.6706	0.1669	0.0219	4.6315	15
16	0.1146	6.1063	29.4693	8.7275	53.2928	0.1638	0.0188	4.8260	16
17	0.1001	6.2064	31.0704	9.9929	62.0203	0.1611	0.0161	5.0062	17
18	0.0874	6.2938	32.5562	11.4419	72.0132	0.1589	0.0139	5.1727	18
19	0.0763	6.3701	33.9301	13.1010	83.4551	0.1570	0.0120	5.3264	19
20	0.0667	6.4368	35.1967	15.0006	96.5561	0.1554	0.0104	5.4680	20
21	0.0582	6.4950	36.3612	17.1757	111.5568	0.1540	0.0090	5.5983	21
22	0.0508	6.5459	37.4290	19.6662	128.7325	0.1528	0.0078	5.7180	22
23	0.0444	6.5903	38.4060	22.5178	148.3987	0.1517	0.0067	5.8277	23
24	0.0388	6.6291	39.2980	25.7829	170.9165	0.1509	0.0059	5.9281	24
25	0.0339	6.6629	40.1110	29.5214	196.6994	0.1501	0.0051	6.0200	25
26	0.0296	6.6925	40.8506	33.8020	226.2208	0.1494	0.0044	6.1039	26
27	0.0258	6.7184	41.5224	38.7033	260.0228	0.1488	0.0038	6.1804	27
28	0.0226	6.7409	42.1317	44.3153	298.7262	0.1483	0.0033	6.2501	28
29	0.0197	6.7606	42.6835	50.7410	343.0415	0.1479	0.0029	6.3135	29
30	0.0172	6.7778	43.1826	58.0985	393.7825	0.1475	0.0025	6.3711	30
31	0.0150	6.7929	43.6336	66.5227	451.8809	0.1472	0.0022	6.4234	31
32	0.0131	6.8060	44.0406	76.1685	518.4037	0.1469	0.0019	6.4708	32
33	0.0115	6.8175	44.4075	87.2130	594.5722	0.1467	0.0017	6.5138	33
34	0.0100	6.8275	44.7380	99.8588	681.7852	0.1465	0.0015	6.5526	34
35	0.0087	6.8362	45.0354	114.3384	781.6440	0.1463	0.0013	6.5877	35
36	0.0076	6.8439	45.3027	130.9174	895.9824	0.1461	0.0011	6.6195	36
37	0.0067	6.8505	45.5429	149.9005	1026.8998	0.1460	0.0010	6.6481	37
38	0.0058	6.8564	45.7584	171.6360	1176.8003	0.1458	0.0008	6.6739	38
39	0.0051	6.8615	45.9518	196.5233	1348.4363	0.1457	0.0007	6.6971	39
40	0.0044	6.8659	46.1251	225.0191	1544.9596	0.1456	0.0006	6.7180	40
41	0.0039	6.8698	46.2804	257.6469	1769.9788	0.1456	0.0006	6.7368	41
42	0.0034	6.8732	46.4193	295.0057	2027.6257	0.1455	0.0005	6.7537	42
43	0.0030	6.8761	46.5437	337.7816	2322.6314	0.1454	0.0004	6.7689	43
44	0.0026	6.8787	46.6549	386.7599	2660.4129	0.1454	0.0004	6.7825	44
45	0.0023	6.8810	46.7542	442.8401	3047.1728	0.1453	0.0003	6.7947	45
46	0.0020	6.8830	46.8430	507.0519	3490.0129	0.1453	0.0003	6.8057	46
47	0.0017	6.8847	46.9222	580.5744	3997.0648	0.1453	0.0003	6.8155	47
48	0.0015	6.8862	46.9929	664.7577	4577.6391	0.1452	0.0002	6.8242	48
49	0.0013	6.8875	47.0560	761.1475	5242.3968	0.1452	0.0002	6.8321	49
50	0.0011	6.8886	47.1122	871.5139	6003.5444	0.1452	0.0002	6.8391	50
51	0.0010	6.8896	47.1623	997.8835	6875.0583	0.1451	0.0001	6.8454	51
52	0.0009	6.8905	47.2069	1142.5766	7872.9417	0.1451	0.0001	6.8510	52
53	0.0008	6.8913	47.2467	1308.2502	9015.5183	0.1451	0.0001	6.8560	53
54	0.0007	6.8919	47.2821	1497.9464	10323.7684	0.1451	0.0001	6.8605	54
55	0.0006	6.8925	47.3135	1715.1487	11821.7149	0.1451	0.0001	6.8645	55
60	0.0003	6.8945	47.4257	3375.4307	23271.9361	0.1450	0.0000	6.8788	60
65	0.0002	6.8955	47.4878	6642.8835	45806.0929	0.1450	0.0000	6.8868	65
70	0.0001	6.8960	47.5219	13073.2651	90153.5523	0.1450	0.0000	6.8912	70
75	0.0000	6.8963	47.5405	25728.3243	177429.8225	0.1450	0.0000	6.8936	75
80	0.0000	6.8964	47.5506	50633.6149	349190.4475	0.1450	0.0000	6.8950	80
85	0.0000	6.8965	47.5561	99647.4908	687217.1782	0.1450	0.0000	6.8957	85
90	0.0000	6.8965	47.5590	196107.3183	1352457.3674	0.1450	0.0000	6.8961	90
95	0.0000	6.8965	47.5606	385941.2812	2661657.1116	0.1450	0.0000	6.8963	95
100	0.0000	6.8965	47.5615	759536.5325	5238176.0864	0.1450	0.0000	6.8964	100

$I = 14.75\%$

n	(P/F)	(P/A)	(P/G)	(F/P)	(F/A)	(A/P)	(A/F)	(A/G)	n
1	0.8715	0.8715	0.0000	1.1475	1.0000	1.1475	1.0000	0.0000	1
2	0.7594	1.6309	0.7594	1.3168	2.1475	0.6132	0.4657	0.4657	2
3	0.6618	2.2927	2.0831	1.5110	3.4643	0.4362	0.2887	0.9086	3
4	0.5768	2.8695	3.8133	1.7338	4.9752	0.3485	0.2010	1.3289	4
5	0.5026	3.3721	5.8238	1.9896	6.7091	0.2966	0.1491	1.7271	5
6	0.4380	3.8101	8.0139	2.2831	8.6987	0.2625	0.1150	2.1033	6
7	0.3817	4.1918	10.3041	2.6198	10.9817	0.2386	0.0911	2.4582	7
8	0.3326	4.5245	12.6326	3.0062	13.6015	0.2210	0.0735	2.7921	8
9	0.2899	4.8143	14.9517	3.4496	16.6078	0.2077	0.0602	3.1057	9
10	0.2526	5.0670	17.2253	3.9585	20.0574	0.1974	0.0499	3.3995	10
11	0.2202	5.2871	19.4268	4.5423	24.0159	0.1891	0.0416	3.6744	11
12	0.1919	5.4790	21.5372	5.2123	28.5582	0.1825	0.0350	3.9309	12
13	0.1672	5.6462	23.5435	5.9812	33.7705	0.1771	0.0296	4.1698	13
14	0.1457	5.7919	25.4376	6.8634	39.7517	0.1727	0.0252	4.3920	14
15	0.1270	5.9188	27.2152	7.8757	46.6151	0.1690	0.0215	4.5981	15
16	0.1107	6.0295	28.8750	9.0374	54.4908	0.1659	0.0184	4.7890	16
17	0.0964	6.1259	30.4178	10.3704	63.5282	0.1632	0.0157	4.9654	17
18	0.0840	6.2099	31.8464	11.9000	73.8986	0.1610	0.0135	5.1283	18
19	0.0732	6.2832	33.1646	13.6553	85.7986	0.1592	0.0117	5.2783	19
20	0.0638	6.3470	34.3771	15.6695	99.4539	0.1576	0.0101	5.4163	20
21	0.0556	6.4026	35.4894	17.9807	115.1234	0.1562	0.0087	5.5430	21
22	0.0485	6.4511	36.5072	20.6329	133.1041	0.1550	0.0075	5.6591	22
23	0.0422	6.4933	37.4364	23.6762	153.7369	0.1540	0.0065	5.7654	23
24	0.0368	6.5301	38.2830	27.1684	177.4131	0.1531	0.0056	5.8625	24
25	0.0321	6.5622	39.0528	31.1758	204.5816	0.1524	0.0049	5.9512	25
26	0.0280	6.5901	39.7516	35.7742	235.7574	0.1517	0.0042	6.0320	26
27	0.0244	6.6145	40.3850	41.0509	271.5316	0.1512	0.0037	6.1055	27
28	0.0212	6.6357	40.9582	47.1059	312.5825	0.1507	0.0032	6.1724	28
29	0.0185	6.6542	41.4762	54.0540	359.6884	0.1503	0.0028	6.2330	29
30	0.0161	6.6704	41.9437	62.0270	413.7424	0.1499	0.0024	6.2881	30
31	0.0140	6.6844	42.3652	71.1760	475.7694	0.1496	0.0021	6.3379	31
32	0.0122	6.6967	42.7448	81.6744	546.9454	0.1493	0.0018	6.3830	32
33	0.0107	6.7073	43.0862	93.7214	628.6199	0.1491	0.0016	6.4238	33
34	0.0093	6.7166	43.3931	107.5453	722.3413	0.1489	0.0014	6.4605	34
35	0.0081	6.7247	43.6686	123.4083	829.8866	0.1487	0.0012	6.4937	35
36	0.0071	6.7318	43.9157	141.6110	953.2949	0.1485	0.0010	6.5236	36
37	0.0062	6.7379	44.1373	162.4986	1094.9059	0.1484	0.0009	6.5506	37
38	0.0054	6.7433	44.3357	186.4672	1257.4045	0.1483	0.0008	6.5748	38
39	0.0047	6.7480	44.5133	213.9711	1443.8717	0.1482	0.0007	6.5965	39
40	0.0041	6.7520	44.6721	245.5318	1657.8428	0.1481	0.0006	6.6161	40
41	0.0035	6.7556	44.8141	281.7478	1903.3746	0.1480	0.0005	6.6336	41
42	0.0031	6.7587	44.9409	323.3055	2185.1224	0.1480	0.0005	6.6493	42
43	0.0027	6.7614	45.0541	370.9931	2508.4279	0.1479	0.0004	6.6634	43
44	0.0023	6.7637	45.1551	425.7146	2879.4210	0.1478	0.0003	6.6761	44
45	0.0020	6.7658	45.2452	488.5075	3305.1356	0.1478	0.0003	6.6874	45
46	0.0018	6.7676	45.3255	560.5624	3793.6431	0.1478	0.0003	6.6975	46
47	0.0016	6.7691	45.3970	643.2453	4354.2055	0.1477	0.0002	6.7065	47
48	0.0014	6.7705	45.4607	738.1240	4997.4508	0.1477	0.0002	6.7145	48
49	0.0012	6.7717	45.5173	846.9973	5735.5748	0.1477	0.0002	6.7217	49
50	0.0010	6.7727	45.5677	971.9294	6582.5721	0.1477	0.0002	6.7282	50
51	0.0009	6.7736	45.6126	1115.2890	7554.5014	0.1476	0.0001	6.7339	51
52	0.0008	6.7744	45.6524	1279.7941	8669.7904	0.1476	0.0001	6.7390	52
53	0.0007	6.7750	45.6878	1468.5637	9949.5845	0.1476	0.0001	6.7435	53
54	0.0006	6.7756	45.7193	1685.1769	11418.1482	0.1476	0.0001	6.7476	54
55	0.0005	6.7762	45.7472	1933.7404	13103.3251	0.1476	0.0001	6.7512	55
60	0.0003	6.7779	45.8461	3847.3496	26076.9465	0.1475	0.0000	6.7641	60
65	0.0001	6.7788	45.9002	7654.6462	51889.1269	0.1475	0.0000	6.7712	65
70	0.0001	6.7792	45.9296	15229.6034	103244.7690	0.1475	0.0000	6.7751	70
75	0.0000	6.7794	45.9455	30300.6584	205421.4130	0.1475	0.0000	6.7772	75
80	0.0000	6.7795	45.9540	60285.8706	408710.9869	0.1475	0.0000	6.7783	80
85	0.0000	6.7796	45.9586	119944.1325	813173.7798	0.1475	0.0000	6.7790	85
90	0.0000	6.7796	45.9611	238639.5816	1617888.6885	0.1475	0.0000	6.7793	90
95	0.0000	6.7796	45.9624	474794.7956	3218940.9874	0.1475	0.0000	6.7795	95
100	0.0000	6.7797	45.9630	944646.7199	6404377.7624	0.1475	0.0000	6.7796	100

$I = 15.00\%$

n	(P/F)	(P/A)	(P/G)	(F/P)	(F/A)	(A/P)	(A/F)	(A/G)	n
1	0.8696	0.8696	0.0000	1.1500	1.0000	1.1500	1.0000	0.0000	1
2	0.7561	1.6257	0.7561	1.3225	2.1500	0.6151	0.4651	0.4651	2
3	0.6575	2.2832	2.0712	1.5209	3.4725	0.4380	0.2880	0.9071	3
4	0.5718	2.8550	3.7864	1.7490	4.9934	0.3503	0.2003	1.3263	4
5	0.4972	3.3522	5.7751	2.0114	6.7424	0.2983	0.1483	1.7228	5
6	0.4323	3.7845	7.9368	2.3131	8.7537	0.2642	0.1142	2.0972	6
7	0.3759	4.1604	10.1924	2.6600	11.0668	0.2404	0.0904	2.4498	7
8	0.3269	4.4873	12.4807	3.0590	13.7268	0.2229	0.0729	2.7813	8
9	0.2843	4.7716	14.7548	3.5179	16.7858	0.2096	0.0596	3.0922	9
10	0.2472	5.0188	16.9795	4.0456	20.3037	0.1993	0.0493	3.3832	10
11	0.2149	5.2337	19.1289	4.6524	24.3493	0.1911	0.0411	3.6549	11
12	0.1869	5.4206	21.1849	5.3503	29.0017	0.1845	0.0345	3.9082	12
13	0.1625	5.5831	23.1352	6.1528	34.3519	0.1791	0.0291	4.1438	13
14	0.1413	5.7245	24.9725	7.0757	40.5047	0.1747	0.0247	4.3624	14
15	0.1229	5.8474	26.6930	8.1371	47.5804	0.1710	0.0210	4.5650	15
16	0.1069	5.9542	28.2960	9.3576	55.7175	0.1679	0.0179	4.7522	16
17	0.0929	6.0472	29.7828	10.7613	65.0751	0.1654	0.0154	4.9251	17
18	0.0808	6.1280	31.1565	12.3755	75.8364	0.1632	0.0132	5.0843	18
19	0.0703	6.1982	32.4213	14.2318	88.2118	0.1613	0.0113	5.2307	19
20	0.0611	6.2593	33.5822	16.3665	102.4436	0.1598	0.0098	5.3651	20
21	0.0531	6.3125	34.6448	18.8215	118.8101	0.1584	0.0084	5.4883	21
22	0.0462	6.3587	35.6150	21.6447	137.6316	0.1573	0.0073	5.6010	22
23	0.0402	6.3988	36.4988	24.8915	159.2764	0.1563	0.0063	5.7040	23
24	0.0349	6.4338	37.3023	28.6252	184.1678	0.1554	0.0054	5.7979	24
25	0.0304	6.4641	38.0314	32.9190	212.7930	0.1547	0.0047	5.8834	25
26	0.0264	6.4906	38.6918	37.8568	245.7120	0.1541	0.0041	5.9612	26
27	0.0230	6.5135	39.2890	43.5353	283.5688	0.1535	0.0035	6.0319	27
28	0.0200	6.5335	39.8283	50.0656	327.1041	0.1531	0.0031	6.0960	28
29	0.0174	6.5509	40.3146	57.5755	377.1697	0.1527	0.0027	6.1541	29
30	0.0151	6.5660	40.7526	66.2118	434.7451	0.1523	0.0023	6.2066	30
31	0.0131	6.5791	41.1466	76.1435	500.9569	0.1520	0.0020	6.2541	31
32	0.0114	6.5905	41.5006	87.5651	577.1005	0.1517	0.0017	6.2970	32
33	0.0099	6.6005	41.8184	100.6998	664.6655	0.1515	0.0015	6.3357	33
34	0.0086	6.6091	42.1033	115.8048	765.3654	0.1513	0.0013	6.3705	34
35	0.0075	6.6166	42.3586	133.1755	881.1702	0.1511	0.0011	6.4019	35
36	0.0065	6.6231	42.5872	153.1519	1014.3457	0.1510	0.0010	6.4301	36
37	0.0057	6.6288	42.7916	176.1246	1167.4975	0.1509	0.0009	6.4554	37
38	0.0049	6.6338	42.9743	202.5433	1343.6222	0.1507	0.0007	6.4781	38
39	0.0043	6.6380	43.1374	232.9248	1546.1655	0.1506	0.0006	6.4985	39
40	0.0037	6.6418	43.2830	267.8635	1779.0903	0.1506	0.0006	6.5168	40
41	0.0032	6.6450	43.4128	308.0431	2046.9539	0.1505	0.0005	6.5331	41
42	0.0028	6.6478	43.5286	354.2495	2354.9969	0.1504	0.0004	6.5478	42
43	0.0025	6.6503	43.6317	407.3870	2709.2465	0.1504	0.0004	6.5609	43
44	0.0021	6.6524	43.7235	468.4950	3116.6334	0.1503	0.0003	6.5725	44
45	0.0019	6.6543	43.8051	538.7693	3585.1285	0.1503	0.0003	6.5830	45
46	0.0016	6.6559	43.8778	619.5847	4123.8977	0.1502	0.0002	6.5923	46
47	0.0014	6.6573	43.9423	712.5224	4743.4824	0.1502	0.0002	6.6006	47
48	0.0012	6.6585	43.9997	819.4007	5456.0047	0.1502	0.0002	6.6080	48
49	0.0011	6.6596	44.0506	942.3108	6275.4055	0.1502	0.0002	6.6146	49
50	0.0009	6.6605	44.0958	1083.6574	7217.7163	0.1501	0.0001	6.6205	50
51	0.0008	6.6613	44.1360	1246.2061	8301.3737	0.1501	0.0001	6.6257	51
52	0.0007	6.6620	44.1715	1433.1370	9547.5798	0.1501	0.0001	6.6304	52
53	0.0006	6.6626	44.2031	1648.1075	10980.7167	0.1501	0.0001	6.6345	53
54	0.0005	6.6631	44.2311	1895.3236	12628.8243	0.1501	0.0001	6.6382	54
55	0.0005	6.6636	44.2558	2179.6222	14524.1479	0.1501	0.0001	6.6414	55
60	0.0002	6.6651	44.3431	4383.9987	29219.9916	0.1500	0.0000	6.6530	60
65	0.0001	6.6659	44.3903	8817.7874	58778.5826	0.1500	0.0000	6.6593	65
70	0.0001	6.6663	44.4156	17735.7200	118231.4669	0.1500	0.0000	6.6627	70
75	0.0000	6.6665	44.4292	35672.8680	237812.4532	0.1500	0.0000	6.6646	75
80	0.0000	6.6666	44.4364	71750.8794	478332.5293	0.1500	0.0000	6.6656	80
85	0.0000	6.6666	44.4402	144316.6470	962104.3133	0.1500	0.0000	6.6661	85
90	0.0000	6.6666	44.4422	290272.3252	1935142.1680	0.1500	0.0000	6.6664	90
95	0.0000	6.6667	44.4433	583841.3276	3892268.8509	0.1500	0.0000	6.6665	95
100	0.0000	6.6667	44.4438	1174313.4507	7828749.6713	0.1500	0.0000	6.6666	100

$I = 15.25\%$

n	(P/F)	(P/A)	(P/G)	(F/P)	(F/A)	(A/P)	(A/F)	(A/G)	n
1	0.8677	0.8677	0.0000	1.1525	1.0000	1.1525	1.0000	0.0000	1
2	0.7529	1.6205	0.7529	1.3283	2.1525	0.6171	0.4646	0.4646	2
3	0.6532	2.2738	2.0594	1.5308	3.4808	0.4398	0.2873	0.9057	3
4	0.5668	2.8406	3.7598	1.7643	5.0116	0.3520	0.1995	1.3236	4
5	0.4918	3.3324	5.7270	2.0333	6.7758	0.3001	0.1476	1.7186	5
6	0.4267	3.7591	7.8607	2.3434	8.8092	0.2660	0.1135	2.0911	6
7	0.3703	4.1294	10.0823	2.7008	11.1525	0.2422	0.0897	2.4416	7
8	0.3213	4.4507	12.3312	3.1126	13.8533	0.2247	0.0722	2.7706	8
9	0.2788	4.7294	14.5613	3.5873	16.9659	0.2114	0.0589	3.0789	9
10	0.2419	4.9713	16.7381	4.1344	20.5532	0.2012	0.0487	3.3669	10
11	0.2099	5.1812	18.8368	4.7649	24.6876	0.1930	0.0405	3.6356	11
12	0.1821	5.3633	20.8399	5.4915	29.4525	0.1865	0.0340	3.8857	12
13	0.1580	5.5213	22.7360	6.3290	34.9440	0.1811	0.0286	4.1179	13
14	0.1371	5.6584	24.5182	7.2941	41.2729	0.1767	0.0242	4.3331	14
15	0.1190	5.7773	26.1836	8.4065	48.5671	0.1731	0.0206	4.5321	15
16	0.1032	5.8806	27.7318	9.6885	56.9735	0.1701	0.0176	4.7159	16
17	0.0896	5.9701	29.1648	11.1660	66.6620	0.1675	0.0150	4.8851	17
18	0.0777	6.0478	30.4858	12.8688	77.8280	0.1653	0.0128	5.0408	18
19	0.0674	6.1152	31.6994	14.8312	90.6967	0.1635	0.0110	5.1837	19
20	0.0585	6.1737	32.8110	17.0930	105.5280	0.1620	0.0095	5.3146	20
21	0.0508	6.2245	33.8263	19.6997	122.6210	0.1607	0.0082	5.4344	21
22	0.0440	6.2686	34.7512	22.7039	142.3207	0.1595	0.0070	5.5437	22
23	0.0382	6.3068	35.5920	26.1662	165.0246	0.1586	0.0061	5.6435	23
24	0.0332	6.3399	36.3547	30.1566	191.1908	0.1577	0.0052	5.7342	24
25	0.0288	6.3687	37.0452	34.7555	221.3474	0.1570	0.0045	5.8168	25
26	0.0250	6.3937	37.6693	40.0557	256.1029	0.1564	0.0039	5.8917	26
27	0.0217	6.4153	38.2325	46.1642	296.1586	0.1559	0.0034	5.9596	27
28	0.0188	6.4341	38.7400	53.2042	342.3228	0.1554	0.0029	6.0210	28
29	0.0163	6.4504	39.1967	61.3179	395.5270	0.1550	0.0025	6.0766	29
30	0.0142	6.4646	39.6070	70.6688	456.8449	0.1547	0.0022	6.1268	30
31	0.0123	6.4769	39.9754	81.4458	527.5138	0.1544	0.0019	6.1720	31
32	0.0107	6.4875	40.3056	93.8663	608.9596	0.1541	0.0016	6.2128	32
33	0.0092	6.4968	40.6014	108.1810	702.8260	0.1539	0.0014	6.2495	33
34	0.0080	6.5048	40.8661	124.6786	811.0069	0.1537	0.0012	6.2825	34
35	0.0070	6.5117	41.1027	143.6920	935.6855	0.1536	0.0011	6.3121	35
36	0.0060	6.5178	41.3141	165.6051	1079.3775	0.1534	0.0009	6.3387	36
37	0.0052	6.5230	41.5027	190.8598	1244.9826	0.1533	0.0008	6.3625	37
38	0.0045	6.5276	41.6709	219.9660	1435.8424	0.1532	0.0007	6.3838	38
39	0.0039	6.5315	41.8208	253.5108	1655.8084	0.1531	0.0006	6.4029	39
40	0.0034	6.5349	41.9543	292.1712	1909.3192	0.1530	0.0005	6.4200	40
41	0.0030	6.5379	42.0731	336.7273	2201.4903	0.1530	0.0005	6.4353	41
42	0.0026	6.5405	42.1787	388.0782	2538.2176	0.1529	0.0004	6.4489	42
43	0.0022	6.5427	42.2726	447.2601	2926.2958	0.1528	0.0003	6.4610	43
44	0.0019	6.5447	42.3560	515.4673	3373.5559	0.1528	0.0003	6.4719	44
45	0.0017	6.5463	42.4301	594.0760	3889.0232	0.1528	0.0003	6.4815	45
46	0.0015	6.5478	42.4958	684.6726	4483.0992	0.1527	0.0002	6.4901	46
47	0.0013	6.5491	42.5541	789.0852	5167.7718	0.1527	0.0002	6.4977	47
48	0.0011	6.5502	42.6058	909.4207	5956.8570	0.1527	0.0002	6.5045	48
49	0.0010	6.5511	42.6516	1048.1074	6866.2777	0.1526	0.0001	6.5106	49
50	0.0008	6.5519	42.6922	1207.9437	7914.3851	0.1526	0.0001	6.5160	50
51	0.0007	6.5527	42.7281	1392.1551	9122.3288	0.1526	0.0001	6.5207	51
52	0.0006	6.5533	42.7599	1604.4588	10514.4840	0.1526	0.0001	6.5249	52
53	0.0005	6.5538	42.7880	1849.1388	12118.9428	0.1526	0.0001	6.5287	53
54	0.0005	6.5543	42.8129	2131.1324	13968.0816	0.1526	0.0001	6.5320	54
55	0.0004	6.5547	42.8348	2456.1301	16099.2140	0.1526	0.0001	6.5350	55
60	0.0002	6.5561	42.9118	4994.0863	32741.5496	0.1525	0.0000	6.5454	60
65	0.0001	6.5567	42.9530	10154.5507	66580.6606	0.1525	0.0000	6.5510	65
70	0.0000	6.5571	42.9749	20647.4006	135386.3335	0.1525	0.0000	6.5540	70
75	0.0000	6.5572	42.9865	41982.6699	275289.6390	0.1525	0.0000	6.5556	75
80	0.0000	6.5573	42.9925	85363.9937	559757.3355	0.1525	0.0000	6.5564	80
85	0.0000	6.5573	42.9957	173571.8911	1138169.7778	0.1525	0.0000	6.5569	85
90	0.0000	6.5574	42.9974	352926.3345	2314264.4885	0.1525	0.0000	6.5571	90
95	0.0000	6.5574	42.9983	717610.4194	4705635.5371	0.1525	0.0000	6.5572	95
100	0.0000	6.5574	42.9987	1459128.0494	9568046.2259	0.1525	0.0000	6.5573	100

$I = 15.50\%$

n	(P/F)	(P/A)	(P/G)	(F/P)	(F/A)	(A/P)	(A/F)	(A/G)	n
1	0.8658	0.8658	0.0000	1.1550	1.0000	1.1550	1.0000	0.0000	1
2	0.7496	1.6154	0.7496	1.3340	2.1550	0.6190	0.4640	0.4640	2
3	0.6490	2.2644	2.0476	1.5408	3.4890	0.4416	0.2866	0.9043	3
4	0.5619	2.8263	3.7334	1.7796	5.0298	0.3538	0.1988	1.3209	4
5	0.4865	3.3129	5.6794	2.0555	6.8094	0.3019	0.1469	1.7144	5
6	0.4212	3.7341	7.7855	2.3741	8.8649	0.2678	0.1128	2.0850	6
7	0.3647	4.0988	9.9737	2.7420	11.2390	0.2440	0.0890	2.4333	7
8	0.3158	4.4145	12.1839	3.1671	13.9810	0.2265	0.0715	2.7600	8
9	0.2734	4.6879	14.3709	3.6580	17.1481	0.2133	0.0583	3.0655	9
10	0.2367	4.9246	16.5012	4.2249	20.8060	0.2031	0.0481	3.3508	10
11	0.2049	5.1295	18.5504	4.8798	25.0310	0.1950	0.0400	3.6164	11
12	0.1774	5.3069	20.5021	5.6362	29.9108	0.1884	0.0334	3.8633	12
13	0.1536	5.4605	22.3455	6.5098	35.5469	0.1831	0.0281	4.0922	13
14	0.1330	5.5935	24.0745	7.5188	42.0567	0.1788	0.0238	4.3040	14
15	0.1152	5.7087	25.6866	8.6842	49.5755	0.1752	0.0202	4.4996	15
16	0.0997	5.8084	27.1821	10.0302	58.2597	0.1722	0.0172	4.6798	16
17	0.0863	5.8947	28.5632	11.5849	68.2899	0.1696	0.0146	4.8456	17
18	0.0747	5.9695	29.8337	13.3806	79.8749	0.1675	0.0125	4.9977	18
19	0.0647	6.0342	30.9984	15.4546	93.2555	0.1657	0.0107	5.1372	19
20	0.0560	6.0902	32.0628	17.8501	108.7101	0.1642	0.0092	5.2647	20
21	0.0485	6.1387	33.0329	20.6168	126.5601	0.1629	0.0079	5.3811	21
22	0.0420	6.1807	33.9148	23.8124	147.1769	0.1618	0.0068	5.4872	22
23	0.0364	6.2170	34.7147	27.5034	170.9894	0.1608	0.0058	5.5838	23
24	0.0315	6.2485	35.4387	31.7664	198.4927	0.1600	0.0050	5.6715	24
25	0.0273	6.2758	36.0928	36.6902	230.2591	0.1593	0.0043	5.7511	25
26	0.0236	6.2994	36.6828	42.3771	266.9493	0.1587	0.0037	5.8232	26
27	0.0204	6.3198	37.2140	48.9456	309.3264	0.1582	0.0032	5.8885	27
28	0.0177	6.3375	37.6916	56.5322	358.2720	0.1578	0.0028	5.9474	28
29	0.0153	6.3528	38.1204	65.2946	414.8041	0.1574	0.0024	6.0006	29
30	0.0133	6.3661	38.5050	75.4153	480.0988	0.1571	0.0021	6.0485	30
31	0.0115	6.3775	38.8494	87.1047	555.5141	0.1568	0.0018	6.0916	31
32	0.0099	6.3875	39.1575	100.6059	642.6188	0.1566	0.0016	6.1303	32
33	0.0086	6.3961	39.4329	116.1998	743.2247	0.1563	0.0013	6.1652	33
34	0.0075	6.4035	39.6788	134.2108	859.4245	0.1562	0.0012	6.1964	34
35	0.0065	6.4100	39.8981	155.0135	993.6353	0.1560	0.0010	6.2244	35
36	0.0056	6.4156	40.0936	179.0406	1148.6488	0.1559	0.0009	6.2494	36
37	0.0048	6.4204	40.2677	206.7918	1327.6893	0.1558	0.0008	6.2718	37
38	0.0042	6.4246	40.4226	238.8446	1534.4812	0.1557	0.0007	6.2918	38
39	0.0036	6.4282	40.5603	275.8655	1773.3258	0.1556	0.0006	6.3097	39
40	0.0031	6.4314	40.6827	318.6246	2049.1913	0.1555	0.0005	6.3257	40
41	0.0027	6.4341	40.7914	368.0115	2367.8159	0.1554	0.0004	6.3399	41
42	0.0024	6.4364	40.8879	425.0532	2735.8274	0.1554	0.0004	6.3526	42
43	0.0020	6.4385	40.9734	490.9365	3160.8806	0.1553	0.0003	6.3638	43
44	0.0018	6.4402	41.0493	567.0317	3651.8171	0.1553	0.0003	6.3739	44
45	0.0015	6.4418	41.1165	654.9216	4218.8488	0.1552	0.0002	6.3828	45
46	0.0013	6.4431	41.1760	756.4344	4873.7704	0.1552	0.0002	6.3907	46
47	0.0011	6.4442	41.2286	873.6817	5630.2048	0.1552	0.0002	6.3978	47
48	0.0010	6.4452	41.2752	1009.1024	6503.8865	0.1552	0.0002	6.4040	48
49	0.0009	6.4461	41.3164	1165.5133	7512.9889	0.1551	0.0001	6.4095	49
50	0.0007	6.4468	41.3528	1346.1678	8678.5022	0.1551	0.0001	6.4144	50
51	0.0006	6.4475	41.3849	1554.8239	10024.6700	0.1551	0.0001	6.4188	51
52	0.0006	6.4480	41.4133	1795.8216	11579.4939	0.1551	0.0001	6.4226	52
53	0.0005	6.4485	41.4384	2074.1739	13375.3155	0.1551	0.0001	6.4260	53
54	0.0004	6.4489	41.4605	2395.6708	15449.4893	0.1551	0.0001	6.4291	54
55	0.0004	6.4493	41.4800	2766.9998	17845.1602	0.1551	0.0001	6.4317	55
60	0.0002	6.4505	41.5479	5687.4691	36686.8977	0.1550	0.0000	6.4411	60
65	0.0001	6.4511	41.5839	11690.3893	75415.4150	0.1550	0.0000	6.4461	65
70	0.0000	6.4513	41.6028	24029.1770	155020.4966	0.1550	0.0000	6.4487	70
75	0.0000	6.4515	41.6127	49391.1135	318645.8935	0.1550	0.0000	6.4501	75
80	0.0000	6.4515	41.6178	101521.6665	654972.0420	0.1550	0.0000	6.4508	80
85	0.0000	6.4516	41.6205	208674.1530	1346278.4065	0.1550	0.0000	6.4512	85
90	0.0000	6.4516	41.6219	428922.2550	2767233.9030	0.1550	0.0000	6.4514	90
95	0.0000	6.4516	41.6226	881634.3479	5687957.0833	0.1550	0.0000	6.4515	95
100	0.0000	6.4516	41.6229	1812167.8566	************	0.1550	0.0000	6.4516	100

$I = 15.75\%$

n	(P/F)	(P/A)	(P/G)	(F/P)	(F/A)	(A/P)	(A/F)	(A/G)	n
1	0.8639	0.8639	0.0000	1.1575	1.0000	1.1575	1.0000	0.0000	1
2	0.7464	1.6103	0.7464	1.3398	2.1575	0.6210	0.4635	0.4635	2
3	0.6448	2.2551	2.0360	1.5508	3.4973	0.4434	0.2859	0.9028	3
4	0.5571	2.8122	3.7072	1.7951	5.0481	0.3556	0.1981	1.3183	4
5	0.4813	3.2935	5.6324	2.0778	6.8432	0.3036	0.1461	1.7102	5
6	0.4158	3.7093	7.7113	2.4051	8.9210	0.2696	0.1121	2.0789	6
7	0.3592	4.0685	9.8666	2.7839	11.3261	0.2458	0.0883	2.4251	7
8	0.3103	4.3788	12.0389	3.2223	14.1099	0.2284	0.0709	2.7494	8
9	0.2681	4.6469	14.1838	3.7298	17.3323	0.2152	0.0577	3.0523	9
10	0.2316	4.8786	16.2685	4.3173	21.0621	0.2050	0.0475	3.3347	10
11	0.2001	5.0787	18.2696	4.9972	25.3794	0.1969	0.0394	3.5973	11
12	0.1729	5.2515	20.1712	5.7843	30.3766	0.1904	0.0329	3.8410	12
13	0.1494	5.4009	21.9635	6.6953	36.1609	0.1852	0.0277	4.0666	13
14	0.1290	5.5299	23.6410	7.7499	42.8563	0.1808	0.0233	4.2751	14
15	0.1115	5.6414	25.2017	8.9705	50.6061	0.1773	0.0198	4.4673	15
16	0.0963	5.7377	26.6463	10.3833	59.5766	0.1743	0.0168	4.6441	16
17	0.0832	5.8209	27.9776	12.0187	69.9599	0.1718	0.0143	4.8064	17
18	0.0719	5.8928	29.1996	13.9116	81.9786	0.1697	0.0122	4.9551	18
19	0.0621	5.9549	30.3174	16.1027	95.8902	0.1679	0.0104	5.0912	19
20	0.0537	6.0086	31.3367	18.6389	111.9929	0.1664	0.0089	5.2153	20
21	0.0464	6.0549	32.2638	21.5745	130.6318	0.1652	0.0077	5.3285	21
22	0.0400	6.0950	33.1047	24.9725	152.2063	0.1641	0.0066	5.4315	22
23	0.0346	6.1296	33.8658	28.9057	177.1788	0.1631	0.0056	5.5250	23
24	0.0299	6.1594	34.5532	33.4583	206.0845	0.1624	0.0049	5.6098	24
25	0.0258	6.1853	35.1729	38.7280	239.5428	0.1617	0.0042	5.6866	25
26	0.0223	6.2076	35.7306	44.8277	278.2708	0.1611	0.0036	5.7560	26
27	0.0193	6.2268	36.2317	51.8880	323.0985	0.1606	0.0031	5.8186	27
28	0.0166	6.2435	36.6812	60.0604	374.9865	0.1602	0.0027	5.8751	28
29	0.0144	6.2579	37.0840	69.5199	435.0468	0.1598	0.0023	5.9260	29
30	0.0124	6.2703	37.4444	80.4693	504.5667	0.1595	0.0020	5.9717	30
31	0.0107	6.2810	37.7665	93.1432	585.0360	0.1592	0.0017	6.0128	31
32	0.0093	6.2903	38.0540	107.8132	678.1791	0.1590	0.0015	6.0496	32
33	0.0080	6.2983	38.3104	124.7938	785.9923	0.1588	0.0013	6.0826	33
34	0.0069	6.3053	38.5389	144.4488	910.7861	0.1586	0.0011	6.1122	34
35	0.0060	6.3112	38.7422	167.1995	1055.2350	0.1584	0.0009	6.1386	35
36	0.0052	6.3164	38.9231	193.5334	1222.4345	0.1583	0.0008	6.1622	36
37	0.0045	6.3209	39.0838	224.0149	1415.9679	0.1582	0.0007	6.1833	37
38	0.0039	6.3247	39.2265	259.2973	1639.9828	0.1581	0.0006	6.2021	38
39	0.0033	6.3281	39.3531	300.1366	1899.2801	0.1580	0.0005	6.2188	39
40	0.0029	6.3309	39.4653	347.4081	2199.4167	0.1580	0.0005	6.2337	40
41	0.0025	6.3334	39.5648	402.1249	2546.8249	0.1579	0.0004	6.2470	41
42	0.0021	6.3356	39.6529	465.4596	2948.9498	0.1578	0.0003	6.2588	42
43	0.0019	6.3374	39.7309	538.7695	3414.4094	0.1578	0.0003	6.2692	43
44	0.0016	6.3390	39.7998	623.6257	3953.1789	0.1578	0.0003	6.2785	44
45	0.0014	6.3404	39.8608	721.8467	4576.8046	0.1577	0.0002	6.2868	45
46	0.0012	6.3416	39.9146	835.5376	5298.6513	0.1577	0.0002	6.2941	46
47	0.0010	6.3426	39.9622	967.1347	6134.1889	0.1577	0.0002	6.3006	47
48	0.0009	6.3435	40.0042	1119.4585	7101.3236	0.1576	0.0001	6.3063	48
49	0.0008	6.3443	40.0412	1295.7732	8220.7821	0.1576	0.0001	6.3114	49
50	0.0007	6.3450	40.0739	1499.8573	9516.5552	0.1576	0.0001	6.3158	50
51	0.0006	6.3455	40.1027	1736.0850	11016.4127	0.1576	0.0001	6.3198	51
52	0.0005	6.3460	40.1281	2009.5184	12752.4977	0.1576	0.0001	6.3233	52
53	0.0004	6.3465	40.1504	2326.0175	14762.0161	0.1576	0.0001	6.3264	53
54	0.0004	6.3468	40.1701	2692.3653	17088.0336	0.1576	0.0001	6.3291	54
55	0.0003	6.3472	40.1874	3116.4128	19780.3989	0.1576	0.0001	6.3316	55
60	0.0002	6.3482	40.2474	6475.3013	41106.6749	0.1575	0.0000	6.3399	60
65	0.0001	6.3487	40.2788	13454.4200	85418.5395	0.1575	0.0000	6.3444	65
70	0.0000	6.3490	40.2951	27955.6747	177489.9981	0.1575	0.0000	6.3467	70
75	0.0000	6.3491	40.3035	58086.4690	368796.6283	0.1575	0.0000	6.3479	75
80	0.0000	6.3492	40.3079	120692.4144	766294.6945	0.1575	0.0000	6.3485	80
85	0.0000	6.3492	40.3101	250775.4242	1592218.5661	0.1575	0.0000	6.3489	85
90	0.0000	6.3492	40.3112	521062.6839	3308328.1516	0.1575	0.0000	6.3490	90
95	0.0000	6.3492	40.3118	1082667.1770	6874070.9649	0.1575	0.0000	6.3491	95
100	0.0000	6.3492	40.3121	2249572.3688	************	0.1575	0.0000	6.3492	100

$I = 16.00\%$

n	(P/F)	(P/A)	(P/G)	(F/P)	(F/A)	(A/P)	(A/F)	(A/G)	n
1	0.8621	0.8621	0.0000	1.1600	1.0000	1.1600	1.0000	0.0000	1
2	0.7432	1.6052	0.7432	1.3456	2.1600	0.6230	0.4630	0.4630	2
3	0.6407	2.2459	2.0245	1.5609	3.5056	0.4453	0.2853	0.9014	3
4	0.5523	2.7982	3.6814	1.8106	5.0665	0.3574	0.1974	1.3156	4
5	0.4761	3.2743	5.5858	2.1003	6.8771	0.3054	0.1454	1.7060	5
6	0.4104	3.6847	7.6380	2.4364	8.9775	0.2714	0.1114	2.0729	6
7	0.3538	4.0386	9.7610	2.8262	11.4139	0.2476	0.0876	2.4169	7
8	0.3050	4.3436	11.8962	3.2784	14.2401	0.2302	0.0702	2.7388	8
9	0.2630	4.6065	13.9998	3.8030	17.5185	0.2171	0.0571	3.0391	9
10	0.2267	4.8332	16.0399	4.4114	21.3215	0.2069	0.0469	3.3187	10
11	0.1954	5.0286	17.9941	5.1173	25.7329	0.1989	0.0389	3.5783	11
12	0.1685	5.1971	19.8472	5.9360	30.8502	0.1924	0.0324	3.8189	12
13	0.1452	5.3423	21.5899	6.8858	36.7862	0.1872	0.0272	4.0413	13
14	0.1252	5.4675	23.2175	7.9875	43.6720	0.1829	0.0229	4.2464	14
15	0.1079	5.5755	24.7284	9.2655	51.6595	0.1794	0.0194	4.4352	15
16	0.0930	5.6685	26.1241	10.7480	60.9250	0.1764	0.0164	4.6086	16
17	0.0802	5.7487	27.4074	12.4677	71.6730	0.1740	0.0140	4.7676	17
18	0.0691	5.8178	28.5828	14.4625	84.1407	0.1719	0.0119	4.9130	18
19	0.0596	5.8775	29.6557	16.7765	98.6032	0.1701	0.0101	5.0457	19
20	0.0514	5.9288	30.6321	19.4608	115.3797	0.1687	0.0087	5.1666	20
21	0.0443	5.9731	31.5180	22.5745	134.8405	0.1674	0.0074	5.2766	21
22	0.0382	6.0113	32.3200	26.1864	157.4150	0.1664	0.0064	5.3765	22
23	0.0329	6.0442	33.0442	30.3762	183.6014	0.1654	0.0054	5.4671	23
24	0.0284	6.0726	33.6970	35.2364	213.9776	0.1647	0.0047	5.5490	24
25	0.0245	6.0971	34.2841	40.8742	249.2140	0.1640	0.0040	5.6230	25
26	0.0211	6.1182	34.8114	47.4141	290.0883	0.1634	0.0034	5.6898	26
27	0.0182	6.1364	35.2841	55.0004	337.5024	0.1630	0.0030	5.7500	27
28	0.0157	6.1520	35.7073	63.8004	392.5028	0.1625	0.0025	5.8041	28
29	0.0135	6.1656	36.0856	74.0085	456.3032	0.1622	0.0022	5.8528	29
30	0.0116	6.1772	36.4234	85.8499	530.3117	0.1619	0.0019	5.8964	30
31	0.0100	6.1872	36.7247	99.5859	616.1616	0.1616	0.0016	5.9356	31
32	0.0087	6.1959	36.9930	115.5196	715.7475	0.1614	0.0014	5.9706	32
33	0.0075	6.2034	37.2318	134.0027	831.2671	0.1612	0.0012	6.0019	33
34	0.0064	6.2098	37.4441	155.4432	965.2698	0.1610	0.0010	6.0299	34
35	0.0055	6.2153	37.6327	180.3141	1120.7130	0.1609	0.0009	6.0548	35
36	0.0048	6.2201	37.8000	209.1643	1301.0270	0.1608	0.0008	6.0771	36
37	0.0041	6.2242	37.9484	242.6306	1510.1914	0.1607	0.0007	6.0969	37
38	0.0036	6.2278	38.0799	281.4515	1752.8220	0.1606	0.0006	6.1145	38
39	0.0031	6.2309	38.1963	326.4838	2034.2735	0.1605	0.0005	6.1302	39
40	0.0026	6.2335	38.2992	378.7212	2360.7572	0.1604	0.0004	6.1441	40
41	0.0023	6.2358	38.3903	439.3165	2739.4784	0.1604	0.0004	6.1565	41
42	0.0020	6.2377	38.4707	509.6072	3178.7949	0.1603	0.0003	6.1674	42
43	0.0017	6.2394	38.5418	591.1443	3688.4021	0.1603	0.0003	6.1771	43
44	0.0015	6.2409	38.6045	685.7274	4279.5465	0.1602	0.0002	6.1857	44
45	0.0013	6.2421	38.6598	795.4438	4965.2739	0.1602	0.0002	6.1934	45
46	0.0011	6.2432	38.7086	922.7148	5760.7177	0.1602	0.0002	6.2001	46
47	0.0009	6.2442	38.7516	1070.3492	6683.4326	0.1601	0.0001	6.2060	47
48	0.0008	6.2450	38.7894	1241.6051	7753.7818	0.1601	0.0001	6.2113	48
49	0.0007	6.2457	38.8227	1440.2619	8995.3869	0.1601	0.0001	6.2160	49
50	0.0006	6.2463	38.8521	1670.7038	10435.6488	0.1601	0.0001	6.2201	50
51	0.0005	6.2468	38.8779	1938.0164	12106.3526	0.1601	0.0001	6.2237	51
52	0.0004	6.2472	38.9006	2248.0990	14044.3690	0.1601	0.0001	6.2269	52
53	0.0004	6.2476	38.9205	2607.7949	16292.4680	0.1601	0.0001	6.2297	53
54	0.0003	6.2479	38.9380	3025.0421	18900.2629	0.1601	0.0001	6.2321	54
55	0.0003	6.2482	38.9534	3509.0488	21925.3050	0.1600	0.0000	6.2343	55
60	0.0001	6.2492	39.0063	7370.2014	46057.5085	0.1600	0.0000	6.2419	60
65	0.0001	6.2496	39.0337	15479.9410	96743.3810	0.1600	0.0000	6.2458	65
70	0.0000	6.2498	39.0478	32513.1648	203201.0302	0.1600	0.0000	6.2478	70
75	0.0000	6.2499	39.0551	68288.7545	426798.4658	0.1600	0.0000	6.2489	75
80	0.0000	6.2500	39.0587	143429.7159	896429.4743	0.1600	0.0000	6.2494	80
85	0.0000	6.2500	39.0606	301251.4072	1882815.0451	0.1600	0.0000	6.2497	85
90	0.0000	6.2500	39.0615	632730.8800	3954561.7500	0.1600	0.0000	6.2499	90
95	0.0000	6.2500	39.0620	1328951.0253	8305937.6582	0.1600	0.0000	6.2499	95
100	0.0000	6.2500	39.0623	2791251.1994	************	0.1600	0.0000	6.2500	100

$I = 16.25\%$

n	(P/F)	(P/A)	(P/G)	(F/P)	(F/A)	(A/P)	(A/F)	(A/G)	n
1	0.8602	0.8602	0.0000	1.1625	1.0000	1.1625	1.0000	0.0000	1
2	0.7400	1.6002	0.7400	1.3514	2.1625	0.6249	0.4624	0.4624	2
3	0.6365	2.2367	2.0130	1.5710	3.5139	0.4471	0.2846	0.9000	3
4	0.5476	2.7843	3.6557	1.8263	5.0849	0.3592	0.1967	1.3130	4
5	0.4710	3.2553	5.5398	2.1231	6.9112	0.3072	0.1447	1.7018	5
6	0.4052	3.6605	7.5656	2.4681	9.0343	0.2732	0.1107	2.0669	6
7	0.3485	4.0090	9.6569	2.8691	11.5024	0.2494	0.0869	2.4088	7
8	0.2998	4.3088	11.7556	3.3354	14.3715	0.2321	0.0696	2.7283	8
9	0.2579	4.5667	13.8188	3.8774	17.7069	0.2190	0.0565	3.0260	9
10	0.2219	4.7886	15.8155	4.5074	21.5842	0.2088	0.0463	3.3028	10
11	0.1908	4.9794	17.7240	5.2399	26.0917	0.2008	0.0383	3.5594	11
12	0.1642	5.1436	19.5298	6.0914	31.3316	0.1944	0.0319	3.7969	12
13	0.1412	5.2848	21.2244	7.0812	37.4229	0.1892	0.0267	4.0161	13
14	0.1215	5.4063	22.8036	8.2319	44.5042	0.1850	0.0225	4.2180	14
15	0.1045	5.5108	24.2666	9.5696	52.7361	0.1815	0.0190	4.4035	15
16	0.0899	5.6007	25.6150	11.1247	62.3057	0.1785	0.0160	4.5735	16
17	0.0773	5.6780	26.8522	12.9324	73.4304	0.1761	0.0136	4.7292	17
18	0.0665	5.7445	27.9829	15.0340	86.3628	0.1741	0.0116	4.8712	18
19	0.0572	5.8017	29.0129	17.4770	101.3968	0.1724	0.0099	5.0007	19
20	0.0492	5.8510	29.9480	20.3170	118.8737	0.1709	0.0084	5.1185	20
21	0.0423	5.8933	30.7948	23.6185	139.1907	0.1697	0.0072	5.2254	21
22	0.0364	5.9297	31.5597	27.4565	162.8092	0.1686	0.0061	5.3223	22
23	0.0313	5.9610	32.2489	31.9182	190.2657	0.1678	0.0053	5.4099	23
24	0.0270	5.9880	32.8688	37.1049	222.1839	0.1670	0.0045	5.4891	24
25	0.0232	6.0112	33.4252	43.1344	259.2888	0.1664	0.0039	5.5605	25
26	0.0199	6.0311	33.9238	50.1438	302.4232	0.1658	0.0033	5.6248	26
27	0.0172	6.0483	34.3698	58.2921	352.5670	0.1653	0.0028	5.6826	27
28	0.0148	6.0630	34.7682	67.7646	410.8591	0.1649	0.0024	5.7345	28
29	0.0127	6.0757	35.1237	78.7764	478.6237	0.1646	0.0021	5.7810	29
30	0.0109	6.0866	35.4403	91.5775	557.4001	0.1643	0.0018	5.8226	30
31	0.0094	6.0960	35.7221	106.4589	648.9776	0.1640	0.0015	5.8599	31
32	0.0081	6.1041	35.9726	123.7584	755.4365	0.1638	0.0013	5.8932	32
33	0.0070	6.1111	36.1951	143.8692	879.1949	0.1636	0.0011	5.9229	33
34	0.0060	6.1171	36.3924	167.2479	1023.0640	0.1635	0.0010	5.9493	34
35	0.0051	6.1222	36.5672	194.4257	1190.3120	0.1633	0.0008	5.9729	35
36	0.0044	6.1266	36.7221	226.0199	1384.7376	0.1632	0.0007	5.9939	36
37	0.0038	6.1304	36.8591	262.7481	1610.7575	0.1631	0.0006	6.0125	37
38	0.0033	6.1337	36.9802	305.4447	1873.5056	0.1630	0.0005	6.0290	38
39	0.0028	6.1365	37.0873	355.0794	2178.9503	0.1630	0.0005	6.0437	39
40	0.0024	6.1389	37.1817	412.7798	2534.0297	0.1629	0.0004	6.0567	40
41	0.0021	6.1410	37.2651	479.8565	2946.8095	0.1628	0.0003	6.0682	41
42	0.0018	6.1428	37.3386	557.8332	3426.6661	0.1628	0.0003	6.0784	42
43	0.0015	6.1444	37.4034	648.4811	3984.4993	0.1628	0.0003	6.0874	43
44	0.0013	6.1457	37.4604	753.8593	4632.9804	0.1627	0.0002	6.0954	44
45	0.0011	6.1468	37.5106	876.3615	5386.8398	0.1627	0.0002	6.1024	45
46	0.0010	6.1478	37.5548	1018.7702	6263.2012	0.1627	0.0002	6.1086	46
47	0.0008	6.1487	37.5936	1184.3204	7281.9714	0.1626	0.0001	6.1141	47
48	0.0007	6.1494	37.6278	1376.7724	8466.2918	0.1626	0.0001	6.1190	48
49	0.0006	6.1500	37.6578	1600.4979	9843.0642	0.1626	0.0001	6.1232	49
50	0.0005	6.1505	37.6841	1860.5788	11443.5621	0.1626	0.0001	6.1270	50
51	0.0005	6.1510	37.7072	2162.9229	13304.1410	0.1626	0.0001	6.1303	51
52	0.0004	6.1514	37.7275	2514.3979	15467.0639	0.1626	0.0001	6.1332	52
53	0.0003	6.1517	37.7453	2922.9875	17981.4617	0.1626	0.0001	6.1357	53
54	0.0003	6.1520	37.7609	3397.9730	20904.4493	0.1625	0.0000	6.1379	54
55	0.0003	6.1523	37.7746	3950.1436	24302.4223	0.1625	0.0000	6.1399	55
60	0.0001	6.1531	37.8213	8386.4410	51602.7136	0.1625	0.0000	6.1467	60
65	0.0001	6.1535	37.8452	17805.0215	109563.2090	0.1625	0.0000	6.1502	65
70	0.0000	6.1537	37.8574	37801.3499	232617.5381	0.1625	0.0000	6.1520	70
75	0.0000	6.1538	37.8636	80255.0033	493870.7898	0.1625	0.0000	6.1529	75
80	0.0000	6.1538	37.8667	170387.1838	1048530.3621	0.1625	0.0000	6.1534	80
85	0.0000	6.1538	37.8683	361744.3301	2226112.8004	0.1625	0.0000	6.1536	85
90	0.0000	6.1538	37.8691	768009.4088	4726205.5928	0.1625	0.0000	6.1537	90
95	0.0000	6.1538	37.8694	1630539.5912	************	0.1625	0.0000	6.1538	95
100	0.0000	6.1538	37.8696	3461753.6294	************	0.1625	0.0000	6.1538	100

ENGINEERING ECONOMIC ANALYSIS

$I = 16.50\%$

n	(P/F)	(P/A)	(P/G)	(F/P)	(F/A)	(A/P)	(A/F)	(A/G)	n
1	0.8584	0.8584	0.0000	1.1650	1.0000	1.1650	1.0000	0.0000	1
2	0.7368	1.5952	0.7368	1.3572	2.1650	0.6269	0.4619	0.4619	2
3	0.6324	2.2276	2.0017	1.5812	3.5222	0.4489	0.2839	0.8986	3
4	0.5429	2.7705	3.6303	1.8421	5.1034	0.3609	0.1959	1.3103	4
5	0.4660	3.2365	5.4942	2.1460	6.9455	0.3090	0.1440	1.6976	5
6	0.4000	3.6365	7.4942	2.5001	9.0915	0.2750	0.1100	2.0608	6
7	0.3433	3.9798	9.5542	2.9126	11.5915	0.2513	0.0863	2.4007	7
8	0.2947	4.2745	11.6171	3.3932	14.5041	0.2339	0.0689	2.7178	8
9	0.2530	4.5275	13.6409	3.9531	17.8973	0.2209	0.0559	3.0129	9
10	0.2171	4.7446	15.5951	4.6053	21.8504	0.2108	0.0458	3.2869	10
11	0.1864	4.9310	17.4590	5.3652	26.4557	0.2028	0.0378	3.5407	11
12	0.1600	5.0910	19.2189	6.2504	31.8209	0.1964	0.0314	3.7751	12
13	0.1373	5.2283	20.8668	7.2818	38.0713	0.1913	0.0263	3.9911	13
14	0.1179	5.3462	22.3993	8.4833	45.3531	0.1870	0.0220	4.1898	14
15	0.1012	5.4474	23.8158	9.8830	53.8364	0.1836	0.0186	4.3720	15
16	0.0869	5.5342	25.1186	11.5137	63.7194	0.1807	0.0157	4.5388	16
17	0.0746	5.6088	26.3115	13.4135	75.2331	0.1783	0.0133	4.6911	17
18	0.0640	5.6728	27.3993	15.6267	88.6465	0.1763	0.0113	4.8300	18
19	0.0549	5.7277	28.3881	18.2051	104.2732	0.1746	0.0096	4.9563	19
20	0.0471	5.7748	29.2839	21.2089	122.4783	0.1732	0.0082	5.0709	20
21	0.0405	5.8153	30.0934	24.7084	143.6872	0.1720	0.0070	5.1748	21
22	0.0347	5.8501	30.8229	28.7853	168.3956	0.1709	0.0059	5.2688	22
23	0.0298	5.8799	31.4789	33.5348	197.1809	0.1701	0.0051	5.3537	23
24	0.0256	5.9055	32.0677	39.0681	230.7157	0.1693	0.0043	5.4302	24
25	0.0220	5.9274	32.5950	45.5143	269.7838	0.1687	0.0037	5.4990	25
26	0.0189	5.9463	33.0665	53.0242	315.2981	0.1682	0.0032	5.5608	26
27	0.0162	5.9625	33.4873	61.7732	368.3223	0.1677	0.0027	5.6163	27
28	0.0139	5.9764	33.8625	71.9658	430.0955	0.1673	0.0023	5.6660	28
29	0.0119	5.9883	34.1965	83.8401	502.0613	0.1670	0.0020	5.7105	29
30	0.0102	5.9986	34.4934	97.6737	585.9014	0.1667	0.0017	5.7503	30
31	0.0088	6.0073	34.7570	113.7899	683.5751	0.1665	0.0015	5.7858	31
32	0.0075	6.0149	34.9909	132.5652	797.3650	0.1663	0.0013	5.8174	32
33	0.0065	6.0214	35.1981	154.4385	929.9302	0.1661	0.0011	5.8455	33
34	0.0056	6.0269	35.3815	179.9208	1084.3687	0.1659	0.0009	5.8706	34
35	0.0048	6.0317	35.5437	209.6078	1264.2895	0.1658	0.0008	5.8928	35
36	0.0041	6.0358	35.6870	244.1931	1473.8973	0.1657	0.0007	5.9126	36
37	0.0035	6.0393	35.8136	284.4849	1718.0904	0.1656	0.0006	5.9301	37
38	0.0030	6.0423	35.9252	331.4249	2002.5753	0.1655	0.0005	5.9456	38
39	0.0026	6.0449	36.0236	386.1100	2334.0002	0.1654	0.0004	5.9593	39
40	0.0022	6.0471	36.1104	449.8182	2720.1102	0.1654	0.0004	5.9715	40
41	0.0019	6.0490	36.1867	524.0382	3169.9284	0.1653	0.0003	5.9822	41
42	0.0016	6.0507	36.2538	610.5045	3693.9666	0.1653	0.0003	5.9917	42
43	0.0014	6.0521	36.3129	711.2377	4304.4711	0.1652	0.0002	6.0001	43
44	0.0012	6.0533	36.3648	828.5920	5015.7089	0.1652	0.0002	6.0074	44
45	0.0010	6.0543	36.4104	965.3096	5844.3008	0.1652	0.0002	6.0139	45
46	0.0009	6.0552	36.4504	1124.5857	6809.6105	0.1651	0.0001	6.0197	46
47	0.0008	6.0560	36.4855	1310.1424	7934.1962	0.1651	0.0001	6.0247	47
48	0.0007	6.0566	36.5163	1526.3159	9244.3386	0.1651	0.0001	6.0291	48
49	0.0006	6.0572	36.5433	1778.1580	10770.6544	0.1651	0.0001	6.0330	49
50	0.0005	6.0577	36.5669	2071.5540	12548.8124	0.1651	0.0001	6.0365	50
51	0.0004	6.0581	36.5877	2413.3605	14620.3664	0.1651	0.0001	6.0395	51
52	0.0004	6.0585	36.6058	2811.5649	17033.7269	0.1651	0.0001	6.0421	52
53	0.0003	6.0588	36.6217	3275.4732	19845.2918	0.1651	0.0001	6.0444	53
54	0.0003	6.0590	36.6356	3815.9262	23120.7650	0.1650	0.0000	6.0465	54
55	0.0002	6.0592	36.6477	4445.5541	26936.6912	0.1650	0.0000	6.0482	55
60	0.0001	6.0600	36.6890	9540.1570	57813.0727	0.1650	0.0000	6.0543	60
65	0.0000	6.0603	36.7099	20473.1726	124073.7735	0.1650	0.0000	6.0574	65
70	0.0000	6.0605	36.7205	43935.4193	266269.2078	0.1650	0.0000	6.0590	70
75	0.0000	6.0605	36.7257	94285.3901	571420.5459	0.1650	0.0000	6.0598	75
80	0.0000	6.0606	36.7284	202336.4048	1226275.1805	0.1650	0.0000	6.0602	80
85	0.0000	6.0606	36.7297	434213.8339	2631592.9325	0.1650	0.0000	6.0604	85
90	0.0000	6.0606	36.7303	931822.6926	5647404.1976	0.1650	0.0000	6.0605	90
95	0.0000	6.0606	36.7306	1999691.0802	************	0.1650	0.0000	6.0606	95
100	0.0000	6.0606	36.7308	4291336.1606	************	0.1650	0.0000	6.0606	100

$I = 16.75\%$

n	(P/F)	(P/A)	(P/G)	(F/P)	(F/A)	(A/P)	(A/F)	(A/G)	n
1	0.8565	0.8565	0.0000	1.1675	1.0000	1.1675	1.0000	0.0000	1
2	0.7336	1.5902	0.7336	1.3631	2.1675	0.6289	0.4614	0.4614	2
3	0.6284	2.2186	1.9904	1.5914	3.5306	0.4507	0.2832	0.8972	3
4	0.5382	2.7568	3.6051	1.8579	5.1219	0.3627	0.1952	1.3077	4
5	0.4610	3.2178	5.4492	2.1691	6.9798	0.3108	0.1433	1.6934	5
6	0.3949	3.6127	7.4236	2.5325	9.1490	0.2768	0.1093	2.0549	6
7	0.3382	3.9509	9.4529	2.9566	11.6814	0.2531	0.0856	2.3926	7
8	0.2897	4.2406	11.4808	3.4519	14.6381	0.2358	0.0683	2.7073	8
9	0.2481	4.4887	13.4659	4.0301	18.0899	0.2228	0.0553	2.9999	9
10	0.2125	4.7013	15.3787	4.7051	22.1200	0.2127	0.0452	3.2712	10
11	0.1820	4.8833	17.1991	5.4932	26.8251	0.2048	0.0373	3.5220	11
12	0.1559	5.0393	18.9143	6.4133	32.3183	0.1984	0.0309	3.7534	12
13	0.1336	5.1728	20.5170	7.4875	38.7316	0.1933	0.0258	3.9663	13
14	0.1144	5.2872	22.0041	8.7417	46.2192	0.1891	0.0216	4.1618	14
15	0.0980	5.3852	23.3758	10.2059	54.9609	0.1857	0.0182	4.3408	15
16	0.0839	5.4691	24.6347	11.9154	65.1668	0.1828	0.0153	4.5043	16
17	0.0719	5.5410	25.7848	13.9113	77.0823	0.1805	0.0130	4.6535	17
18	0.0616	5.6026	26.8315	16.2414	90.9936	0.1785	0.0110	4.7892	18
19	0.0527	5.6553	27.7808	18.9619	107.2350	0.1768	0.0093	4.9124	19
20	0.0452	5.7005	28.6391	22.1380	126.1968	0.1754	0.0079	5.0240	20
21	0.0387	5.7392	29.4129	25.8461	148.3348	0.1742	0.0067	5.1249	21
22	0.0331	5.7723	30.1088	30.1753	174.1809	0.1732	0.0057	5.2161	22
23	0.0284	5.8007	30.7333	35.2297	204.3562	0.1724	0.0049	5.2982	23
24	0.0243	5.8250	31.2925	41.1306	239.5858	0.1717	0.0042	5.3721	24
25	0.0208	5.8458	31.7923	48.0200	280.7165	0.1711	0.0036	5.4385	25
26	0.0178	5.8637	32.2382	56.0634	328.7365	0.1705	0.0030	5.4980	26
27	0.0153	5.8789	32.6354	65.4540	384.7998	0.1701	0.0026	5.5512	27
28	0.0131	5.8920	32.9888	76.4175	450.2538	0.1697	0.0022	5.5989	28
29	0.0112	5.9032	33.3026	89.2174	526.6713	0.1694	0.0019	5.6414	29
30	0.0096	5.9128	33.5810	104.1614	615.8888	0.1691	0.0016	5.6793	30
31	0.0082	5.9211	33.8277	121.6084	720.0501	0.1689	0.0014	5.7131	31
32	0.0070	5.9281	34.0460	141.9778	841.6585	0.1687	0.0012	5.7432	32
33	0.0060	5.9341	34.2391	165.7591	983.6363	0.1685	0.0010	5.7699	33
34	0.0052	5.9393	34.4096	193.5237	1149.3954	0.1684	0.0009	5.7935	34
35	0.0044	5.9437	34.5601	225.9390	1342.9192	0.1682	0.0007	5.8146	35
36	0.0038	5.9475	34.6928	263.7837	1568.8581	0.1681	0.0006	5.8332	36
37	0.0032	5.9508	34.8097	307.9675	1832.6419	0.1680	0.0005	5.8496	37
38	0.0028	5.9535	34.9126	359.5521	2140.6094	0.1680	0.0005	5.8642	38
39	0.0024	5.9559	35.0031	419.7770	2500.1614	0.1679	0.0004	5.8770	39
40	0.0020	5.9580	35.0827	490.0897	2919.9385	0.1678	0.0003	5.8884	40
41	0.0017	5.9597	35.1526	572.1797	3410.0282	0.1678	0.0003	5.8984	41
42	0.0015	5.9612	35.2140	668.0198	3982.2079	0.1678	0.0003	5.9072	42
43	0.0013	5.9625	35.2678	779.9131	4650.2277	0.1677	0.0002	5.9149	43
44	0.0011	5.9636	35.3150	910.5486	5430.1408	0.1677	0.0002	5.9218	44
45	0.0009	5.9645	35.3564	1063.0655	6340.6894	0.1677	0.0002	5.9278	45
46	0.0008	5.9653	35.3927	1241.1289	7403.7549	0.1676	0.0001	5.9331	46
47	0.0007	5.9660	35.4244	1449.0180	8644.8839	0.1676	0.0001	5.9377	47
48	0.0006	5.9666	35.4522	1691.7286	10093.9019	0.1676	0.0001	5.9418	48
49	0.0005	5.9671	35.4765	1975.0931	11785.6305	0.1676	0.0001	5.9453	49
50	0.0004	5.9676	35.4978	2305.9212	13760.7236	0.1676	0.0001	5.9485	50
51	0.0004	5.9679	35.5163	2692.1630	16066.6448	0.1676	0.0001	5.9512	51
52	0.0003	5.9682	35.5326	3143.1003	18758.8078	0.1676	0.0001	5.9536	52
53	0.0003	5.9685	35.5467	3669.5696	21901.9081	0.1675	0.0000	5.9557	53
54	0.0002	5.9688	35.5591	4284.2225	25571.4777	0.1675	0.0000	5.9575	54
55	0.0002	5.9690	35.5699	5001.8298	29855.7002	0.1675	0.0000	5.9592	55
60	0.0001	5.9696	35.6064	10849.5907	64767.7056	0.1675	0.0000	5.9646	60
65	0.0000	5.9699	35.6247	23534.1112	140496.1860	0.1675	0.0000	5.9674	65
70	0.0000	5.9700	35.6338	51048.4132	304760.6756	0.1675	0.0000	5.9688	70
75	0.0000	5.9701	35.6383	110730.3552	661070.7770	0.1675	0.0000	5.9695	75
80	0.0000	5.9701	35.6405	240187.9077	1433951.6880	0.1675	0.0000	5.9698	80
85	0.0000	5.9701	35.6416	520997.4351	3110426.4782	0.1675	0.0000	5.9700	85
90	0.0000	5.9701	35.6422	1130108.2137	6746908.7388	0.1675	0.0000	5.9701	90
95	0.0000	5.9701	35.6424	2451345.2251	************	0.1675	0.0000	5.9701	95
100	0.0000	5.9701	35.6426	5317272.5756	************	0.1675	0.0000	5.9701	100

$I = 17.00\%$

n	(P/F)	(P/A)	(P/G)	(F/P)	(F/A)	(A/P)	(A/F)	(A/G)	n
1	0.8547	0.8547	0.0000	1.1700	1.0000	1.1700	1.0000	0.0000	1
2	0.7305	1.5852	0.7305	1.3689	2.1700	0.6308	0.4608	0.4608	2
3	0.6244	2.2096	1.9793	1.6016	3.5389	0.4526	0.2826	0.8958	3
4	0.5337	2.7432	3.5802	1.8739	5.1405	0.3645	0.1945	1.3051	4
5	0.4561	3.1993	5.4046	2.1924	7.0144	0.3126	0.1426	1.6893	5
6	0.3898	3.5892	7.3538	2.5652	9.2068	0.2786	0.1086	2.0489	6
7	0.3332	3.9224	9.3530	3.0012	11.7720	0.2549	0.0849	2.3845	7
8	0.2848	4.2072	11.3465	3.5115	14.7733	0.2377	0.0677	2.6969	8
9	0.2434	4.4506	13.2937	4.1084	18.2847	0.2247	0.0547	2.9870	9
10	0.2080	4.6586	15.1661	4.8068	22.3931	0.2147	0.0447	3.2555	10
11	0.1778	4.8364	16.9442	5.6240	27.1999	0.2068	0.0368	3.5035	11
12	0.1520	4.9884	18.6159	6.5801	32.8239	0.2005	0.0305	3.7318	12
13	0.1299	5.1183	20.1746	7.6987	39.4040	0.1954	0.0254	3.9417	13
14	0.1110	5.2293	21.6178	9.0075	47.1027	0.1912	0.0212	4.1340	14
15	0.0949	5.3242	22.9463	10.5387	56.1101	0.1878	0.0178	4.3098	15
16	0.0811	5.4053	24.1628	12.3303	66.6488	0.1850	0.0150	4.4702	16
17	0.0693	5.4746	25.2719	14.4265	78.9792	0.1827	0.0127	4.6162	17
18	0.0592	5.5339	26.2790	16.8790	93.4056	0.1807	0.0107	4.7488	18
19	0.0506	5.5845	27.1905	19.7484	110.2846	0.1791	0.0091	4.8689	19
20	0.0433	5.6278	28.0128	23.1056	130.0329	0.1777	0.0077	4.9776	20
21	0.0370	5.6648	28.7526	27.0336	153.1385	0.1765	0.0065	5.0757	21
22	0.0316	5.6964	29.4166	31.6293	180.1721	0.1756	0.0056	5.1641	22
23	0.0270	5.7234	30.0111	37.0062	211.8013	0.1747	0.0047	5.2436	23
24	0.0231	5.7465	30.5423	43.2973	248.8076	0.1740	0.0040	5.3149	24
25	0.0197	5.7662	31.0160	50.6578	292.1049	0.1734	0.0034	5.3789	25
26	0.0169	5.7831	31.4378	59.2697	342.7627	0.1729	0.0029	5.4362	26
27	0.0144	5.7975	31.8128	69.3455	402.0323	0.1725	0.0025	5.4873	27
28	0.0123	5.8099	32.1456	81.1342	471.3778	0.1721	0.0021	5.5329	28
29	0.0105	5.8204	32.4405	94.9271	552.5121	0.1718	0.0018	5.5736	29
30	0.0090	5.8294	32.7016	111.0647	647.4391	0.1715	0.0015	5.6098	30
31	0.0077	5.8371	32.9325	129.9456	758.5038	0.1713	0.0013	5.6419	31
32	0.0066	5.8437	33.1364	152.0364	888.4494	0.1711	0.0011	5.6705	32
33	0.0056	5.8493	33.3163	177.8826	1040.4858	0.1710	0.0010	5.6958	33
34	0.0048	5.8541	33.4748	208.1226	1218.3684	0.1708	0.0008	5.7182	34
35	0.0041	5.8582	33.6145	243.5035	1426.4910	0.1707	0.0007	5.7380	35
36	0.0035	5.8617	33.7373	284.8991	1669.9945	0.1706	0.0006	5.7555	36
37	0.0030	5.8647	33.8453	333.3319	1954.8936	0.1705	0.0005	5.7710	37
38	0.0026	5.8673	33.9402	389.9983	2288.2255	0.1704	0.0004	5.7847	38
39	0.0022	5.8695	34.0235	456.2980	2678.2238	0.1704	0.0004	5.7967	39
40	0.0019	5.8713	34.0965	533.8687	3134.5218	0.1703	0.0003	5.8073	40
41	0.0016	5.8729	34.1606	624.6264	3668.3906	0.1703	0.0003	5.8166	41
42	0.0014	5.8743	34.2167	730.8129	4293.0169	0.1702	0.0002	5.8248	42
43	0.0012	5.8755	34.2658	855.0511	5023.8298	0.1702	0.0002	5.8320	43
44	0.0010	5.8765	34.3088	1000.4098	5878.8809	0.1702	0.0002	5.8383	44
45	0.0009	5.8773	34.3464	1170.4794	6879.2907	0.1701	0.0001	5.8439	45
46	0.0007	5.8781	34.3792	1369.4609	8049.7701	0.1701	0.0001	5.8487	46
47	0.0006	5.8787	34.4079	1602.2693	9419.2310	0.1701	0.0001	5.8530	47
48	0.0005	5.8792	34.4330	1874.6550	11021.5002	0.1701	0.0001	5.8567	48
49	0.0005	5.8797	34.4549	2193.3464	12896.1553	0.1701	0.0001	5.8600	49
50	0.0004	5.8801	34.4740	2566.2153	15089.5017	0.1701	0.0001	5.8629	50
51	0.0003	5.8804	34.4906	3002.4719	17655.7170	0.1701	0.0001	5.8654	51
52	0.0003	5.8807	34.5052	3512.8921	20658.1888	0.1700	0.0000	5.8675	52
53	0.0002	5.8809	34.5178	4110.0838	24171.0809	0.1700	0.0000	5.8695	53
54	0.0002	5.8811	34.5288	4808.7980	28281.1647	0.1700	0.0000	5.8711	54
55	0.0002	5.8813	34.5384	5626.2937	33089.9627	0.1700	0.0000	5.8726	55
60	0.0001	5.8819	34.5707	12335.3565	72555.0381	0.1700	0.0000	5.8775	60
65	0.0000	5.8821	34.5867	27044.6281	159080.1652	0.1700	0.0000	5.8799	65
70	0.0000	5.8823	34.5945	59293.9417	348780.0102	0.1700	0.0000	5.8812	70
75	0.0000	5.8823	34.5984	129998.8861	764693.4475	0.1700	0.0000	5.8818	75
80	0.0000	5.8823	34.6003	285015.8024	1676557.6612	0.1700	0.0000	5.8821	80
85	0.0000	5.8823	34.6012	624882.3361	3675772.5655	0.1700	0.0000	5.8822	85
90	0.0000	5.8823	34.6017	1370022.0504	8058947.3554	0.1700	0.0000	5.8823	90
95	0.0000	5.8824	34.6019	3003702.1533	************	0.1700	0.0000	5.8823	95
100	0.0000	5.8824	34.6020	6585460.8858	************	0.1700	0.0000	5.8823	100

$I = 17.25\%$

n	(P/F)	(P/A)	(P/G)	(F/P)	(F/A)	(A/P)	(A/F)	(A/G)	n
1	0.8529	0.8529	0.0000	1.1725	1.0000	1.1725	1.0000	0.0000	1
2	0.7274	1.5803	0.7274	1.3748	2.1725	0.6328	0.4603	0.4603	2
3	0.6204	2.2007	1.9682	1.6119	3.5473	0.4544	0.2819	0.8944	3
4	0.5291	2.7298	3.5555	1.8900	5.1592	0.3663	0.1938	1.3025	4
5	0.4513	3.1810	5.3606	2.2160	7.0491	0.3144	0.1419	1.6852	5
6	0.3849	3.5659	7.2850	2.5982	9.2651	0.2804	0.1079	2.0429	6
7	0.3283	3.8942	9.2545	3.0464	11.8633	0.2568	0.0843	2.3765	7
8	0.2800	4.1741	11.2142	3.5719	14.9097	0.2396	0.0671	2.6866	8
9	0.2388	4.4129	13.1244	4.1881	18.4817	0.2266	0.0541	2.9741	9
10	0.2036	4.6166	14.9572	4.9105	22.6697	0.2166	0.0441	3.2399	10
11	0.1737	4.7902	16.6940	5.7576	27.5803	0.2088	0.0363	3.4850	11
12	0.1481	4.9384	18.3235	6.7508	33.3379	0.2025	0.0300	3.7104	12
13	0.1263	5.0647	19.8395	7.9153	40.0887	0.1974	0.0249	3.9172	13
14	0.1078	5.1725	21.2403	9.2807	48.0040	0.1933	0.0208	4.1064	14
15	0.0919	5.2644	22.5269	10.8816	57.2846	0.1900	0.0175	4.2791	15
16	0.0784	5.3427	23.7025	12.7587	68.1662	0.1872	0.0147	4.4364	16
17	0.0668	5.4096	24.7721	14.9595	80.9249	0.1849	0.0124	4.5793	17
18	0.0570	5.4666	25.7413	17.5401	95.8845	0.1829	0.0104	4.7088	18
19	0.0486	5.5152	26.6165	20.5657	113.4245	0.1813	0.0088	4.8260	19
20	0.0415	5.5567	27.4045	24.1133	133.9903	0.1800	0.0075	4.9318	20
21	0.0354	5.5921	28.1119	28.2729	158.1036	0.1788	0.0063	5.0271	21
22	0.0302	5.6222	28.7454	33.1499	186.3765	0.1779	0.0054	5.1128	22
23	0.0257	5.6480	29.3114	38.8683	219.5264	0.1771	0.0046	5.1897	23
24	0.0219	5.6699	29.8161	45.5731	258.3947	0.1764	0.0039	5.2587	24
25	0.0187	5.6886	30.2652	53.4344	303.9678	0.1758	0.0033	5.3203	25
26	0.0160	5.7046	30.6642	62.6519	357.4022	0.1753	0.0028	5.3754	26
27	0.0136	5.7182	31.0182	73.4593	420.0541	0.1749	0.0024	5.4245	27
28	0.0116	5.7298	31.3317	86.1311	493.5134	0.1745	0.0020	5.4682	28
29	0.0099	5.7397	31.6089	100.9887	579.6445	0.1742	0.0017	5.5071	29
30	0.0084	5.7481	31.8538	118.4092	680.6332	0.1740	0.0015	5.5416	30
31	0.0072	5.7553	32.0699	138.8348	799.0424	0.1738	0.0013	5.5722	31
32	0.0061	5.7615	32.2603	162.7838	937.8772	0.1736	0.0011	5.5993	32
33	0.0052	5.7667	32.4280	190.8640	1100.6611	0.1734	0.0009	5.6233	33
34	0.0045	5.7712	32.5755	223.7881	1291.5251	0.1733	0.0008	5.6445	34
35	0.0038	5.7750	32.7050	262.3915	1515.3132	0.1732	0.0007	5.6632	35
36	0.0033	5.7783	32.8188	307.6541	1777.7047	0.1731	0.0006	5.6797	36
37	0.0028	5.7810	32.9186	360.7244	2085.3588	0.1730	0.0005	5.6942	37
38	0.0024	5.7834	33.0061	422.9493	2446.0831	0.1729	0.0004	5.7070	38
39	0.0020	5.7854	33.0827	495.9081	2869.0325	0.1728	0.0003	5.7183	39
40	0.0017	5.7871	33.1498	581.4523	3364.9406	0.1728	0.0003	5.7282	40
41	0.0015	5.7886	33.2085	681.7528	3946.3928	0.1728	0.0003	5.7369	41
42	0.0013	5.7898	33.2597	799.3551	4628.1456	0.1727	0.0002	5.7445	42
43	0.0011	5.7909	33.3046	937.2439	5427.5007	0.1727	0.0002	5.7512	43
44	0.0009	5.7918	33.3437	1098.9184	6364.7446	0.1727	0.0002	5.7570	44
45	0.0008	5.7926	33.3778	1288.4819	7463.6630	0.1726	0.0001	5.7621	45
46	0.0007	5.7933	33.4076	1510.7450	8752.1449	0.1726	0.0001	5.7666	46
47	0.0006	5.7938	33.4336	1771.3485	10262.8899	0.1726	0.0001	5.7706	47
48	0.0005	5.7943	33.4562	2076.9061	12034.2384	0.1726	0.0001	5.7740	48
49	0.0004	5.7947	33.4759	2435.1724	14111.1445	0.1726	0.0001	5.7770	49
50	0.0004	5.7951	33.4931	2855.2397	16546.3169	0.1726	0.0001	5.7796	50
51	0.0003	5.7954	33.5080	3347.7685	19401.5566	0.1726	0.0001	5.7819	51
52	0.0003	5.7956	33.5210	3925.2586	22749.3251	0.1725	0.0000	5.7839	52
53	0.0002	5.7958	33.5323	4602.3657	26674.5837	0.1725	0.0000	5.7856	53
54	0.0002	5.7960	33.5421	5396.2738	31276.9494	0.1725	0.0000	5.7871	54
55	0.0002	5.7962	33.5507	6327.1310	36673.2232	0.1725	0.0000	5.7884	55
60	0.0001	5.7967	33.5792	14020.7448	81273.8827	0.1725	0.0000	5.7928	60
65	0.0000	5.7969	33.5932	31069.5770	180107.6926	0.1725	0.0000	5.7950	65
70	0.0000	5.7970	33.6000	68849.3108	399120.6422	0.1725	0.0000	5.7961	70
75	0.0000	5.7971	33.6033	152568.1408	884447.1932	0.1725	0.0000	5.7966	75
80	0.0000	5.7971	33.6049	338086.7191	1959917.2124	0.1725	0.0000	5.7969	80
85	0.0000	5.7971	33.6057	749190.6832	4343128.5980	0.1725	0.0000	5.7970	85
90	0.0000	5.7971	33.6061	1660185.5322	9624258.1574	0.1725	0.0000	5.7970	90
95	0.0000	5.7971	33.6062	3678924.5557	************	0.1725	0.0000	5.7971	95
100	0.0000	5.7971	33.6063	8152393.5876	************	0.1725	0.0000	5.7971	100

I = 17.50%

n	(P/F)	(P/A)	(P/G)	(F/P)	(F/A)	(A/P)	(A/F)	(A/G)	n
1	0.8511	0.8511	0.0000	1.1750	1.0000	1.1750	1.0000	0.0000	1
2	0.7243	1.5754	0.7243	1.3806	2.1750	0.6348	0.4598	0.4598	2
3	0.6164	2.1918	1.9572	1.6222	3.5556	0.4562	0.2812	0.8930	3
4	0.5246	2.7164	3.5311	1.9061	5.1779	0.3681	0.1931	1.2999	4
5	0.4465	3.1629	5.3170	2.2397	7.0840	0.3162	0.1412	1.6810	5
6	0.3800	3.5429	7.2170	2.6316	9.3237	0.2823	0.1073	2.0370	6
7	0.3234	3.8663	9.1573	3.0922	11.9553	0.2586	0.0836	2.3685	7
8	0.2752	4.1415	11.0840	3.6333	15.0475	0.2415	0.0665	2.6763	8
9	0.2342	4.3758	12.9579	4.2691	18.6808	0.2285	0.0535	2.9613	9
10	0.1994	4.5751	14.7520	5.0162	22.9500	0.2186	0.0436	3.2244	10
11	0.1697	4.7448	16.4487	5.8941	27.9662	0.2108	0.0358	3.4667	11
12	0.1444	4.8892	18.0370	6.9256	33.8603	0.2045	0.0295	3.6892	12
13	0.1229	5.0121	19.5116	8.1375	40.7858	0.1995	0.0245	3.8929	13
14	0.1046	5.1167	20.8712	9.5616	48.9234	0.1954	0.0204	4.0791	14
15	0.0890	5.2057	22.1173	11.2349	58.4850	0.1921	0.0171	4.2487	15
16	0.0758	5.2814	23.2536	13.2010	69.7198	0.1893	0.0143	4.4029	16
17	0.0645	5.3459	24.2851	15.5111	82.9208	0.1871	0.0121	4.5428	17
18	0.0549	5.4008	25.2179	18.2256	98.4319	0.1852	0.0102	4.6693	18
19	0.0467	5.4475	26.0584	21.4151	116.6575	0.1836	0.0086	4.7836	19
20	0.0397	5.4872	26.8135	25.1627	138.0726	0.1822	0.0072	4.8866	20
21	0.0338	5.5210	27.4900	29.5662	163.2353	0.1811	0.0061	4.9792	21
22	0.0288	5.5498	28.0944	34.7403	192.8015	0.1802	0.0052	5.0622	22
23	0.0245	5.5743	28.6334	40.8198	227.5417	0.1794	0.0044	5.1367	23
24	0.0208	5.5951	29.1129	47.9633	268.3616	0.1787	0.0037	5.2032	24
25	0.0177	5.6129	29.5388	56.3568	316.3248	0.1782	0.0032	5.2627	25
26	0.0151	5.6280	29.9163	66.2193	372.6817	0.1777	0.0027	5.3156	26
27	0.0129	5.6408	30.2505	77.8077	438.9010	0.1773	0.0023	5.3628	27
28	0.0109	5.6518	30.5458	91.4240	516.7086	0.1769	0.0019	5.4046	28
29	0.0093	5.6611	30.8065	107.4232	608.1326	0.1766	0.0016	5.4418	29
30	0.0079	5.6690	31.0362	126.2223	715.5558	0.1764	0.0014	5.4747	30
31	0.0067	5.6758	31.2385	148.3112	841.7781	0.1762	0.0012	5.5038	31
32	0.0057	5.6815	31.4164	174.2656	990.0893	0.1760	0.0010	5.5296	32
33	0.0049	5.6864	31.5727	204.7621	1164.3549	0.1759	0.0009	5.5523	33
34	0.0042	5.6905	31.7098	240.5955	1369.1170	0.1757	0.0007	5.5724	34
35	0.0035	5.6941	31.8301	282.6997	1609.7125	0.1756	0.0006	5.5900	35
36	0.0030	5.6971	31.9355	332.1721	1892.4122	0.1755	0.0005	5.6056	36
37	0.0026	5.6996	32.0277	390.3023	2224.5843	0.1754	0.0004	5.6192	37
38	0.0022	5.7018	32.1084	458.6052	2614.8866	0.1754	0.0004	5.6312	38
39	0.0019	5.7037	32.1789	538.8611	3073.4917	0.1753	0.0003	5.6418	39
40	0.0016	5.7053	32.2405	633.1617	3612.3528	0.1753	0.0003	5.6510	40
41	0.0013	5.7066	32.2943	743.9650	4245.5145	0.1752	0.0002	5.6591	41
42	0.0011	5.7077	32.3412	874.1589	4989.4796	0.1752	0.0002	5.6662	42
43	0.0010	5.7087	32.3820	1027.1367	5863.6385	0.1752	0.0002	5.6724	43
44	0.0008	5.7096	32.4177	1206.8857	6890.7753	0.1751	0.0001	5.6778	44
45	0.0007	5.7103	32.4487	1418.0907	8097.6609	0.1751	0.0001	5.6825	45
46	0.0006	5.7109	32.4757	1666.2565	9515.7516	0.1751	0.0001	5.6867	46
47	0.0005	5.7114	32.4992	1957.8514	11182.0081	0.1751	0.0001	5.6903	47
48	0.0004	5.7118	32.5196	2300.4754	13139.8595	0.1751	0.0001	5.6934	48
49	0.0004	5.7122	32.5374	2703.0586	15440.3350	0.1751	0.0001	5.6962	49
50	0.0003	5.7125	32.5528	3176.0939	18143.3936	0.1751	0.0001	5.6985	50
51	0.0003	5.7128	32.5662	3731.9103	21319.4875	0.1750	0.0000	5.7006	51
52	0.0002	5.7130	32.5779	4384.9946	25051.3978	0.1750	0.0000	5.7024	52
53	0.0002	5.7132	32.5879	5152.3687	29436.3924	0.1750	0.0000	5.7040	53
54	0.0002	5.7133	32.5967	6054.0332	34588.7610	0.1750	0.0000	5.7054	54
55	0.0001	5.7135	32.6043	7113.4890	40642.7942	0.1750	0.0000	5.7066	55
60	0.0001	5.7139	32.6295	15932.0623	91034.6418	0.1750	0.0000	5.7105	60
65	0.0000	5.7141	32.6417	35682.9975	203897.1286	0.1750	0.0000	5.7125	65
70	0.0000	5.7142	32.6476	79919.1144	456674.9392	0.1750	0.0000	5.7134	70
75	0.0000	5.7143	32.6505	178994.6274	1022820.7279	0.1750	0.0000	5.7139	75
80	0.0000	5.7143	32.6518	400893.7897	2290815.9413	0.1750	0.0000	5.7141	80
85	0.0000	5.7143	32.6525	897880.7521	5130741.4404	0.1750	0.0000	5.7142	85
90	0.0000	5.7143	32.6528	2010981.1266	************	0.1750	0.0000	5.7142	90
95	0.0000	5.7143	32.6529	4503989.0680	************	0.1750	0.0000	5.7143	95
100	0.0000	5.7143	32.6530	************	************	0.1750	0.0000	5.7143	100

$I = 17.75\%$

n	(P/F)	(P/A)	(P/G)	(F/P)	(F/A)	(A/P)	(A/F)	(A/G)	n
1	0.8493	0.8493	0.0000	1.1775	1.0000	1.1775	1.0000	0.0000	1
2	0.7212	1.5705	0.7212	1.3865	2.1775	0.6367	0.4592	0.4592	2
3	0.6125	2.1830	1.9463	1.6326	3.5640	0.4581	0.2806	0.8916	3
4	0.5202	2.7032	3.5068	1.9224	5.1966	0.3699	0.1924	1.2973	4
5	0.4418	3.1450	5.2739	2.2636	7.1190	0.3180	0.1405	1.6769	5
6	0.3752	3.5201	7.1498	2.6654	9.3826	0.2841	0.1066	2.0311	6
7	0.3186	3.8388	9.0615	3.1385	12.0481	0.2605	0.0830	2.3605	7
8	0.2706	4.1093	10.9556	3.6956	15.1866	0.2433	0.0658	2.6660	8
9	0.2298	4.3391	12.7940	4.3516	18.8822	0.2305	0.0530	2.9485	9
10	0.1952	4.5343	14.5505	5.1240	23.2338	0.2205	0.0430	3.2090	10
11	0.1657	4.7001	16.2079	6.0335	28.3578	0.2128	0.0353	3.4484	11
12	0.1408	4.8408	17.7562	7.1045	34.3913	0.2066	0.0291	3.6680	12
13	0.1195	4.9603	19.1907	8.3655	41.4958	0.2016	0.0241	3.8688	13
14	0.1015	5.0619	20.5104	9.8504	49.8613	0.1976	0.0201	4.0519	14
15	0.0862	5.1481	21.7174	11.5988	59.7117	0.1942	0.0167	4.2186	15
16	0.0732	5.2213	22.8157	13.6576	71.3105	0.1915	0.0140	4.3697	16
17	0.0622	5.2835	23.8106	16.0818	84.9681	0.1893	0.0118	4.5066	17
18	0.0528	5.3363	24.7084	18.9364	101.0499	0.1874	0.0099	4.6303	18
19	0.0448	5.3811	25.5156	22.2976	119.9863	0.1858	0.0083	4.7417	19
20	0.0381	5.4192	26.2393	26.2554	142.2838	0.1845	0.0070	4.8419	20
21	0.0323	5.4516	26.8862	30.9157	168.5392	0.1834	0.0059	4.9318	21
22	0.0275	5.4790	27.4631	36.4032	199.4549	0.1825	0.0050	5.0124	22
23	0.0233	5.5024	27.9763	42.8648	235.8582	0.1817	0.0042	5.0844	23
24	0.0198	5.5222	28.4320	50.4733	278.7230	0.1811	0.0036	5.1487	24
25	0.0168	5.5390	28.8358	59.4323	329.1963	0.1805	0.0030	5.2060	25
26	0.0143	5.5533	29.1931	69.9816	388.6287	0.1801	0.0026	5.2569	26
27	0.0121	5.5654	29.5086	82.4033	458.6103	0.1797	0.0022	5.3021	27
28	0.0103	5.5757	29.7869	97.0299	541.0136	0.1793	0.0018	5.3422	28
29	0.0088	5.5845	30.0319	114.2527	638.0435	0.1791	0.0016	5.3777	29
30	0.0074	5.5919	30.2475	134.5326	752.2962	0.1788	0.0013	5.4091	30
31	0.0063	5.5982	30.4369	158.4121	886.8288	0.1786	0.0011	5.4369	31
32	0.0054	5.6036	30.6031	186.5303	1045.2409	0.1785	0.0010	5.4613	32
33	0.0046	5.6082	30.7488	219.6394	1231.7712	0.1783	0.0008	5.4829	33
34	0.0039	5.6120	30.8764	258.6254	1451.4106	0.1782	0.0007	5.5018	34
35	0.0033	5.6153	30.9880	304.5314	1710.0360	0.1781	0.0006	5.5185	35
36	0.0028	5.6181	31.0856	358.5857	2014.5673	0.1780	0.0005	5.5331	36
37	0.0024	5.6205	31.1709	422.2347	2373.1530	0.1779	0.0004	5.5460	37
38	0.0020	5.6225	31.2453	497.1813	2795.3877	0.1779	0.0004	5.5572	38
39	0.0017	5.6242	31.3102	585.4310	3292.5690	0.1778	0.0003	5.5671	39
40	0.0015	5.6256	31.3668	689.3450	3878.0000	0.1778	0.0003	5.5757	40
41	0.0012	5.6269	31.4161	811.7037	4567.3450	0.1777	0.0002	5.5832	41
42	0.0010	5.6279	31.4590	955.7812	5379.0488	0.1777	0.0002	5.5898	42
43	0.0009	5.6288	31.4963	1125.4323	6334.8299	0.1777	0.0002	5.5956	43
44	0.0008	5.6296	31.5287	1325.1965	7460.2622	0.1776	0.0001	5.6006	44
45	0.0006	5.6302	31.5569	1560.4189	8785.4588	0.1776	0.0001	5.6049	45
46	0.0005	5.6307	31.5814	1837.3933	10345.8777	0.1776	0.0001	5.6088	46
47	0.0005	5.6312	31.6027	2163.5306	12183.2710	0.1776	0.0001	5.6121	47
48	0.0004	5.6316	31.6211	2547.5573	14346.8016	0.1776	0.0001	5.6150	48
49	0.0003	5.6319	31.6371	2999.7487	16894.3589	0.1776	0.0001	5.6175	49
50	0.0003	5.6322	31.6510	3532.2041	19894.1076	0.1776	0.0001	5.6196	50
51	0.0002	5.6324	31.6630	4159.1703	23426.3117	0.1775	0.0000	5.6215	51
52	0.0002	5.6327	31.6734	4897.4231	27585.4820	0.1775	0.0000	5.6232	52
53	0.0002	5.6328	31.6825	5766.7157	32482.9051	0.1775	0.0000	5.6246	53
54	0.0001	5.6330	31.6903	6790.3077	38249.6207	0.1775	0.0000	5.6258	54
55	0.0001	5.6331	31.6970	7995.5873	45039.9284	0.1775	0.0000	5.6269	55
60	0.0001	5.6335	31.7193	18099.0153	101960.6498	0.1775	0.0000	5.6305	60
65	0.0000	5.6337	31.7300	40969.3928	230807.8466	0.1775	0.0000	5.6322	65
70	0.0000	5.6337	31.7351	92739.3625	522469.6477	0.1775	0.0000	5.6330	70
75	0.0000	5.6338	31.7376	209927.1864	1182682.7403	0.1775	0.0000	5.6334	75
80	0.0000	5.6338	31.7387	475196.5337	2677157.9365	0.1775	0.0000	5.6336	80
85	0.0000	5.6338	31.7393	1075666.9946	6060090.1103	0.1775	0.0000	5.6337	85
90	0.0000	5.6338	31.7395	2434907.2459	************	0.1775	0.0000	5.6338	90
95	0.0000	5.6338	31.7396	5511718.1490	************	0.1775	0.0000	5.6338	95
100	0.0000	5.6338	31.7397	************	************	0.1775	0.0000	5.6338	100

$I = 18.00\%$

n	(P/F)	(P/A)	(P/G)	(F/P)	(F/A)	(A/P)	(A/F)	(A/G)	n
1	0.8475	0.8475	0.0000	1.1800	1.0000	1.1800	1.0000	0.0000	1
2	0.7182	1.5656	0.7182	1.3924	2.1800	0.6387	0.4587	0.4587	2
3	0.6086	2.1743	1.9354	1.6430	3.5724	0.4599	0.2799	0.8902	3
4	0.5158	2.6901	3.4828	1.9388	5.2154	0.3717	0.1917	1.2947	4
5	0.4371	3.1272	5.2312	2.2878	7.1542	0.3198	0.1398	1.6728	5
6	0.3704	3.4976	7.0834	2.6996	9.4420	0.2859	0.1059	2.0252	6
7	0.3139	3.8115	8.9670	3.1855	12.1415	0.2624	0.0824	2.3526	7
8	0.2660	4.0776	10.8292	3.7589	15.3270	0.2452	0.0652	2.6558	8
9	0.2255	4.3030	12.6329	4.4355	19.0859	0.2324	0.0524	2.9358	9
10	0.1911	4.4941	14.3525	5.2338	23.5213	0.2225	0.0425	3.1936	10
11	0.1619	4.6560	15.9716	6.1759	28.7551	0.2148	0.0348	3.4303	11
12	0.1372	4.7932	17.4811	7.2876	34.9311	0.2086	0.0286	3.6470	12
13	0.1163	4.9095	18.8765	8.5994	42.2187	0.2037	0.0237	3.8449	13
14	0.0985	5.0081	20.1576	10.1472	50.8180	0.1997	0.0197	4.0250	14
15	0.0835	5.0916	21.3269	11.9737	60.9653	0.1964	0.0164	4.1887	15
16	0.0708	5.1624	22.3885	14.1290	72.9390	0.1937	0.0137	4.3369	16
17	0.0600	5.2223	23.3482	16.6722	87.0680	0.1915	0.0115	4.4708	17
18	0.0508	5.2732	24.2123	19.6733	103.7403	0.1896	0.0096	4.5916	18
19	0.0431	5.3162	24.9877	23.2144	123.4135	0.1881	0.0081	4.7003	19
20	0.0365	5.3527	25.6813	27.3930	146.6280	0.1868	0.0068	4.7978	20
21	0.0309	5.3837	26.3000	32.3238	174.0210	0.1857	0.0057	4.8851	21
22	0.0262	5.4099	26.8506	38.1421	206.3448	0.1848	0.0048	4.9632	22
23	0.0222	5.4321	27.3394	45.0076	244.4868	0.1841	0.0041	5.0329	23
24	0.0188	5.4509	27.7725	53.1090	289.4945	0.1835	0.0035	5.0950	24
25	0.0160	5.4669	28.1555	62.6686	342.6035	0.1829	0.0029	5.1502	25
26	0.0135	5.4804	28.4935	73.9490	405.2721	0.1825	0.0025	5.1991	26
27	0.0115	5.4919	28.7915	87.2598	479.2211	0.1821	0.0021	5.2425	27
28	0.0097	5.5016	29.0537	102.9666	566.4809	0.1818	0.0018	5.2810	28
29	0.0082	5.5098	29.2842	121.5005	669.4475	0.1815	0.0015	5.3149	29
30	0.0070	5.5168	29.4864	143.3706	790.9480	0.1813	0.0013	5.3448	30
31	0.0059	5.5227	29.6638	169.1774	934.3186	0.1811	0.0011	5.3712	31
32	0.0050	5.5277	29.8191	199.6293	1103.4960	0.1809	0.0009	5.3945	32
33	0.0042	5.5320	29.9549	235.5625	1303.1253	0.1808	0.0008	5.4149	33
34	0.0036	5.5356	30.0736	277.9638	1538.6878	0.1806	0.0006	5.4328	34
35	0.0030	5.5386	30.1773	327.9973	1816.6516	0.1806	0.0006	5.4485	35
36	0.0026	5.5412	30.2677	387.0368	2144.6489	0.1805	0.0005	5.4623	36
37	0.0022	5.5434	30.3465	456.7034	2531.6857	0.1804	0.0004	5.4744	37
38	0.0019	5.5452	30.4152	538.9100	2988.3891	0.1803	0.0003	5.4849	38
39	0.0016	5.5468	30.4749	635.9139	3527.2992	0.1803	0.0003	5.4941	39
40	0.0013	5.5482	30.5269	750.3783	4163.2130	0.1802	0.0002	5.5022	40
41	0.0011	5.5493	30.5721	885.4464	4913.5914	0.1802	0.0002	5.5092	41
42	0.0010	5.5502	30.6113	1044.8268	5799.0378	0.1802	0.0002	5.5153	42
43	0.0008	5.5510	30.6454	1232.8956	6843.8646	0.1801	0.0001	5.5207	43
44	0.0007	5.5517	30.6750	1454.8168	8076.7603	0.1801	0.0001	5.5253	44
45	0.0006	5.5523	30.7006	1716.6839	9531.5771	0.1801	0.0001	5.5293	45
46	0.0005	5.5528	30.7228	2025.6870	11248.2610	0.1801	0.0001	5.5328	46
47	0.0004	5.5532	30.7420	2390.3106	13273.9480	0.1801	0.0001	5.5359	47
48	0.0004	5.5536	30.7587	2820.5665	15664.2586	0.1801	0.0001	5.5385	48
49	0.0003	5.5539	30.7731	3328.2685	18484.8251	0.1801	0.0001	5.5408	49
50	0.0003	5.5541	30.7856	3927.3569	21813.0937	0.1800	0.0000	5.5428	50
51	0.0002	5.5544	30.7964	4634.2811	25740.4505	0.1800	0.0000	5.5445	51
52	0.0002	5.5545	30.8057	5468.4517	30374.7316	0.1800	0.0000	5.5460	52
53	0.0002	5.5547	30.8138	6452.7730	35843.1833	0.1800	0.0000	5.5473	53
54	0.0001	5.5548	30.8207	7614.2721	42295.9563	0.1800	0.0000	5.5485	54
55	0.0001	5.5549	30.8268	8984.8411	49910.2284	0.1800	0.0000	5.5494	55
60	0.0000	5.5553	30.8465	20555.1400	114189.6665	0.1800	0.0000	5.5526	60
65	0.0000	5.5554	30.8559	47025.1809	261245.4494	0.1800	0.0000	5.5542	65
70	0.0000	5.5555	30.8603	107582.2224	597673.4576	0.1800	0.0000	5.5549	70
75	0.0000	5.5555	30.8624	246122.0637	1367339.2429	0.1800	0.0000	5.5553	75
80	0.0000	5.5555	30.8634	563067.6604	3128148.1133	0.1800	0.0000	5.5554	80
85	0.0000	5.5556	30.8638	1288162.4077	7156452.2647	0.1800	0.0000	5.5555	85
90	0.0000	5.5556	30.8640	2947003.5401	************	0.1800	0.0000	5.5555	90
95	0.0000	5.5556	30.8641	6742030.2082	************	0.1800	0.0000	5.5555	95
100	0.0000	5.5556	30.8642	************	************	0.1800	0.0000	5.5555	100

$I = 18.25\%$

n	(P/F)	(P/A)	(P/G)	(F/P)	(F/A)	(A/P)	(A/F)	(A/G)	n
1	0.8457	0.8457	0.0000	1.1825	1.0000	1.1825	1.0000	0.0000	1
2	0.7152	1.5608	0.7152	1.3983	2.1825	0.6407	0.4582	0.4582	2
3	0.6048	2.1656	1.9247	1.6535	3.5808	0.4618	0.2793	0.8888	3
4	0.5114	2.6770	3.4590	1.9553	5.2343	0.3735	0.1910	1.2921	4
5	0.4325	3.1095	5.1891	2.3121	7.1896	0.3216	0.1391	1.6688	5
6	0.3658	3.4753	7.0179	2.7341	9.5017	0.2877	0.1052	2.0193	6
7	0.3093	3.7846	8.8737	3.2330	12.2357	0.2642	0.0817	2.3447	7
8	0.2616	4.0462	10.7047	3.8230	15.4687	0.2471	0.0646	2.6456	8
9	0.2212	4.2674	12.4743	4.5207	19.2918	0.2343	0.0518	2.9232	9
10	0.1871	4.4544	14.1579	5.3458	23.8125	0.2245	0.0420	3.1784	10
11	0.1582	4.6126	15.7398	6.3214	29.1583	0.2168	0.0343	3.4123	11
12	0.1338	4.7464	17.2114	7.4750	35.4797	0.2107	0.0282	3.6262	12
13	0.1131	4.8596	18.5690	8.8392	42.9547	0.2058	0.0233	3.8211	13
14	0.0957	4.9552	19.8127	10.4524	51.7940	0.2018	0.0193	3.9983	14
15	0.0809	5.0361	20.9454	12.3600	62.2464	0.1986	0.0161	4.1590	15
16	0.0684	5.1045	21.9717	14.6157	74.6063	0.1959	0.0134	4.3043	16
17	0.0579	5.1624	22.8975	17.2830	89.2220	0.1937	0.0112	4.4354	17
18	0.0489	5.2113	23.7293	20.4372	106.5050	0.1919	0.0094	4.5534	18
19	0.0414	5.2527	24.4741	24.1669	126.9422	0.1904	0.0079	4.6593	19
20	0.0350	5.2877	25.1389	28.5774	151.1091	0.1891	0.0066	4.7542	20
21	0.0296	5.3173	25.7308	33.7928	179.6866	0.1881	0.0056	4.8391	21
22	0.0250	5.3423	26.2563	39.9600	213.4794	0.1872	0.0047	4.9148	22
23	0.0212	5.3635	26.7219	47.2527	253.4393	0.1864	0.0039	4.9822	23
24	0.0179	5.3814	27.1335	55.8763	300.6920	0.1858	0.0033	5.0421	24
25	0.0151	5.3965	27.4968	66.0737	356.5683	0.1853	0.0028	5.0953	25
26	0.0128	5.4093	27.8167	78.1322	422.6420	0.1849	0.0024	5.1424	26
27	0.0108	5.4201	28.0981	92.3913	500.7742	0.1845	0.0020	5.1840	27
28	0.0092	5.4293	28.3453	109.2527	593.1655	0.1842	0.0017	5.2208	28
29	0.0077	5.4370	28.5620	129.1913	702.4182	0.1839	0.0014	5.2532	29
30	0.0065	5.4436	28.7518	152.7687	831.6095	0.1837	0.0012	5.2818	30
31	0.0055	5.4491	28.9179	180.6490	984.3782	0.1835	0.0010	5.3069	31
32	0.0047	5.4538	29.0630	213.6175	1165.0273	0.1834	0.0009	5.3289	32
33	0.0040	5.4578	29.1897	252.6027	1378.6447	0.1832	0.0007	5.3483	33
34	0.0033	5.4611	29.3002	298.7027	1631.2474	0.1831	0.0006	5.3652	34
35	0.0028	5.4639	29.3964	353.2159	1929.9501	0.1830	0.0005	5.3801	35
36	0.0024	5.4663	29.4802	417.6778	2283.1660	0.1829	0.0004	5.3931	36
37	0.0020	5.4684	29.5531	493.9040	2700.8437	0.1829	0.0004	5.4044	37
38	0.0017	5.4701	29.6165	584.0415	3194.7477	0.1828	0.0003	5.4143	38
39	0.0014	5.4715	29.6715	690.6290	3778.7892	0.1828	0.0003	5.4229	39
40	0.0012	5.4727	29.7192	816.6688	4469.4182	0.1827	0.0002	5.4304	40
41	0.0010	5.4738	29.7607	965.7109	5286.0870	0.1827	0.0002	5.4370	41
42	0.0009	5.4747	29.7966	1141.9531	6251.7979	0.1827	0.0002	5.4426	42
43	0.0007	5.4754	29.8277	1350.3596	7393.7510	0.1826	0.0001	5.4476	43
44	0.0006	5.4760	29.8546	1596.8002	8744.1106	0.1826	0.0001	5.4519	44
45	0.0005	5.4766	29.8779	1888.2162	10340.9108	0.1826	0.0001	5.4556	45
46	0.0004	5.4770	29.8981	2232.8157	12229.1270	0.1826	0.0001	5.4588	46
47	0.0004	5.4774	29.9155	2640.3045	14461.9427	0.1826	0.0001	5.4616	47
48	0.0003	5.4777	29.9305	3122.1601	17102.2472	0.1826	0.0001	5.4641	48
49	0.0003	5.4780	29.9435	3691.9543	20224.4073	0.1825	0.0000	5.4662	49
50	0.0002	5.4782	29.9548	4365.7360	23916.3617	0.1825	0.0000	5.4680	50
51	0.0002	5.4784	29.9644	5162.4828	28282.0977	0.1825	0.0000	5.4696	51
52	0.0002	5.4786	29.9728	6104.6359	33444.5805	0.1825	0.0000	5.4709	52
53	0.0001	5.4787	29.9800	7218.7320	39549.2164	0.1825	0.0000	5.4721	53
54	0.0001	5.4788	29.9862	8536.1506	46767.9484	0.1825	0.0000	5.4731	54
55	0.0001	5.4789	29.9916	10093.9981	55304.0990	0.1825	0.0000	5.4740	55
60	0.0000	5.4792	30.0090	23338.2864	127875.5418	0.1825	0.0000	5.4769	60
65	0.0000	5.4794	30.0172	53960.3442	295667.6397	0.1825	0.0000	5.4782	65
70	0.0000	5.4794	30.0211	124761.4630	683618.9751	0.1825	0.0000	5.4789	70
75	0.0000	5.4794	30.0229	288460.4029	1580599.4678	0.1825	0.0000	5.4792	75
80	0.0000	5.4794	30.0237	666947.9666	3654503.9264	0.1825	0.0000	5.4793	80
85	0.0000	5.4794	30.0241	1542047.3163	8449568.8563	0.1825	0.0000	5.4794	85
90	0.0000	5.4795	30.0242	3565360.4851	************	0.1825	0.0000	5.4794	90
95	0.0000	5.4795	30.0243	8243453.5273	************	0.1825	0.0000	5.4794	95
100	0.0000	5.4795	30.0244	************	************	0.1825	0.0000	5.4794	100

$I = 18.50\%$

n	(P/F)	(P/A)	(P/G)	(F/P)	(F/A)	(A/P)	(A/F)	(A/G)	n
1	0.8439	0.8439	0.0000	1.1850	1.0000	1.1850	1.0000	0.0000	1
2	0.7121	1.5560	0.7121	1.4042	2.1850	0.6427	0.4577	0.4577	2
3	0.6010	2.1570	1.9141	1.6640	3.5892	0.4636	0.2786	0.8874	3
4	0.5071	2.6641	3.4355	1.9718	5.2532	0.3754	0.1904	1.2895	4
5	0.4280	3.0921	5.1473	2.3366	7.2251	0.3234	0.1384	1.6647	5
6	0.3612	3.4532	6.9531	2.7689	9.5617	0.2896	0.1046	2.0135	6
7	0.3048	3.7580	8.7817	3.2812	12.3306	0.2661	0.0811	2.3368	7
8	0.2572	4.0152	10.5820	3.8882	15.6118	0.2491	0.0641	2.6355	8
9	0.2170	4.2322	12.3183	4.6075	19.5000	0.2363	0.0513	2.9106	9
10	0.1832	4.4154	13.9667	5.4599	24.1075	0.2265	0.0415	3.1632	10
11	0.1546	4.5699	15.5123	6.4700	29.5674	0.2188	0.0338	3.3944	11
12	0.1304	4.7004	16.9471	7.6669	36.0373	0.2127	0.0277	3.6055	12
13	0.1101	4.8104	18.2679	9.0853	43.7042	0.2079	0.0229	3.7975	13
14	0.0929	4.9033	19.4754	10.7661	52.7895	0.2039	0.0189	3.9719	14
15	0.0784	4.9817	20.5727	12.7578	63.5556	0.2007	0.0157	4.1297	15
16	0.0661	5.0479	21.5649	15.1180	76.3134	0.1981	0.0131	4.2721	16
17	0.0558	5.1037	22.4581	17.9148	91.4313	0.1959	0.0109	4.4004	17
18	0.0471	5.1508	23.2588	21.2290	109.3461	0.1941	0.0091	4.5156	18
19	0.0398	5.1905	23.9744	25.1564	130.5752	0.1927	0.0077	4.6189	19
20	0.0335	5.2241	24.6117	29.8103	155.7316	0.1914	0.0064	4.7112	20
21	0.0283	5.2524	25.1779	35.3253	185.5419	0.1904	0.0054	4.7936	21
22	0.0239	5.2763	25.6796	41.8604	220.8672	0.1895	0.0045	4.8670	22
23	0.0202	5.2964	26.1231	49.6046	262.7276	0.1888	0.0038	4.9322	23
24	0.0170	5.3134	26.5144	58.7815	312.3322	0.1882	0.0032	4.9900	24
25	0.0144	5.3278	26.8589	69.6560	371.1137	0.1877	0.0027	5.0413	25
26	0.0121	5.3399	27.1618	82.5424	440.7697	0.1873	0.0023	5.0866	26
27	0.0102	5.3501	27.4276	97.8127	523.3122	0.1869	0.0019	5.1265	27
28	0.0086	5.3588	27.6605	115.9081	621.1249	0.1866	0.0016	5.1617	28
29	0.0073	5.3661	27.8644	137.3511	737.0330	0.1864	0.0014	5.1927	29
30	0.0061	5.3722	28.0426	162.7611	874.3841	0.1861	0.0011	5.2199	30
31	0.0052	5.3774	28.1981	192.8719	1037.1452	0.1860	0.0010	5.2438	31
32	0.0044	5.3818	28.3337	228.5531	1230.0170	0.1858	0.0008	5.2648	32
33	0.0037	5.3854	28.4519	270.8355	1458.5702	0.1857	0.0007	5.2831	33
34	0.0031	5.3886	28.5547	320.9400	1729.4057	0.1856	0.0006	5.2991	34
35	0.0026	5.3912	28.6441	380.3140	2050.3457	0.1855	0.0005	5.3131	35
36	0.0022	5.3934	28.7218	450.6720	2430.6597	0.1854	0.0004	5.3253	36
37	0.0019	5.3953	28.7892	534.0464	2881.3317	0.1853	0.0003	5.3360	37
38	0.0016	5.3969	28.8477	632.8449	3415.3781	0.1853	0.0003	5.3453	38
39	0.0013	5.3982	28.8983	749.9213	4048.2230	0.1852	0.0002	5.3533	39
40	0.0011	5.3993	28.9422	888.6567	4798.1443	0.1852	0.0002	5.3603	40
41	0.0009	5.4003	28.9802	1053.0582	5686.8009	0.1852	0.0002	5.3664	41
42	0.0008	5.4011	29.0131	1247.8739	6739.8591	0.1851	0.0001	5.3717	42
43	0.0007	5.4017	29.0415	1478.7306	7987.7331	0.1851	0.0001	5.3763	43
44	0.0006	5.4023	29.0660	1752.2958	9466.4637	0.1851	0.0001	5.3803	44
45	0.0005	5.4028	29.0872	2076.4705	11218.7595	0.1851	0.0001	5.3837	45
46	0.0004	5.4032	29.1055	2460.6175	13295.2300	0.1851	0.0001	5.3867	46
47	0.0003	5.4036	29.1213	2915.8318	15755.8475	0.1851	0.0001	5.3893	47
48	0.0003	5.4038	29.1349	3455.2607	18671.6793	0.1851	0.0001	5.3915	48
49	0.0002	5.4041	29.1466	4094.4839	22126.9400	0.1850	0.0000	5.3934	49
50	0.0002	5.4043	29.1567	4851.9634	26221.4239	0.1850	0.0000	5.3951	50
51	0.0002	5.4045	29.1654	5749.5766	31073.3873	0.1850	0.0000	5.3965	51
52	0.0001	5.4046	29.1729	6813.2483	36822.9639	0.1850	0.0000	5.3978	52
53	0.0001	5.4047	29.1793	8073.6993	43636.2122	0.1850	0.0000	5.3988	53
54	0.0001	5.4048	29.1848	9567.3336	51709.9115	0.1850	0.0000	5.3998	54
55	0.0001	5.4049	29.1896	11337.2903	61277.2451	0.1850	0.0000	5.4006	55
60	0.0000	5.4052	29.2051	26491.1628	143190.0689	0.1850	0.0000	5.4031	60
65	0.0000	5.4053	29.2123	61900.3027	334590.8252	0.1850	0.0000	5.4044	65
70	0.0000	5.4054	29.2156	144638.7048	781825.4315	0.1850	0.0000	5.4049	70
75	0.0000	5.4054	29.2171	337968.5403	1826851.5691	0.1850	0.0000	5.4052	75
80	0.0000	5.4054	29.2178	789710.7096	4268701.1328	0.1850	0.0000	5.4053	80
85	0.0000	5.4054	29.2181	1845269.3979	9974423.7723	0.1850	0.0000	5.4054	85
90	0.0000	5.4054	29.2183	4311729.7378	***********	0.1850	0.0000	5.4054	90
95	0.0000	5.4054	29.2184	***********	***********	0.1850	0.0000	5.4054	95
100	0.0000	5.4054	29.2184	***********	***********	0.1850	0.0000	5.4054	100

$I = 18.75\%$

n	(P/F)	(P/A)	(P/G)	(F/P)	(F/A)	(A/P)	(A/F)	(A/G)	n
1	0.8421	0.8421	0.0000	1.1875	1.0000	1.1875	1.0000	0.0000	1
2	0.7091	1.5512	0.7091	1.4102	2.1875	0.6446	0.4571	0.4571	2
3	0.5972	2.1484	1.9035	1.6746	3.5977	0.4655	0.2780	0.8860	3
4	0.5029	2.6513	3.4121	1.9885	5.2722	0.3772	0.1897	1.2870	4
5	0.4235	3.0748	5.1060	2.3614	7.2608	0.3252	0.1377	1.6606	5
6	0.3566	3.4314	6.8891	2.8042	9.6221	0.2914	0.1039	2.0077	6
7	0.3003	3.7317	8.6910	3.3299	12.4263	0.2680	0.0805	2.3290	7
8	0.2529	3.9846	10.4612	3.9543	15.7562	0.2510	0.0635	2.6254	8
9	0.2130	4.1975	12.1649	4.6957	19.7105	0.2382	0.0507	2.8981	9
10	0.1793	4.3769	13.7789	5.5762	24.4063	0.2285	0.0410	3.1481	10
11	0.1510	4.5279	15.2891	6.6217	29.9824	0.2209	0.0334	3.3766	11
12	0.1272	4.6551	16.6880	7.8633	36.6041	0.2148	0.0273	3.5849	12
13	0.1071	4.7622	17.9731	9.3376	44.4674	0.2100	0.0225	3.7741	13
14	0.0902	4.8524	19.1455	11.0884	53.8050	0.2061	0.0186	3.9456	14
15	0.0759	4.9283	20.2087	13.1675	64.8935	0.2029	0.0154	4.1005	15
16	0.0640	4.9922	21.1680	15.6364	78.0610	0.2003	0.0128	4.2402	16
17	0.0539	5.0461	22.0297	18.5683	93.6975	0.1982	0.0107	4.3657	17
18	0.0454	5.0915	22.8007	22.0498	112.2657	0.1964	0.0089	4.4782	18
19	0.0382	5.1296	23.4881	26.1842	134.3156	0.1949	0.0074	4.5789	19
20	0.0322	5.1618	24.0992	31.0937	160.4997	0.1937	0.0062	4.6687	20
21	0.0271	5.1889	24.6408	36.9238	191.5934	0.1927	0.0052	4.7488	21
22	0.0228	5.2117	25.1198	43.8470	228.5172	0.1919	0.0044	4.8199	22
23	0.0192	5.2309	25.5423	52.0683	272.3642	0.1912	0.0037	4.8830	23
24	0.0162	5.2471	25.9143	61.8311	324.4324	0.1906	0.0031	4.9388	24
25	0.0136	5.2607	26.2411	73.4244	386.2635	0.1901	0.0026	4.9881	25
26	0.0115	5.2722	26.5278	87.1915	459.6879	0.1897	0.0022	5.0317	26
27	0.0097	5.2818	26.7790	103.5399	546.8794	0.1893	0.0018	5.0700	27
28	0.0081	5.2900	26.9986	122.9536	650.4193	0.1890	0.0015	5.1037	28
29	0.0068	5.2968	27.1903	146.0074	773.3729	0.1888	0.0013	5.1333	29
30	0.0058	5.3026	27.3576	173.3838	919.3803	0.1886	0.0011	5.1593	30
31	0.0049	5.3074	27.5033	205.8933	1092.7641	0.1884	0.0009	5.1820	31
32	0.0041	5.3115	27.6301	244.4983	1298.6574	0.1883	0.0008	5.2019	32
33	0.0034	5.3150	27.7403	290.3417	1543.1557	0.1881	0.0006	5.2193	33
34	0.0029	5.3179	27.8360	344.7808	1833.4974	0.1880	0.0005	5.2344	34
35	0.0024	5.3203	27.9190	409.4272	2178.2781	0.1880	0.0005	5.2476	35
36	0.0021	5.3224	27.9910	486.1947	2587.7053	0.1879	0.0004	5.2591	36
37	0.0017	5.3241	28.0534	577.3563	3073.9000	0.1878	0.0003	5.2691	37
38	0.0015	5.3256	28.1074	685.6106	3651.2563	0.1878	0.0003	5.2778	38
39	0.0012	5.3268	28.1540	814.1625	4336.8669	0.1877	0.0002	5.2854	39
40	0.0010	5.3278	28.1944	966.8180	5151.0294	0.1877	0.0002	5.2919	40
41	0.0009	5.3287	28.2292	1148.0964	6117.8474	0.1877	0.0002	5.2976	41
42	0.0007	5.3294	28.2593	1363.3645	7265.9438	0.1876	0.0001	5.3025	42
43	0.0006	5.3300	28.2852	1618.9953	8629.3082	0.1876	0.0001	5.3068	43
44	0.0005	5.3306	28.3076	1922.5569	10248.3035	0.1876	0.0001	5.3104	44
45	0.0004	5.3310	28.3269	2283.0363	12170.8605	0.1876	0.0001	5.3136	45
46	0.0004	5.3314	28.3435	2711.1056	14453.8968	0.1876	0.0001	5.3164	46
47	0.0003	5.3317	28.3577	3219.4380	17165.0024	0.1876	0.0001	5.3187	47
48	0.0003	5.3319	28.3700	3823.0826	20384.4404	0.1875	0.0000	5.3208	48
49	0.0002	5.3322	28.3806	4539.9106	24207.5230	0.1875	0.0000	5.3225	49
50	0.0002	5.3323	28.3897	5391.1438	28747.4335	0.1875	0.0000	5.3241	50
51	0.0002	5.3325	28.3975	6401.9832	34138.5773	0.1875	0.0000	5.3254	51
52	0.0001	5.3326	28.4042	7602.3551	40540.5605	0.1875	0.0000	5.3265	52
53	0.0001	5.3327	28.4100	9027.7967	48142.9156	0.1875	0.0000	5.3275	53
54	0.0001	5.3328	28.4149	10720.5086	57170.7123	0.1875	0.0000	5.3283	54
55	0.0001	5.3329	28.4192	12730.6039	67891.2209	0.1875	0.0000	5.3290	55
60	0.0000	5.3332	28.4329	30061.9465	160325.0481	0.1875	0.0000	5.3313	60
65	0.0000	5.3333	28.4392	70988.0406	378597.5501	0.1875	0.0000	5.3324	65
70	0.0000	5.3333	28.4420	167630.5928	894024.4948	0.1875	0.0000	5.3329	70
75	0.0000	5.3333	28.4434	395841.5443	2111149.5696	0.1875	0.0000	5.3331	75
80	0.0000	5.3333	28.4440	934737.0644	4985259.0099	0.1875	0.0000	5.3332	80
85	0.0000	5.3333	28.4442	2207280.6457	************	0.1875	0.0000	5.3333	85
90	0.0000	5.3333	28.4443	5212254.9052	************	0.1875	0.0000	5.3333	90
95	0.0000	5.3333	28.4444	************	************	0.1875	0.0000	5.3333	95
100	0.0000	5.3333	28.4444	************	************	0.1875	0.0000	5.3333	100

$I = 19.00\%$

n	(P/F)	(P/A)	(P/G)	(F/P)	(F/A)	(A/P)	(A/F)	(A/G)	n
1	0.8403	0.8403	0.0000	1.1900	1.0000	1.1900	1.0000	0.0000	1
2	0.7062	1.5465	0.7062	1.4161	2.1900	0.6466	0.4566	0.4566	2
3	0.5934	2.1399	1.8930	1.6852	3.6061	0.4673	0.2773	0.8846	3
4	0.4987	2.6386	3.3890	2.0053	5.2913	0.3790	0.1890	1.2844	4
5	0.4190	3.0576	5.0652	2.3864	7.2966	0.3271	0.1371	1.6566	5
6	0.3521	3.4098	6.8259	2.8398	9.6830	0.2933	0.1033	2.0019	6
7	0.2959	3.7057	8.6014	3.3793	12.5227	0.2699	0.0799	2.3211	7
8	0.2487	3.9544	10.3421	4.0214	15.9020	0.2529	0.0629	2.6154	8
9	0.2090	4.1633	12.0138	4.7854	19.9234	0.2402	0.0502	2.8856	9
10	0.1756	4.3389	13.5943	5.6947	24.7089	0.2305	0.0405	3.1331	10
11	0.1476	4.4865	15.0699	6.7767	30.4035	0.2229	0.0329	3.3589	11
12	0.1240	4.6105	16.4340	8.0642	37.1802	0.2169	0.0269	3.5645	12
13	0.1042	4.7147	17.6844	9.5964	45.2445	0.2121	0.0221	3.7509	13
14	0.0876	4.8023	18.8228	11.4198	54.8409	0.2082	0.0182	3.9196	14
15	0.0736	4.8759	19.8530	13.5895	66.2607	0.2051	0.0151	4.0717	15
16	0.0618	4.9377	20.7806	16.1715	79.8502	0.2025	0.0125	4.2086	16
17	0.0520	4.9897	21.6120	19.2441	96.0218	0.2004	0.0104	4.3314	17
18	0.0437	5.0333	22.3543	22.9005	115.2659	0.1987	0.0087	4.4413	18
19	0.0367	5.0700	23.0148	27.2516	138.1664	0.1972	0.0072	4.5394	19
20	0.0308	5.1009	23.6007	32.4294	165.4180	0.1960	0.0060	4.6268	20
21	0.0259	5.1268	24.1190	38.5910	197.8474	0.1951	0.0051	4.7045	21
22	0.0218	5.1486	24.5763	45.9233	236.4385	0.1942	0.0042	4.7734	22
23	0.0183	5.1668	24.9788	54.6487	282.3618	0.1935	0.0035	4.8344	23
24	0.0154	5.1822	25.3325	65.0320	337.0105	0.1930	0.0030	4.8883	24
25	0.0129	5.1951	25.6426	77.3881	402.0425	0.1925	0.0025	4.9359	25
26	0.0109	5.2060	25.9141	92.0918	479.4306	0.1921	0.0021	4.9777	26
27	0.0091	5.2151	26.1514	109.5893	571.5224	0.1917	0.0017	5.0145	27
28	0.0077	5.2228	26.3584	130.4112	681.1116	0.1915	0.0015	5.0468	28
29	0.0064	5.2292	26.5388	155.1893	811.5228	0.1912	0.0012	5.0751	29
30	0.0054	5.2347	26.6958	184.6753	966.7122	0.1910	0.0010	5.0998	30
31	0.0046	5.2392	26.8324	219.7636	1151.3875	0.1909	0.0009	5.1215	31
32	0.0038	5.2430	26.9509	261.5187	1371.1511	0.1907	0.0007	5.1403	32
33	0.0032	5.2462	27.0537	311.2073	1632.6698	0.1906	0.0006	5.1568	33
34	0.0027	5.2489	27.1428	370.3366	1943.8771	0.1905	0.0005	5.1711	34
35	0.0023	5.2512	27.2200	440.7006	2314.2137	0.1904	0.0004	5.1836	35
36	0.0019	5.2531	27.2867	524.4337	2754.9143	0.1904	0.0004	5.1944	36
37	0.0016	5.2547	27.3444	624.0761	3279.3481	0.1903	0.0003	5.2038	37
38	0.0013	5.2561	27.3942	742.6506	3903.4242	0.1903	0.0003	5.2119	38
39	0.0011	5.2572	27.4372	883.7542	4646.0748	0.1902	0.0002	5.2190	39
40	0.0010	5.2582	27.4743	1051.6675	5529.8290	0.1902	0.0002	5.2251	40
41	0.0008	5.2590	27.5063	1251.4843	6581.4965	0.1902	0.0002	5.2304	41
42	0.0007	5.2596	27.5338	1489.2664	7832.9808	0.1901	0.0001	5.2349	42
43	0.0006	5.2602	27.5575	1772.2270	9322.2472	0.1901	-0.0001	5.2389	43
44	0.0005	5.2607	27.5779	2108.9501	11094.4741	0.1901	0.0001	5.2423	44
45	0.0004	5.2611	27.5954	2509.6506	13203.4242	0.1901	0.0001	5.2452	45
46	0.0003	5.2614	27.6105	2986.4842	15713.0748	0.1901	0.0001	5.2478	46
47	0.0003	5.2617	27.6234	3553.9162	18699.5590	0.1901	0.0001	5.2499	47
48	0.0002	5.2619	27.6345	4229.1603	22253.4753	0.1900	0.0000	5.2518	48
49	0.0002	5.2621	27.6441	5032.7008	26482.6356	0.1900	0.0000	5.2534	49
50	0.0002	5.2623	27.6523	5988.9139	31515.3363	0.1900	0.0000	5.2548	50
51	0.0001	5.2624	27.6593	7126.8075	37504.2502	0.1900	0.0000	5.2560	51
52	0.0001	5.2625	27.6653	8480.9010	44631.0578	0.1900	0.0000	5.2570	52
53	0.0001	5.2626	27.6704	10092.2722	53111.9588	0.1900	0.0000	5.2579	53
54	0.0001	5.2627	27.6749	12009.8039	63204.2309	0.1900	0.0000	5.2587	54
55	0.0001	5.2628	27.6786	14291.6666	75214.0348	0.1900	0.0000	5.2593	55
60	0.0000	5.2630	27.6908	34104.9709	179494.5838	0.1900	0.0000	5.2614	60
65	0.0000	5.2631	27.6963	81386.5222	428344.8535	0.1900	0.0000	5.2624	65
70	0.0000	5.2631	27.6988	194217.0251	1022189.6056	0.1900	0.0000	5.2628	70
75	0.0000	5.2631	27.6999	463470.5086	2439313.2029	0.1900	0.0000	5.2630	75
80	0.0000	5.2632	27.7004	1106004.5444	5821071.2861	0.1900	0.0000	5.2631	80
85	0.0000	5.2632	27.7007	2639317.9923	************	0.1900	0.0000	5.2631	85
90	0.0000	5.2632	27.7008	6298346.1505	************	0.1900	0.0000	5.2631	90
95	0.0000	5.2632	27.7008	************	************	0.1900	0.0000	5.2632	95
100	0.0000	5.2632	27.7008	************	************	0.1900	0.0000	5.2632	100

$I = 19.25\%$

n	(P/F)	(P/A)	(P/G)	(F/P)	(F/A)	(A/P)	(A/F)	(A/G)	n
1	0.8386	0.8386	0.0000	1.1925	1.0000	1.1925	1.0000	0.0000	1
2	0.7032	1.5418	0.7032	1.4221	2.1925	0.6486	0.4561	0.4561	2
3	0.5897	2.1315	1.8826	1.6958	3.6146	0.4692	0.2767	0.8832	3
4	0.4945	2.6260	3.3661	2.0222	5.3104	0.3808	0.1883	1.2818	4
5	0.4147	3.0406	5.0248	2.4115	7.3326	0.3289	0.1364	1.6525	5
6	0.3477	3.3884	6.7635	2.8757	9.7441	0.2951	0.1026	1.9961	6
7	0.2916	3.6800	8.5131	3.4293	12.6199	0.2717	0.0792	2.3133	7
8	0.2445	3.9245	10.2248	4.0895	16.0492	0.2548	0.0623	2.6054	8
9	0.2051	4.1296	11.8653	4.8767	20.1387	0.2422	0.0497	2.8732	9
10	0.1720	4.3015	13.4129	5.8155	25.0154	0.2325	0.0400	3.1182	10
11	0.1442	4.4457	14.8548	6.9349	30.8308	0.2249	0.0324	3.3414	11
12	0.1209	4.5666	16.1850	8.2699	37.7658	0.2190	0.0265	3.5442	12
13	0.1014	4.6680	17.4018	9.8619	46.0357	0.2142	0.0217	3.7278	13
14	0.0850	4.7531	18.5072	11.7603	55.8975	0.2104	0.0179	3.8937	14
15	0.0713	4.8244	19.5055	14.0241	67.6578	0.2073	0.0148	4.0431	15
16	0.0598	4.8842	20.4024	16.7238	81.6819	0.2047	0.0122	4.1772	16
17	0.0501	4.9343	21.2047	19.9431	98.4057	0.2027	0.0102	4.2974	17
18	0.0420	4.9764	21.9195	23.7821	118.3488	0.2009	0.0084	4.4047	18
19	0.0353	5.0116	22.5542	28.3602	142.1309	0.1995	0.0070	4.5004	19
20	0.0296	5.0412	23.1160	33.8195	170.4911	0.1984	0.0059	4.5854	20
21	0.0248	5.0660	23.6119	40.3298	204.3107	0.1974	0.0049	4.6609	21
22	0.0208	5.0868	24.0485	48.0933	244.6405	0.1966	0.0041	4.7276	22
23	0.0174	5.1042	24.4322	57.3513	292.7338	0.1959	0.0034	4.7867	23
24	0.0146	5.1188	24.7685	68.3914	350.0850	0.1954	0.0029	4.8387	24
25	0.0123	5.1311	25.0627	81.5567	418.4764	0.1949	0.0024	4.8845	25
26	0.0103	5.1414	25.3198	97.2564	500.0331	0.1945	0.0020	4.9247	26
27	0.0086	5.1500	25.5440	115.9782	597.2895	0.1942	0.0017	4.9600	27
28	0.0072	5.1572	25.7392	138.3040	713.2677	0.1939	0.0014	4.9909	28
29	0.0061	5.1633	25.9090	164.9276	851.5718	0.1937	0.0012	5.0179	29
30	0.0051	5.1684	26.0564	196.6761	1016.4993	0.1935	0.0010	5.0415	30
31	0.0043	5.1727	26.1843	234.5363	1213.1754	0.1933	0.0008	5.0621	31
32	0.0036	5.1762	26.2952	279.6845	1447.7117	0.1932	0.0007	5.0800	32
33	0.0030	5.1792	26.3911	333.5238	1727.3962	0.1931	0.0006	5.0956	33
34	0.0025	5.1817	26.4741	397.7271	2060.9200	0.1930	0.0005	5.1091	34
35	0.0021	5.1839	26.5458	474.2896	2458.6471	0.1929	0.0004	5.1209	35
36	0.0018	5.1856	26.6076	565.5903	2932.9367	0.1928	0.0003	5.1310	36
37	0.0015	5.1871	26.6610	674.4664	3498.5270	0.1928	0.0003	5.1399	37
38	0.0012	5.1883	26.7070	804.3012	4172.9934	0.1927	0.0002	5.1475	38
39	0.0010	5.1894	26.7466	959.1292	4977.2946	0.1927	0.0002	5.1541	39
40	0.0009	5.1903	26.7807	1143.7616	5936.4239	0.1927	0.0002	5.1598	40
41	0.0007	5.1910	26.8101	1363.9357	7080.1854	0.1926	0.0001	5.1647	41
42	0.0006	5.1916	26.8353	1626.4933	8444.1211	0.1926	0.0001	5.1690	42
43	0.0005	5.1921	26.8569	1939.5933	10070.6145	0.1926	0.0001	5.1726	43
44	0.0004	5.1926	26.8755	2312.9650	12010.2077	0.1926	0.0001	5.1758	44
45	0.0004	5.1929	26.8915	2758.2108	14323.1727	0.1926	0.0001	5.1785	45
46	0.0003	5.1932	26.9051	3289.1663	17081.3835	0.1926	0.0001	5.1808	46
47	0.0003	5.1935	26.9169	3922.3308	20370.5498	0.1925	0.0000	5.1828	47
48	0.0002	5.1937	26.9269	4677.3795	24292.8806	0.1925	0.0000	5.1845	48
49	0.0002	5.1939	26.9355	5577.7751	28970.2602	0.1925	0.0000	5.1860	49
50	0.0002	5.1940	26.9429	6651.4968	34548.0353	0.1925	0.0000	5.1873	50
51	0.0001	5.1942	26.9492	7931.9099	41199.5320	0.1925	0.0000	5.1884	51
52	0.0001	5.1943	26.9546	9458.8026	49131.4420	0.1925	0.0000	5.1893	52
53	0.0001	5.1943	26.9592	11279.6221	58590.2445	0.1925	0.0000	5.1901	53
54	0.0001	5.1944	26.9631	13450.9493	69869.8666	0.1925	0.0000	5.1908	54
55	0.0001	5.1945	26.9665	16040.2571	83320.8159	0.1925	0.0000	5.1914	55
60	0.0000	5.1947	26.9772	38681.4961	200937.6419	0.1925	0.0000	5.1933	60
65	0.0000	5.1947	26.9821	93281.4313	484573.6692	0.1925	0.0000	5.1941	65
70	0.0000	5.1948	26.9843	224950.5917	1168569.3075	0.1925	0.0000	5.1945	70
75	0.0000	5.1948	26.9852	542474.1880	2818042.5350	0.1925	0.0000	5.1947	75
80	0.0000	5.1948	26.9857	1308190.5783	6795790.0170	0.1925	0.0000	5.1947	80
85	0.0000	5.1948	26.9859	3154735.5191	************	0.1925	0.0000	5.1948	85
90	0.0000	5.1948	26.9859	7607726.5508	************	0.1925	0.0000	5.1948	90
95	0.0000	5.1948	26.9860	************	************	0.1925	0.0000	5.1948	95
100	0.0000	5.1948	26.9860	************	************	0.1925	0.0000	5.1948	100

$I = 19.50\%$

n	(P/F)	(P/A)	(P/G)	(F/P)	(F/A)	(A/P)	(A/F)	(A/G)	n
1	0.8368	0.8368	0.0000	1.1950	1.0000	1.1950	1.0000	0.0000	1
2	0.7003	1.5371	0.7003	1.4280	2.1950	0.6506	0.4556	0.4556	2
3	0.5860	2.1231	1.8723	1.7065	3.6230	0.4710	0.2760	0.8819	3
4	0.4904	2.6135	3.3434	2.0393	5.3295	0.3826	0.1876	1.2793	4
5	0.4104	3.0238	4.9848	2.4369	7.3688	0.3307	0.1357	1.6485	5
6	0.3434	3.3672	6.7018	2.9121	9.8057	0.2970	0.1020	1.9903	6
7	0.2874	3.6546	8.4259	3.4800	12.7178	0.2736	0.0786	2.3056	7
8	0.2405	3.8950	10.1092	4.1586	16.1978	0.2567	0.0617	2.5954	8
9	0.2012	4.0963	11.7190	4.9695	20.3563	0.2441	0.0491	2.8609	9
10	0.1684	4.2647	13.2346	5.9385	25.3258	0.2345	0.0395	3.1033	10
11	0.1409	4.4056	14.6437	7.0965	31.2643	0.2270	0.0320	3.3239	11
12	0.1179	4.5235	15.9408	8.4804	38.3609	0.2211	0.0261	3.5240	12
13	0.0987	4.6222	17.1249	10.1340	46.8412	0.2163	0.0213	3.7050	13
14	0.0826	4.7047	18.1984	12.1102	56.9753	0.2126	0.0176	3.8681	14
15	0.0691	4.7738	19.1658	14.4717	69.0855	0.2095	0.0145	4.0148	15
16	0.0578	4.8317	20.0332	17.2936	83.5571	0.2070	0.0120	4.1462	16
17	0.0484	4.8801	20.8074	20.6659	100.8508	0.2049	0.0099	4.2638	17
18	0.0405	4.9205	21.4958	24.6958	121.5167	0.2032	0.0082	4.3686	18
19	0.0339	4.9544	22.1057	29.5114	146.2124	0.2018	0.0068	4.4618	19
20	0.0284	4.9828	22.6445	35.2662	175.7239	0.2007	0.0057	4.5445	20
21	0.0237	5.0065	23.1191	42.1431	210.9900	0.1997	0.0047	4.6176	21
22	0.0199	5.0264	23.5360	50.3610	253.1331	0.1990	0.0040	4.6825	22
23	0.0166	5.0430	23.9016	60.1813	303.4940	0.1983	0.0033	4.7396	23
24	0.0139	5.0569	24.2214	71.9167	363.6754	0.1977	0.0027	4.7898	24
25	0.0116	5.0685	24.5007	85.9405	435.5921	0.1973	0.0023	4.8339	25
26	0.0097	5.0783	24.7441	102.6988	521.5325	0.1969	0.0019	4.8725	26
27	0.0081	5.0864	24.9560	122.7251	624.2314	0.1966	0.0016	4.9064	27
28	0.0068	5.0932	25.1401	146.6565	746.9565	0.1963	0.0013	4.9360	28
29	0.0057	5.0989	25.2998	175.2545	893.6130	0.1961	0.0011	4.9618	29
30	0.0048	5.1037	25.4383	209.4292	1068.8675	0.1959	0.0009	4.9843	30
31	0.0040	5.1077	25.5582	250.2679	1278.2967	0.1958	0.0008	5.0038	31
32	0.0033	5.1111	25.6618	299.0701	1528.5645	0.1957	0.0007	5.0208	32
33	0.0028	5.1139	25.7514	357.3887	1827.6346	0.1955	0.0005	5.0356	33
34	0.0023	5.1162	25.8287	427.0796	2185.0233	0.1955	0.0005	5.0484	34
35	0.0020	5.1182	25.8953	510.3601	2612.1029	0.1954	0.0004	5.0595	35
36	0.0016	5.1198	25.9527	609.8803	3122.4630	0.1953	0.0003	5.0691	36
37	0.0014	5.1212	26.0021	728.8069	3732.3432	0.1953	0.0003	5.0774	37
38	0.0011	5.1223	26.0445	870.9243	4461.1502	0.1952	0.0002	5.0845	38
39	0.0010	5.1233	26.0811	1040.7545	5332.0745	0.1952	0.0002	5.0907	39
40	0.0008	5.1241	26.1124	1243.7017	6372.8290	0.1952	0.0002	5.0960	40
41	0.0007	5.1248	26.1393	1486.2235	7616.5306	0.1951	0.0001	5.1006	41
42	0.0006	5.1253	26.1624	1776.0371	9102.7541	0.1951	0.0001	5.1045	42
43	0.0005	5.1258	26.1822	2122.3643	10878.7912	0.1951	0.0001	5.1079	43
44	0.0004	5.1262	26.1992	2536.2253	13001.1554	0.1951	0.0001	5.1108	44
45	0.0003	5.1265	26.2137	3030.7892	15537.3808	0.1951	0.0001	5.1134	45
46	0.0003	5.1268	26.2261	3621.7932	18568.1700	0.1951	0.0001	5.1155	46
47	0.0002	5.1270	26.2367	4328.0428	22189.9632	0.1950	0.0000	5.1173	47
48	0.0002	5.1272	26.2458	5172.0112	26518.0060	0.1950	0.0000	5.1189	48
49	0.0002	5.1274	26.2536	6180.5533	31690.0171	0.1950	0.0000	5.1203	49
50	0.0001	5.1275	26.2602	7385.7612	37870.5705	0.1950	0.0000	5.1214	50
51	0.0001	5.1276	26.2659	8825.9847	45256.3317	0.1950	0.0000	5.1224	51
52	0.0001	5.1277	26.2707	10547.0517	54082.3164	0.1950	0.0000	5.1233	52
53	0.0001	5.1278	26.2748	12603.7268	64629.3681	0.1950	0.0000	5.1240	53
54	0.0001	5.1279	26.2784	15061.4535	77233.0949	0.1950	0.0000	5.1246	54
55	0.0001	5.1279	26.2814	17998.4369	92294.5484	0.1950	0.0000	5.1251	55
60	0.0000	5.1281	26.2909	43860.5746	224920.8955	0.1950	0.0000	5.1268	60
65	0.0000	5.1282	26.2951	106884.2818	548119.3940	0.1950	0.0000	5.1276	65
70	0.0000	5.1282	26.2970	260467.3970	1335725.1128	0.1950	0.0000	5.1279	70
75	0.0000	5.1282	26.2978	634735.6575	3255049.5255	0.1950	0.0000	5.1281	75
80	0.0000	5.1282	26.2982	1546793.8004	7932270.7712	0.1950	0.0000	5.1282	80
85	0.0000	5.1282	26.2984	3769397.5952	************	0.1950	0.0000	5.1282	85
90	0.0000	5.1282	26.2984	9185683.4618	************	0.1950	0.0000	5.1282	90
95	0.0000	5.1282	26.2985	************	************	0.1950	0.0000	5.1282	95
100	0.0000	5.1282	26.2985	************	************	0.1950	0.0000	5.1282	100

$I = 19.75\%$

n	(P/F)	(P/A)	(P/G)	(F/P)	(F/A)	(A/P)	(A/F)	(A/G)	n
1	0.8351	0.8351	0.0000	1.1975	1.0000	1.1975	1.0000	0.0000	1
2	0.6973	1.5324	0.6973	1.4340	2.1975	0.6526	0.4551	0.4551	2
3	0.5823	2.1148	1.8620	1.7172	3.6315	0.4729	0.2754	0.8805	3
4	0.4863	2.6010	3.3209	2.0564	5.3487	0.3845	0.1870	1.2768	4
5	0.4061	3.0071	4.9453	2.4625	7.4051	0.3325	0.1350	1.6445	5
6	0.3391	3.3463	6.6408	2.9489	9.8676	0.2988	0.1013	1.9846	6
7	0.2832	3.6294	8.3399	3.5313	12.8165	0.2755	0.0780	2.2979	7
8	0.2365	3.8659	9.9953	4.2287	16.3477	0.2587	0.0612	2.5855	8
9	0.1975	4.0634	11.5751	5.0638	20.5764	0.2461	0.0486	2.8486	9
10	0.1649	4.2283	13.0593	6.0639	25.6402	0.2365	0.0390	3.0885	10
11	0.1377	4.3660	14.4364	7.2616	31.7042	0.2290	0.0315	3.3065	11
12	0.1150	4.4810	15.7014	8.6957	38.9657	0.2232	0.0257	3.5040	12
13	0.0960	4.5771	16.8538	10.4131	47.6615	0.2185	0.0210	3.6822	13
14	0.0802	4.6572	17.8963	12.4697	58.0746	0.2147	0.0172	3.8427	14
15	0.0670	4.7242	18.8339	14.9325	70.5444	0.2117	0.0142	3.9867	15
16	0.0559	4.7801	19.6727	17.8817	85.4769	0.2092	0.0117	4.1155	16
17	0.0467	4.8268	20.4199	21.4133	103.3585	0.2072	0.0097	4.2305	17
18	0.0390	4.8658	21.0829	25.6424	124.7719	0.2055	0.0080	4.3328	18
19	0.0326	4.8984	21.6691	30.7068	150.4143	0.2041	0.0066	4.4237	19
20	0.0272	4.9256	22.1858	36.7714	181.1211	0.2030	0.0055	4.5042	20
21	0.0227	4.9483	22.6400	44.0338	217.8926	0.2021	0.0046	4.5753	21
22	0.0190	4.9673	23.0382	52.7305	261.9263	0.2013	0.0038	4.6380	22
23	0.0158	4.9831	23.3866	63.1447	314.6568	0.2007	0.0032	4.6932	23
24	0.0132	4.9963	23.6908	75.6158	377.8015	0.2001	0.0026	4.7416	24
25	0.0110	5.0074	23.9559	90.5499	453.4173	0.1997	0.0022	4.7841	25
26	0.0092	5.0166	24.1864	108.4335	543.9672	0.1993	0.0018	4.8213	26
27	0.0077	5.0243	24.3867	129.8491	652.4007	0.1990	0.0015	4.8537	27
28	0.0064	5.0307	24.5603	155.4944	782.2499	0.1988	0.0013	4.8821	28
29	0.0054	5.0361	24.7107	186.2045	937.7442	0.1986	0.0011	4.9067	29
30	0.0045	5.0406	24.8407	222.9799	1123.9487	0.1984	0.0009	4.9281	30
31	0.0037	5.0443	24.9531	267.0184	1346.9286	0.1982	0.0007	4.9468	31
32	0.0031	5.0475	25.0500	319.7545	1613.9470	0.1981	0.0006	4.9629	32
33	0.0026	5.0501	25.1336	382.9061	1933.7015	0.1980	0.0005	4.9769	33
34	0.0022	5.0522	25.2056	458.5300	2316.6076	0.1979	0.0004	4.9890	34
35	0.0018	5.0541	25.2675	549.0897	2775.1376	0.1979	0.0004	4.9994	35
36	0.0015	5.0556	25.3207	657.5349	3324.2272	0.1978	0.0003	5.0085	36
37	0.0013	5.0569	25.3664	787.3980	3981.7621	0.1978	0.0003	5.0162	37
38	0.0011	5.0579	25.4057	942.9091	4769.1601	0.1977	0.0002	5.0229	38
39	0.0009	5.0588	25.4393	1129.1337	5712.0693	0.1977	0.0002	5.0287	39
40	0.0007	5.0595	25.4682	1352.1376	6841.2029	0.1976	0.0001	5.0337	40
41	0.0006	5.0602	25.4929	1619.1848	8193.3405	0.1976	0.0001	5.0380	41
42	0.0005	5.0607	25.5140	1938.9737	9812.5253	0.1976	0.0001	5.0416	42
43	0.0004	5.0611	25.5321	2321.9211	11751.4990	0.1976	0.0001	5.0448	43
44	0.0004	5.0615	25.5476	2780.5005	14073.4201	0.1976	0.0001	5.0475	44
45	0.0003	5.0618	25.5608	3329.6493	16853.9205	0.1976	0.0001	5.0498	45
46	0.0003	5.0620	25.5721	3987.2550	20183.5698	0.1975	0.0000	5.0518	46
47	0.0002	5.0622	25.5817	4774.7379	24170.8249	0.1975	0.0000	5.0534	47
48	0.0002	5.0624	25.5899	5717.7487	28945.5628	0.1975	0.0000	5.0549	48
49	0.0001	5.0626	25.5969	6847.0040	34663.3114	0.1975	0.0000	5.0561	49
50	0.0001	5.0627	25.6029	8199.2873	41510.3155	0.1975	0.0000	5.0572	50
51	0.0001	5.0628	25.6080	9818.6465	49709.6028	0.1975	0.0000	5.0581	51
52	0.0001	5.0629	25.6123	11757.8292	59528.2493	0.1975	0.0000	5.0589	52
53	0.0001	5.0629	25.6160	14080.0005	71286.0785	0.1975	0.0000	5.0595	53
54	0.0001	5.0630	25.6192	16860.8006	85366.0791	0.1975	0.0000	5.0601	54
55	0.0000	5.0630	25.6219	20190.8087	102226.8797	0.1975	0.0000	5.0606	55
60	0.0000	5.0632	25.6303	49720.0235	251741.8911	0.1975	0.0000	5.0621	60
65	0.0000	5.0632	25.6340	122435.9444	619923.7693	0.1975	0.0000	5.0628	65
70	0.0000	5.0633	25.6357	301499.4652	1526574.5071	0.1975	0.0000	5.0631	70
75	0.0000	5.0633	25.6364	742444.7772	3759208.9984	0.1975	0.0000	5.0632	75
80	0.0000	5.0633	25.6367	1828276.0365	9257088.7923	0.1975	0.0000	5.0632	80
85	0.0000	5.0633	25.6368	4502143.9551	***********	0.1975	0.0000	5.0633	85
90	0.0000	5.0633	25.6369	***********	***********	0.1975	0.0000	5.0633	90
95	0.0000	5.0633	25.6369	***********	***********	0.1975	0.0000	5.0633	95
100	0.0000	5.0633	25.6369	***********	***********	0.1975	0.0000	5.0633	100

$I = 20.00\%$

n	(P/F)	(P/A)	(P/G)	(F/P)	(F/A)	(A/P)	(A/F)	(A/G)	n
1	0.8333	0.8333	0.0000	1.2000	1.0000	1.2000	1.0000	0.0000	1
2	0.6944	1.5278	0.6944	1.4400	2.2000	0.6545	0.4545	0.4545	2
3	0.5787	2.1065	1.8519	1.7280	3.6400	0.4747	0.2747	0.8791	3
4	0.4823	2.5887	3.2986	2.0736	5.3680	0.3863	0.1863	1.2742	4
5	0.4019	2.9906	4.9061	2.4883	7.4416	0.3344	0.1344	1.6405	5
6	0.3349	3.3255	6.5806	2.9860	9.9299	0.3007	0.1007	1.9788	6
7	0.2791	3.6046	8.2551	3.5832	12.9159	0.2774	0.0774	2.2902	7
8	0.2326	3.8372	9.8831	4.2998	16.4991	0.2606	0.0606	2.5756	8
9	0.1938	4.0310	11.4335	5.1598	20.7989	0.2481	0.0481	2.8364	9
10	0.1615	4.1925	12.8871	6.1917	25.9587	0.2385	0.0385	3.0739	10
11	0.1346	4.3271	14.2330	7.4301	32.1504	0.2311	0.0311	3.2893	11
12	0.1122	4.4392	15.4667	8.9161	39.5805	0.2253	0.0253	3.4841	12
13	0.0935	4.5327	16.5883	10.6993	48.4966	0.2206	0.0206	3.6597	13
14	0.0779	4.6106	17.6008	12.8392	59.1959	0.2169	0.0169	3.8175	14
15	0.0649	4.6755	18.5095	15.4070	72.0351	0.2139	0.0139	3.9588	15
16	0.0541	4.7296	19.3208	18.4884	87.4421	0.2114	0.0114	4.0851	16
17	0.0451	4.7746	20.0419	22.1861	105.9306	0.2094	0.0094	4.1976	17
18	0.0376	4.8122	20.6805	26.6233	128.1167	0.2078	0.0078	4.2975	18
19	0.0313	4.8435	21.2439	31.9480	154.7400	0.2065	0.0065	4.3861	19
20	0.0261	4.8696	21.7395	38.3376	186.6880	0.2054	0.0054	4.4643	20
21	0.0217	4.8913	22.1742	46.0051	225.0256	0.2044	0.0044	4.5334	21
22	0.0181	4.9094	22.5546	55.2061	271.0307	0.2037	0.0037	4.5941	22
23	0.0151	4.9245	22.8867	66.2474	326.2369	0.2031	0.0031	4.6475	23
24	0.0126	4.9371	23.1760	79.4968	392.4842	0.2025	0.0025	4.6943	24
25	0.0105	4.9476	23.4276	95.3962	471.9811	0.2021	0.0021	4.7352	25
26	0.0087	4.9563	23.6460	114.4755	567.3773	0.2018	0.0018	4.7709	26
27	0.0073	4.9636	23.8353	137.3706	681.8528	0.2015	0.0015	4.8020	27
28	0.0061	4.9697	23.9991	164.8447	819.2233	0.2012	0.0012	4.8291	28
29	0.0051	4.9747	24.1406	197.8136	984.0680	0.2010	0.0010	4.8527	29
30	0.0042	4.9789	24.2628	237.3763	1181.8816	0.2008	0.0008	4.8731	30
31	0.0035	4.9824	24.3681	284.8516	1419.2579	0.2007	0.0007	4.8908	31
32	0.0029	4.9854	24.4588	341.8219	1704.1095	0.2006	0.0006	4.9061	32
33	0.0024	4.9878	24.5368	410.1863	2045.9314	0.2005	0.0005	4.9194	33
34	0.0020	4.9898	24.6038	492.2235	2456.1176	0.2004	0.0004	4.9308	34
35	0.0017	4.9915	24.6614	590.6682	2948.3411	0.2003	0.0003	4.9406	35
36	0.0014	4.9929	24.7108	708.8019	3539.0094	0.2003	0.0003	4.9491	36
37	0.0012	4.9941	24.7531	850.5622	4247.8112	0.2002	0.0002	4.9564	37
38	0.0010	4.9951	24.7894	1020.6747	5098.3735	0.2002	0.0002	4.9627	38
39	0.0008	4.9959	24.8204	1224.8096	6119.0482	0.2002	0.0002	4.9681	39
40	0.0007	4.9966	24.8469	1469.7716	7343.8578	0.2001	0.0001	4.9728	40
41	0.0006	4.9972	24.8696	1763.7259	8813.6294	0.2001	0.0001	4.9767	41
42	0.0005	4.9976	24.8890	2116.4711	10577.3553	0.2001	0.0001	4.9801	42
43	0.0004	4.9980	24.9055	2539.7653	12693.8263	0.2001	0.0001	4.9831	43
44	0.0003	4.9984	24.9196	3047.7183	15233.5916	0.2001	0.0001	4.9856	44
45	0.0003	4.9986	24.9316	3657.2620	18281.3099	0.2001	0.0001	4.9877	45
46	0.0002	4.9989	24.9419	4388.7144	21938.5719	0.2000	0.0000	4.9895	46
47	0.0002	4.9991	24.9506	5266.4573	26327.2863	0.2000	0.0000	4.9911	47
48	0.0002	4.9992	24.9581	6319.7487	31593.7436	0.2000	0.0000	4.9924	48
49	0.0001	4.9993	24.9644	7583.6985	37913.4923	0.2000	0.0000	4.9935	49
50	0.0001	4.9995	24.9698	9100.4382	45497.1908	0.2000	0.0000	4.9945	50
51	0.0001	4.9995	24.9744	10920.5258	54597.6289	0.2000	0.0000	4.9953	51
52	0.0001	4.9996	24.9783	13104.6309	65518.1547	0.2000	0.0000	4.9960	52
53	0.0001	4.9997	24.9816	15725.5571	78622.7856	0.2000	0.0000	4.9966	53
54	0.0001	4.9997	24.9844	18870.6685	94348.3427	0.2000	0.0000	4.9971	54
55	0.0000	4.9998	24.9868	22644.8023	113219.0113	0.2000	0.0000	4.9976	55
60	0.0000	4.9999	24.9942	56347.5144	281732.5718	0.2000	0.0000	4.9989	60
65	0.0000	5.0000	24.9975	140210.6469	701048.2346	0.2000	0.0000	4.9995	65
70	0.0000	5.0000	24.9989	348888.9569	1744439.7847	0.2000	0.0000	4.9998	70
75	0.0000	5.0000	24.9995	868147.3693	4340731.8466	0.2000	0.0000	4.9999	75
80	0.0000	5.0000	24.9998	2160228.4620	************	0.2000	0.0000	5.0000	80
85	0.0000	5.0000	24.9999	5375339.6866	************	0.2000	0.0000	5.0000	85
90	0.0000	5.0000	25.0000	************	************	0.2000	0.0000	5.0000	90
95	0.0000	5.0000	25.0000	************	************	0.2000	0.0000	5.0000	95
100	0.0000	5.0000	25.0000	************	************	0.2000	0.0000	5.0000	100

$I = 20.25\%$

n	(P/F)	(P/A)	(P/G)	(F/P)	(F/A)	(A/P)	(A/F)	(A/G)	n
1	0.8316	0.8316	0.0000	1.2025	1.0000	1.2025	1.0000	0.0000	1
2	0.6916	1.5232	0.6916	1.4460	2.2025	0.6565	0.4540	0.4540	2
3	0.5751	2.0983	1.8418	1.7388	3.6485	0.4766	0.2741	0.8778	3
4	0.4783	2.5765	3.2765	2.0909	5.3873	0.3881	0.1856	1.2717	4
5	0.3977	2.9742	4.8674	2.5143	7.4783	0.3362	0.1337	1.6365	5
6	0.3307	3.3050	6.5211	3.0235	9.9926	0.3026	0.1001	1.9731	6
7	0.2750	3.5800	8.1714	3.6358	13.0161	0.2793	0.0768	2.2825	7
8	0.2287	3.8088	9.7725	4.3720	16.6519	0.2626	0.0601	2.5658	8
9	0.1902	3.9990	11.2942	5.2573	21.0239	0.2501	0.0476	2.8243	9
10	0.1582	4.1571	12.7178	6.3219	26.2812	0.2405	0.0380	3.0593	10
11	0.1315	4.2887	14.0332	7.6021	32.6032	0.2332	0.0307	3.2721	11
12	0.1094	4.3981	15.2365	9.1416	40.2053	0.2274	0.0249	3.4644	12
13	0.0910	4.4890	16.3281	10.9927	49.3469	0.2228	0.0203	3.6373	13
14	0.0756	4.5647	17.3116	13.2188	60.3396	0.2191	0.0166	3.7925	14
15	0.0629	4.6276	18.1923	15.8956	73.5584	0.2161	0.0136	3.9313	15
16	0.0523	4.6799	18.9771	19.1144	89.4540	0.2137	0.0112	4.0550	16
17	0.0435	4.7234	19.6732	22.9851	108.5684	0.2117	0.0092	4.1650	17
18	0.0362	4.7596	20.2882	27.6396	131.5535	0.2101	0.0076	4.2626	18
19	0.0301	4.7897	20.8298	33.2366	159.1931	0.2088	0.0063	4.3489	19
20	0.0250	4.8147	21.3052	39.9670	192.4297	0.2077	0.0052	4.4250	20
21	0.0208	4.8355	21.7213	48.0603	232.3967	0.2068	0.0043	4.4920	21
22	0.0173	4.8528	22.0847	57.7926	280.4570	0.2061	0.0036	4.5509	22
23	0.0144	4.8672	22.4013	69.4955	338.2496	0.2055	0.0030	4.6025	23
24	0.0120	4.8792	22.6765	83.5684	407.7451	0.2050	0.0025	4.6476	24
25	0.0100	4.8891	22.9153	100.4910	491.3135	0.2045	0.0020	4.6870	25
26	0.0083	4.8974	23.1222	120.8404	591.8045	0.2042	0.0017	4.7213	26
27	0.0069	4.9043	23.3011	145.3106	712.6449	0.2039	0.0014	4.7512	27
28	0.0057	4.9100	23.4556	174.7360	857.9555	0.2037	0.0012	4.7771	28
29	0.0048	4.9148	23.5889	210.1200	1032.6915	0.2035	0.0010	4.7996	29
30	0.0040	4.9187	23.7037	252.6693	1242.8116	0.2033	0.0008	4.8191	30
31	0.0033	4.9220	23.8024	303.8349	1495.4809	0.2032	0.0007	4.8359	31
32	0.0027	4.9248	23.8873	365.3614	1799.3158	0.2031	0.0006	4.8504	32
33	0.0023	4.9270	23.9601	439.3471	2164.6772	0.2030	0.0005	4.8630	33
34	0.0019	4.9289	24.0226	528.3149	2604.0244	0.2029	0.0004	4.8738	34
35	0.0016	4.9305	24.0761	635.2987	3132.3393	0.2028	0.0003	4.8831	35
36	0.0013	4.9318	24.1219	763.9467	3767.6380	0.2028	0.0003	4.8911	36
37	0.0011	4.9329	24.1611	918.6459	4531.5848	0.2027	0.0002	4.8980	37
38	0.0009	4.9338	24.1946	1104.6717	5450.2307	0.2027	0.0002	4.9038	38
39	0.0008	4.9346	24.2232	1328.3677	6554.9024	0.2027	0.0002	4.9089	39
40	0.0006	4.9352	24.2476	1597.3622	7883.2701	0.2026	0.0001	4.9132	40
41	0.0005	4.9357	24.2684	1920.8280	9480.6323	0.2026	0.0001	4.9169	41
42	0.0004	4.9361	24.2862	2309.7957	11401.4603	0.2026	0.0001	4.9201	42
43	0.0004	4.9365	24.3013	2777.5294	13711.2561	0.2026	0.0001	4.9228	43
44	0.0003	4.9368	24.3142	3339.9790	16488.7854	0.2026	0.0001	4.9251	44
45	0.0002	4.9370	24.3251	4016.3248	19828.7645	0.2026	0.0001	4.9271	45
46	0.0002	4.9372	24.3344	4829.6306	23845.0893	0.2025	0.0000	4.9287	46
47	0.0002	4.9374	24.3424	5807.6308	28674.7198	0.2025	0.0000	4.9302	47
48	0.0001	4.9376	24.3491	6983.6760	34482.3506	0.2025	0.0000	4.9314	48
49	0.0001	4.9377	24.3548	8397.8704	41466.0266	0.2025	0.0000	4.9324	49
50	0.0001	4.9378	24.3597	10098.4391	49863.8970	0.2025	0.0000	4.9333	50
51	0.0001	4.9379	24.3638	12143.3731	59962.3361	0.2025	0.0000	4.9341	51
52	0.0001	4.9379	24.3673	14602.4061	72105.7092	0.2025	0.0000	4.9347	52
53	0.0001	4.9380	24.3702	17559.3933	86708.1153	0.2025	0.0000	4.9353	53
54	0.0000	4.9380	24.3727	21115.1705	104267.5086	0.2025	0.0000	4.9357	54
55	0.0000	4.9381	24.3749	25390.9925	125382.6791	0.2025	0.0000	4.9361	55
60	0.0000	4.9382	24.3815	63841.7970	315263.1949	0.2025	0.0000	4.9373	60
65	0.0000	4.9382	24.3844	160520.5089	792688.9330	0.2025	0.0000	4.9379	65
70	0.0000	4.9383	24.3856	403604.4567	1993103.4900	0.2025	0.0000	4.9381	70
75	0.0000	4.9383	24.3861	1014802.1495	5011363.7011	0.2025	0.0000	4.9382	75
80	0.0000	4.9383	24.3864	2551565.9835	***********	0.2025	0.0000	4.9382	80
85	0.0000	4.9383	24.3865	6415525.4022	***********	0.2025	0.0000	4.9383	85
90	0.0000	4.9383	24.3865	***********	***********	0.2025	0.0000	4.9383	90
95	0.0000	4.9383	24.3865	***********	***********	0.2025	0.0000	4.9383	95
100	0.0000	4.9383	24.3865	***********	***********	0.2025	0.0000	4.9383	100

$I = 20.50\%$

n	(P/F)	(P/A)	(P/G)	(F/P)	(F/A)	(A/P)	(A/F)	(A/G)	n
1	0.8299	0.8299	0.0000	1.2050	1.0000	1.2050	1.0000	0.0000	1
2	0.6887	1.5186	0.6887	1.4520	2.2050	0.6585	0.4535	0.4535	2
3	0.5715	2.0901	1.8318	1.7497	3.6570	0.4784	0.2734	0.8764	3
4	0.4743	2.5644	3.2546	2.1084	5.4067	0.3900	0.1850	1.2692	4
5	0.3936	2.9580	4.8291	2.5406	7.5151	0.3381	0.1331	1.6325	5
6	0.3266	3.2847	6.4623	3.0614	10.0557	0.3044	0.0994	1.9674	6
7	0.2711	3.5557	8.0888	3.6890	13.1171	0.2812	0.0762	2.2749	7
8	0.2250	3.7807	9.6635	4.4453	16.8061	0.2645	0.0595	2.5560	8
9	0.1867	3.9674	11.1570	5.3565	21.2514	0.2521	0.0471	2.8122	9
10	0.1549	4.1223	12.5513	6.4546	26.6079	0.2426	0.0376	3.0447	10
11	0.1286	4.2509	13.8370	7.7778	33.0625	0.2352	0.0302	3.2551	11
12	0.1067	4.3576	15.0107	9.3723	40.8403	0.2295	0.0245	3.4447	12
13	0.0885	4.4461	16.0733	11.2936	50.2126	0.2249	0.0199	3.6151	13
14	0.0735	4.5196	17.0285	13.6088	61.5062	0.2213	0.0163	3.7677	14
15	0.0610	4.5806	17.8823	16.3986	75.1149	0.2183	0.0133	3.9039	15
16	0.0506	4.6312	18.6414	19.7603	91.5135	0.2159	0.0109	4.0252	16
17	0.0420	4.6732	19.3133	23.8111	111.2737	0.2140	0.0090	4.1328	17
18	0.0349	4.7080	19.9058	28.6924	135.0849	0.2124	0.0074	4.2281	18
19	0.0289	4.7370	20.4264	34.5743	163.7773	0.2111	0.0061	4.3121	19
20	0.0240	4.7610	20.8825	41.6621	198.3516	0.2100	0.0050	4.3862	20
21	0.0199	4.7809	21.2809	50.2028	240.0137	0.2092	0.0042	4.4512	21
22	0.0165	4.7974	21.6280	60.4944	290.2165	0.2084	0.0034	4.5083	22
23	0.0137	4.8111	21.9298	72.8957	350.7109	0.2079	0.0029	4.5581	23
24	0.0114	4.8225	22.1917	87.8394	423.6066	0.2074	0.0024	4.6017	24
25	0.0094	4.8320	22.4184	105.8464	511.4460	0.2070	0.0020	4.6396	25
26	0.0078	4.8398	22.6144	127.5449	617.2924	0.2066	0.0016	4.6726	26
27	0.0065	4.8463	22.7836	153.6916	744.8373	0.2063	0.0013	4.7012	27
28	0.0054	4.8517	22.9294	185.1984	898.5290	0.2061	0.0011	4.7260	28
29	0.0045	4.8562	23.0548	223.1641	1083.7274	0.2059	0.0009	4.7475	29
30	0.0037	4.8599	23.1627	268.9128	1306.8915	0.2058	0.0008	4.7661	30
31	0.0031	4.8630	23.2553	324.0399	1575.8043	0.2056	0.0006	4.7821	31
32	0.0026	4.8656	23.3346	390.4680	1899.8441	0.2055	0.0005	4.7959	32
33	0.0021	4.8677	23.4027	470.5140	2290.3122	0.2054	0.0004	4.8078	33
34	0.0018	4.8694	23.4609	566.9694	2760.8262	0.2054	0.0004	4.8180	34
35	0.0015	4.8709	23.5106	683.1981	3327.7955	0.2053	0.0003	4.8267	35
36	0.0012	4.8721	23.5531	823.2537	4010.9936	0.2052	0.0002	4.8343	36
37	0.0010	4.8731	23.5894	992.0207	4834.2473	0.2052	0.0002	4.8407	37
38	0.0008	4.8740	23.6204	1195.3849	5826.2680	0.2052	0.0002	4.8462	38
39	0.0007	4.8747	23.6468	1440.4389	7021.6530	0.2051	0.0001	4.8510	39
40	0.0006	4.8752	23.6692	1735.7288	8462.0918	0.2051	0.0001	4.8550	40
41	0.0005	4.8757	23.6884	2091.5532	10197.8207	0.2051	0.0001	4.8584	41
42	0.0004	4.8761	23.7046	2520.3217	12289.3739	0.2051	0.0001	4.8614	42
43	0.0003	4.8764	23.7185	3036.9876	14809.6956	0.2051	0.0001	4.8639	43
44	0.0003	4.8767	23.7302	3659.5700	17846.6831	0.2051	0.0001	4.8660	44
45	0.0002	4.8769	23.7402	4409.7819	21506.2532	0.2050	0.0000	4.8678	45
46	0.0002	4.8771	23.7487	5313.7872	25916.0351	0.2050	0.0000	4.8694	46
47	0.0002	4.8773	23.7558	6403.1136	31229.8223	0.2050	0.0000	4.8707	47
48	0.0001	4.8774	23.7619	7715.7519	37632.9359	0.2050	0.0000	4.8718	48
49	0.0001	4.8775	23.7671	9297.4810	45348.6877	0.2050	0.0000	4.8728	49
50	0.0001	4.8776	23.7715	11203.4646	54646.1687	0.2050	0.0000	4.8736	50
51	0.0001	4.8777	23.7752	13500.1748	65849.6333	0.2050	0.0000	4.8743	51
52	0.0001	4.8777	23.7783	16267.7107	79349.8081	0.2050	0.0000	4.8749	52
53	0.0001	4.8778	23.7810	19602.5913	95617.5188	0.2050	0.0000	4.8753	53
54	0.0000	4.8778	23.7832	23621.1226	115220.1101	0.2050	0.0000	4.8758	54
55	0.0000	4.8779	23.7851	28463.4527	138841.2327	0.2050	0.0000	4.8761	55
60	0.0000	4.8780	23.7910	72314.0716	352746.6905	0.2050	0.0000	4.8772	60
65	0.0000	4.8780	23.7935	183720.6821	896193.5713	0.2050	0.0000	4.8777	65
70	0.0000	4.8780	23.7946	466759.6266	2276871.3494	0.2050	0.0000	4.8779	70
75	0.0000	4.8780	23.7950	1185846.6152	5784612.7573	0.2050	0.0000	4.8780	75
80	0.0000	4.8780	23.7952	3012754.5629	************	0.2050	0.0000	4.8780	80
85	0.0000	4.8780	23.7953	7654185.5747	************	0.2050	0.0000	4.8780	85
90	0.0000	4.8780	23.7953	************	************	0.2050	0.0000	4.8780	90
95	0.0000	4.8780	23.7954	************	************	0.2050	0.0000	4.8780	95
100	0.0000	4.8780	23.7954	************	************	0.2050	0.0000	4.8780	100

$I = 20.75\%$

n	(P/F)	(P/A)	(P/G)	(F/P)	(F/A)	(A/P)	(A/F)	(A/G)	n
1	0.8282	0.8282	0.0000	1.2075	1.0000	1.2075	1.0000	0.0000	1
2	0.6858	1.5140	0.6858	1.4581	2.2075	0.6605	0.4530	0.4530	2
3	0.5680	2.0820	1.8218	1.7606	3.6656	0.4803	0.2728	0.8750	3
4	0.4704	2.5524	3.2330	2.1259	5.4262	0.3918	0.1843	1.2667	4
5	0.3896	2.9419	4.7912	2.5671	7.5521	0.3399	0.1324	1.6286	5
6	0.3226	3.2645	6.4042	3.0997	10.1191	0.3063	0.0988	1.9618	6
7	0.2672	3.5317	8.0072	3.7429	13.2189	0.2831	0.0756	2.2672	7
8	0.2213	3.7530	9.5561	4.5196	16.9618	0.2665	0.0590	2.5463	8
9	0.1832	3.9362	11.0220	5.4574	21.4814	0.2541	0.0466	2.8002	9
10	0.1517	4.0880	12.3877	6.5898	26.9387	0.2446	0.0371	3.0303	10
11	0.1257	4.2136	13.6445	7.9572	33.5285	0.2373	0.0298	3.2382	11
12	0.1041	4.3177	14.7893	9.6083	41.4857	0.2316	0.0241	3.4253	12
13	0.0862	4.4039	15.8236	11.6020	51.0940	0.2271	0.0196	3.5931	13
14	0.0714	4.4753	16.7515	14.0094	62.6960	0.2234	0.0159	3.7431	14
15	0.0591	4.5344	17.5791	16.9164	76.7054	0.2205	0.0130	3.8769	15
16	0.0490	4.5833	18.3135	20.4265	93.6217	0.2182	0.0107	3.9957	16
17	0.0405	4.6239	18.9622	24.6650	114.0483	0.2163	0.0088	4.1009	17
18	0.0336	4.6575	19.5330	29.7830	138.7133	0.2147	0.0072	4.1939	18
19	0.0278	4.6853	20.0335	35.9630	168.4963	0.2134	0.0059	4.2758	19
20	0.0230	4.7083	20.4710	43.4253	204.4593	0.2124	0.0049	4.3479	20
21	0.0191	4.7274	20.8524	52.4360	247.8846	0.2115	0.0040	4.4110	21
22	0.0158	4.7432	21.1841	63.3165	300.3206	0.2108	0.0033	4.4662	22
23	0.0131	4.7562	21.4719	76.4547	363.6371	0.2102	0.0027	4.5145	23
24	0.0108	4.7671	21.7210	92.3191	440.0918	0.2098	0.0023	4.5565	24
25	0.0090	4.7760	21.9363	111.4753	532.4109	0.2094	0.0019	4.5930	25
26	0.0074	4.7835	22.1220	134.6064	643.8861	0.2091	0.0016	4.6247	26
27	0.0062	4.7896	22.2820	162.5372	778.4925	0.2088	0.0013	4.6521	27
28	0.0051	4.7947	22.4196	196.2637	941.0297	0.2086	0.0011	4.6759	28
29	0.0042	4.7989	22.5377	236.9884	1137.2934	0.2084	0.0009	4.6964	29
30	0.0035	4.8024	22.6390	286.1635	1374.2817	0.2082	0.0007	4.7141	30
31	0.0029	4.8053	22.7259	345.5424	1660.4452	0.2081	0.0006	4.7293	31
32	0.0024	4.8077	22.8002	417.2424	2005.9876	0.2080	0.0005	4.7424	32
33	0.0020	4.8097	22.8637	503.8202	2423.2300	0.2079	0.0004	4.7536	33
34	0.0016	4.8114	22.9179	608.3629	2927.0502	0.2078	0.0003	4.7633	34
35	0.0014	4.8127	22.9642	734.5982	3535.4132	0.2078	0.0003	4.7716	35
36	0.0011	4.8138	23.0037	887.0274	4270.0114	0.2077	0.0002	4.7786	36
37	0.0009	4.8148	23.0373	1071.0855	5157.0388	0.2077	0.0002	4.7847	37
38	0.0008	4.8156	23.0659	1293.3358	6228.1243	0.2077	0.0002	4.7899	38
39	0.0006	4.8162	23.0902	1561.7030	7521.4601	0.2076	0.0001	4.7943	39
40	0.0005	4.8167	23.1109	1885.7563	9083.1631	0.2076	0.0001	4.7981	40
41	0.0004	4.8172	23.1285	2277.0508	10968.9194	0.2076	0.0001	4.8013	41
42	0.0004	4.8175	23.1434	2749.5388	13245.9702	0.2076	0.0001	4.8040	42
43	0.0003	4.8178	23.1560	3320.0681	15995.5090	0.2076	0.0001	4.8063	43
44	0.0002	4.8181	23.1667	4008.9823	19315.5772	0.2076	0.0001	4.8083	44
45	0.0002	4.8183	23.1758	4840.8461	23324.5594	0.2075	0.0000	4.8100	45
46	0.0002	4.8185	23.1835	5845.3216	28165.4055	0.2075	0.0000	4.8114	46
47	0.0001	4.8186	23.1901	7058.2259	34010.7271	0.2075	0.0000	4.8126	47
48	0.0001	4.8187	23.1956	8522.8078	41068.9530	0.2075	0.0000	4.8136	48
49	0.0001	4.8188	23.2002	10291.2904	49591.7608	0.2075	0.0000	4.8145	49
50	0.0001	4.8189	23.2042	12426.7331	59883.0511	0.2075	0.0000	4.8153	50
51	0.0001	4.8190	23.2075	15005.2802	72309.7843	0.2075	0.0000	4.8159	51
52	0.0001	4.8190	23.2103	18118.8759	87315.0645	0.2075	0.0000	4.8164	52
53	0.0000	4.8191	23.2127	21878.5426	105433.9404	0.2075	0.0000	4.8169	53
54	0.0000	4.8191	23.2147	26418.3402	127312.4830	0.2075	0.0000	4.8172	54
55	0.0000	4.8191	23.2164	31900.1458	153730.8232	0.2075	0.0000	4.8176	55
60	0.0000	4.8192	23.2216	81889.5275	394643.5062	0.2075	0.0000	4.8185	60
65	0.0000	4.8193	23.2238	210215.1745	1013080.3588	0.2075	0.0000	4.8190	65
70	0.0000	4.8193	23.2248	539634.5651	2600643.6871	0.2075	0.0000	4.8191	70
75	0.0000	4.8193	23.2252	1385273.2781	6676010.9788	0.2075	0.0000	4.8192	75
80	0.0000	4.8193	23.2253	3556076.9809	************	0.2075	0.0000	4.8193	80
85	0.0000	4.8193	23.2254	9128656.1969	************	0.2075	0.0000	4.8193	85
90	0.0000	4.8193	23.2254	************	************	0.2075	0.0000	4.8193	90
95	0.0000	4.8193	23.2254	************	************	0.2075	0.0000	4.8193	95
100	0.0000	4.8193	23.2254	************	************	0.2075	0.0000	4.8193	100

$I = 21.00\%$

n	(P/F)	(P/A)	(P/G)	(F/P)	(F/A)	(A/P)	(A/F)	(A/G)	n
1	0.8264	0.8264	0.0000	1.2100	1.0000	1.2100	1.0000	0.0000	1
2	0.6830	1.5095	0.6830	1.4641	2.2100	0.6625	0.4525	0.4525	2
3	0.5645	2.0739	1.8120	1.7716	3.6741	0.4822	0.2722	0.8737	3
4	0.4665	2.5404	3.2115	2.1436	5.4457	0.3936	0.1836	1.2641	4
5	0.3855	2.9260	4.7537	2.5937	7.5892	0.3418	0.1318	1.6246	5
6	0.3186	3.2446	6.3468	3.1384	10.1830	0.3082	0.0982	1.9561	6
7	0.2633	3.5079	7.9268	3.7975	13.3214	0.2851	0.0751	2.2597	7
8	0.2176	3.7256	9.4502	4.5950	17.1189	0.2684	0.0584	2.5366	8
9	0.1799	3.9054	10.8891	5.5599	21.7139	0.2561	0.0461	2.7882	9
10	0.1486	4.0541	12.2269	6.7275	27.2738	0.2467	0.0367	3.0159	10
11	0.1228	4.1769	13.4553	8.1403	34.0013	0.2394	0.0294	3.2213	11
12	0.1015	4.2784	14.5721	9.8497	42.1416	0.2337	0.0237	3.4059	12
13	0.0839	4.3624	15.5790	11.9182	51.9913	0.2292	0.0192	3.5712	13
14	0.0693	4.4317	16.4804	14.4210	63.9095	0.2256	0.0156	3.7188	14
15	0.0573	4.4890	17.2828	17.4494	78.3305	0.2228	0.0128	3.8500	15
16	0.0474	4.5364	17.9932	21.1138	95.7799	0.2204	0.0104	3.9664	16
17	0.0391	4.5755	18.6195	25.5477	116.8937	0.2186	0.0086	4.0694	17
18	0.0323	4.6079	19.1694	30.9127	142.4413	0.2170	0.0070	4.1602	18
19	0.0267	4.6346	19.6506	37.4043	173.3540	0.2158	0.0058	4.2400	19
20	0.0221	4.6567	20.0704	45.2593	210.7584	0.2147	0.0047	4.3100	20
21	0.0183	4.6750	20.4356	54.7637	256.0176	0.2139	0.0039	4.3713	21
22	0.0151	4.6900	20.7526	66.2641	310.7813	0.2132	0.0032	4.4248	22
23	0.0125	4.7025	21.0269	80.1795	377.0454	0.2127	0.0027	4.4714	23
24	0.0103	4.7128	21.2640	97.0172	457.2249	0.2122	0.0022	4.5119	24
25	0.0085	4.7213	21.4685	117.3909	554.2422	0.2118	0.0018	4.5471	25
26	0.0070	4.7284	21.6445	142.0429	671.6330	0.2115	0.0015	4.5776	26
27	0.0058	4.7342	21.7957	171.8719	813.6759	0.2112	0.0012	4.6039	27
28	0.0048	4.7390	21.9256	207.9651	985.5479	0.2110	0.0010	4.6266	28
29	0.0040	4.7430	22.0368	251.6377	1193.5129	0.2108	0.0008	4.6462	29
30	0.0033	4.7463	22.1321	304.4816	1445.1507	0.2107	0.0007	4.6631	30
31	0.0027	4.7490	22.2135	368.4228	1749.6323	0.2106	0.0006	4.6775	31
32	0.0022	4.7512	22.2830	445.7916	2118.0551	0.2105	0.0005	4.6900	32
33	0.0019	4.7531	22.3424	539.4078	2563.8467	0.2104	0.0004	4.7006	33
34	0.0015	4.7546	22.3929	652.6834	3103.2545	0.2103	0.0003	4.7097	34
35	0.0013	4.7559	22.4360	789.7470	3755.9379	0.2103	0.0003	4.7175	35
36	0.0010	4.7569	22.4726	955.5938	4545.6848	0.2102	0.0002	4.7242	36
37	0.0009	4.7578	22.5037	1156.2685	5501.2787	0.2102	0.0002	4.7299	37
38	0.0007	4.7585	22.5302	1399.0849	6657.5472	0.2102	0.0002	4.7347	38
39	0.0006	4.7591	22.5526	1692.8927	8056.6321	0.2101	0.0001	4.7389	39
40	0.0005	4.7596	22.5717	2048.4002	9749.5248	0.2101	0.0001	4.7424	40
41	0.0004	4.7600	22.5878	2478.5643	11797.9250	0.2101	0.0001	4.7454	41
42	0.0003	4.7603	22.6015	2999.0628	14276.4893	0.2101	0.0001	4.7479	42
43	0.0003	4.7606	22.6131	3628.8659	17275.5521	0.2101	0.0001	4.7501	43
44	0.0002	4.7608	22.6229	4390.9278	20904.4180	0.2100	0.0000	4.7519	44
45	0.0002	4.7610	22.6311	5313.0226	25295.3458	0.2100	0.0000	4.7534	45
46	0.0002	4.7612	22.6381	6428.7574	30608.3684	0.2100	0.0000	4.7547	46
47	0.0001	4.7613	22.6441	7778.7964	37037.1257	0.2100	0.0000	4.7559	47
48	0.0001	4.7614	22.6490	9412.3437	44815.9221	0.2100	0.0000	4.7568	48
49	0.0001	4.7615	22.6533	11388.9358	54228.2658	0.2100	0.0000	4.7576	49
50	0.0001	4.7616	22.6568	13780.6123	65617.2016	0.2100	0.0000	4.7583	50
51	0.0001	4.7616	22.6598	16674.5409	79397.8140	0.2100	0.0000	4.7588	51
52	0.0000	4.7617	22.6623	20176.1945	96072.3549	0.2100	0.0000	4.7593	52
53	0.0000	4.7617	22.6645	24413.1954	116248.5494	0.2100	0.0000	4.7597	53
54	0.0000	4.7617	22.6663	29539.9664	140661.7448	0.2100	0.0000	4.7601	54
55	0.0000	4.7618	22.6678	35743.3594	170201.7112	0.2100	0.0000	4.7604	55
60	0.0000	4.7619	22.6724	92709.0688	441466.9944	0.2100	0.0000	4.7613	60
65	0.0000	4.7619	22.6744	240463.4482	1145059.2773	0.2100	0.0000	4.7616	65
70	0.0000	4.7619	22.6752	623700.2558	2969996.4561	0.2100	0.0000	4.7618	70
75	0.0000	4.7619	22.6755	1617717.8358	7703413.5037	0.2100	0.0000	4.7619	75
80	0.0000	4.7619	22.6756	4195943.4391	************	0.2100	0.0000	4.7619	80
85	0.0000	4.7619	22.6757	************	************	0.2100	0.0000	4.7619	85
90	0.0000	4.7619	22.6757	************	************	0.2100	0.0000	4.7619	90
95	0.0000	4.7619	22.6757	************	************	0.2100	0.0000	4.7619	95
100	0.0000	4.7619	22.6757	************	************	0.2100	0.0000	4.7619	100

$I = 21.25\%$

n	(P/F)	(P/A)	(P/G)	(F/P)	(F/A)	(A/P)	(A/F)	(A/G)	n
1	0.8247	0.8247	0.0000	1.2125	1.0000	1.2125	1.0000	0.0000	1
2	0.6802	1.5049	0.6802	1.4702	2.2125	0.6645	0.4520	0.4520	2
3	0.5610	2.0659	1.8022	1.7826	3.6827	0.4840	0.2715	0.8723	3
4	0.4627	2.5286	3.1902	2.1614	5.4652	0.3955	0.1830	1.2616	4
5	0.3816	2.9102	4.7165	2.6206	7.6266	0.3436	0.1311	1.6207	5
6	0.3147	3.2249	6.2901	3.1775	10.2472	0.3101	0.0976	1.9505	6
7	0.2596	3.4845	7.8474	3.8528	13.4248	0.2870	0.0745	2.2521	7
8	0.2141	3.6985	9.3459	4.6715	17.2775	0.2704	0.0579	2.5269	8
9	0.1765	3.8751	10.7582	5.6642	21.9490	0.2581	0.0456	2.7763	9
10	0.1456	4.0207	12.0687	6.8678	27.6132	0.2487	0.0362	3.0017	10
11	0.1201	4.1408	13.2696	8.3272	34.4810	0.2415	0.0290	3.2046	11
12	0.0990	4.2398	14.3591	10.0967	42.8082	0.2359	0.0234	3.3867	12
13	0.0817	4.3215	15.3393	12.2423	52.9049	0.2314	0.0189	3.5495	13
14	0.0674	4.3889	16.2151	14.8438	65.1472	0.2278	0.0153	3.6946	14
15	0.0556	4.4444	16.9929	17.9981	79.9910	0.2250	0.0125	3.8234	15
16	0.0458	4.4902	17.6803	21.8227	97.9891	0.2227	0.0102	3.9375	16
17	0.0378	4.5280	18.2850	26.4600	119.8117	0.2208	0.0083	4.0382	17
18	0.0312	4.5592	18.8148	32.0827	146.2717	0.2193	0.0068	4.1268	18
19	0.0257	4.5849	19.2776	38.9003	178.3545	0.2181	0.0056	4.2046	19
20	0.0212	4.6061	19.6804	47.1666	217.2548	0.2171	0.0046	4.2727	20
21	0.0175	4.6236	20.0301	57.1896	264.4214	0.2163	0.0038	4.3321	21
22	0.0144	4.6380	20.3329	69.3423	321.6110	0.2156	0.0031	4.3840	22
23	0.0119	4.6499	20.5946	84.0776	390.9533	0.2151	0.0026	4.4290	23
24	0.0098	4.6597	20.8202	101.9441	475.0309	0.2146	0.0021	4.4681	24
25	0.0081	4.6678	21.0144	123.6072	576.9749	0.2142	0.0017	4.5020	25
26	0.0067	4.6745	21.1812	149.8737	700.5821	0.2139	0.0014	4.5312	26
27	0.0055	4.6800	21.3243	181.7219	850.4558	0.2137	0.0012	4.5565	27
28	0.0045	4.6845	21.4468	220.3378	1032.1777	0.2135	0.0010	4.5782	28
29	0.0037	4.6883	21.5516	267.1595	1252.5154	0.2133	0.0008	4.5969	29
30	0.0031	4.6914	21.6411	323.9309	1519.6750	0.2132	0.0007	4.6130	30
31	0.0025	4.6939	21.7175	392.7663	1843.6059	0.2130	0.0005	4.6268	31
32	0.0021	4.6960	21.7826	476.2291	2236.3721	0.2129	0.0004	4.6385	32
33	0.0017	4.6977	21.8380	577.4278	2712.6012	0.2129	0.0004	4.6486	33
34	0.0014	4.6992	21.8852	700.1312	3290.0290	0.2128	0.0003	4.6573	34
35	0.0012	4.7003	21.9252	848.9090	3990.1602	0.2128	0.0003	4.6646	35
36	0.0010	4.7013	21.9592	1029.3022	4839.0692	0.2127	0.0002	4.6709	36
37	0.0008	4.7021	21.9881	1248.0289	5868.3714	0.2127	0.0002	4.6762	37
38	0.0007	4.7028	22.0125	1513.2351	7116.4003	0.2126	0.0001	4.6808	38
39	0.0005	4.7033	22.0332	1834.7975	8629.6354	0.2126	0.0001	4.6846	39
40	0.0004	4.7038	22.0508	2224.6920	10464.4329	0.2126	0.0001	4.6879	40
41	0.0004	4.7041	22.0656	2697.4390	12689.1249	0.2126	0.0001	4.6907	41
42	0.0003	4.7044	22.0781	3270.6448	15386.5639	0.2126	0.0001	4.6930	42
43	0.0003	4.7047	22.0887	3965.6569	18657.2087	0.2126	0.0001	4.6950	43
44	0.0002	4.7049	22.0977	4808.3589	22622.8656	0.2125	0.0000	4.6967	44
45	0.0002	4.7051	22.1052	5830.1352	27431.2245	0.2125	0.0000	4.6982	45
46	0.0001	4.7052	22.1116	7069.0390	33261.3598	0.2125	0.0000	4.6994	46
47	0.0001	4.7053	22.1169	8571.2097	40330.3987	0.2125	0.0000	4.7004	47
48	0.0001	4.7054	22.1215	10392.5918	48901.6084	0.2125	0.0000	4.7013	48
49	0.0001	4.7055	22.1253	12601.0176	59294.2002	0.2125	0.0000	4.7020	49
50	0.0001	4.7056	22.1285	15278.7338	71895.2178	0.2125	0.0000	4.7026	50
51	0.0001	4.7056	22.1312	18525.4647	87173.9516	0.2125	0.0000	4.7031	51
52	0.0000	4.7057	22.1334	22462.1260	105699.4163	0.2125	0.0000	4.7036	52
53	0.0000	4.7057	22.1354	27235.3277	128161.5422	0.2125	0.0000	4.7039	53
54	0.0000	4.7057	22.1370	33022.8349	155396.8700	0.2125	0.0000	4.7042	54
55	0.0000	4.7058	22.1383	40040.1873	188419.7048	0.2125	0.0000	4.7045	55
60	0.0000	4.7058	22.1424	104931.2476	493789.4004	0.2125	0.0000	4.7053	60
65	0.0000	4.7059	22.1441	274987.8926	1294055.9654	0.2125	0.0000	4.7056	65
70	0.0000	4.7059	22.1448	720646.5456	3391273.1558	0.2125	0.0000	4.7058	70
75	0.0000	4.7059	22.1451	1888561.1243	8887341.7615	0.2125	0.0000	4.7058	75
80	0.0000	4.7059	22.1452	4949254.4466	************	0.2125	0.0000	4.7059	80
85	0.0000	4.7059	22.1453	************	************	0.2125	0.0000	4.7059	85
90	0.0000	4.7059	22.1453	************	************	0.2125	0.0000	4.7059	90
95	0.0000	4.7059	22.1453	************	************	0.2125	0.0000	4.7059	95
100	0.0000	4.7059	22.1453	************	************	0.2125	0.0000	4.7059	100

ENGINEERING ECONOMIC ANALYSIS

$I = 21.50\%$

n	(P/F)	(P/A)	(P/G)	(F/P)	(F/A)	(A/P)	(A/F)	(A/G)	n
1	0.8230	0.8230	0.0000	1.2150	1.0000	1.2150	1.0000	0.0000	1
2	0.6774	1.5004	0.6774	1.4762	2.2150	0.6665	0.4515	0.4515	2
3	0.5575	2.0580	1.7925	1.7936	3.6912	0.4859	0.2709	0.8710	3
4	0.4589	2.5169	3.1691	2.1792	5.4848	0.3973	0.1823	1.2591	4
5	0.3777	2.8945	4.6798	2.6478	7.6641	0.3455	0.1305	1.6168	5
6	0.3108	3.2054	6.2340	3.2170	10.3119	0.3120	0.0970	1.9449	6
7	0.2558	3.4612	7.7690	3.9087	13.5289	0.2889	0.0739	2.2446	7
8	0.2106	3.6718	9.2430	4.7491	17.4376	0.2723	0.0573	2.5173	8
9	0.1733	3.8451	10.6295	5.7701	22.1867	0.2601	0.0451	2.7644	9
10	0.1426	3.9877	11.9132	7.0107	27.9568	0.2508	0.0358	2.9875	10
11	0.1174	4.1051	13.0872	8.5180	34.9676	0.2436	0.0286	3.1880	11
12	0.0966	4.2017	14.1501	10.3494	43.4856	0.2380	0.0230	3.3677	12
13	0.0795	4.2813	15.1044	12.5745	53.8350	0.2336	0.0186	3.5280	13
14	0.0655	4.3467	15.9553	15.2780	66.4095	0.2301	0.0151	3.6706	14
15	0.0539	4.4006	16.7095	18.5628	81.6876	0.2272	0.0122	3.7971	15
16	0.0443	4.4449	17.3745	22.5538	100.2504	0.2250	0.0100	3.9088	16
17	0.0365	4.4814	17.9584	27.4029	122.8042	0.2231	0.0081	4.0073	17
18	0.0300	4.5115	18.4690	33.2945	150.2072	0.2217	0.0067	4.0938	18
19	0.0247	4.5362	18.9140	40.4529	183.5017	0.2204	0.0054	4.1696	19
20	0.0203	4.5565	19.3005	49.1502	223.9546	0.2195	0.0045	4.2358	20
21	0.0167	4.5733	19.6354	59.7175	273.1048	0.2187	0.0037	4.2935	21
22	0.0138	4.5871	19.9249	72.5568	332.8223	0.2180	0.0030	4.3437	22
23	0.0113	4.5984	20.1744	88.1565	405.3791	0.2175	0.0025	4.3873	23
24	0.0093	4.6077	20.3892	107.1102	493.5356	0.2170	0.0020	4.4250	24
25	0.0077	4.6154	20.5736	130.1388	600.6458	0.2167	0.0017	4.4576	25
26	0.0063	4.6217	20.7317	158.1187	730.7846	0.2164	0.0014	4.4857	26
27	0.0052	4.6270	20.8670	192.1142	888.9033	0.2161	0.0011	4.5099	27
28	0.0043	4.6312	20.9827	233.4188	1081.0175	0.2159	0.0009	4.5307	28
29	0.0035	4.6348	21.0814	283.6038	1314.4363	0.2158	0.0008	4.5485	29
30	0.0029	4.6377	21.1656	344.5786	1598.0401	0.2156	0.0006	4.5638	30
31	0.0024	4.6401	21.2372	418.6630	1942.6188	0.2155	0.0005	4.5769	31
32	0.0020	4.6420	21.2982	508.6756	2361.2818	0.2154	0.0004	4.5881	32
33	0.0016	4.6436	21.3500	618.0408	2869.9574	0.2153	0.0003	4.5977	33
34	0.0013	4.6450	21.3939	750.9196	3487.9982	0.2153	0.0003	4.6058	34
35	0.0011	4.6461	21.4312	912.3673	4238.9178	0.2152	0.0002	4.6128	35
36	0.0009	4.6470	21.4628	1108.5263	5151.2852	0.2152	0.0002	4.6187	36
37	0.0007	4.6477	21.4895	1346.8595	6259.8115	0.2152	0.0002	4.6237	37
38	0.0006	4.6483	21.5121	1636.4343	7606.6709	0.2151	0.0001	4.6279	38
39	0.0005	4.6488	21.5312	1988.2676	9243.1052	0.2151	0.0001	4.6315	39
40	0.0004	4.6492	21.5473	2415.7452	11231.3728	0.2151	0.0001	4.6346	40
41	0.0003	4.6496	21.5610	2935.1304	13647.1180	0.2151	0.0001	4.6372	41
42	0.0003	4.6499	21.5725	3566.1834	16582.2483	0.2151	0.0001	4.6394	42
43	0.0002	4.6501	21.5822	4332.9128	20148.4317	0.2150	0.0000	4.6412	43
44	0.0002	4.6503	21.5903	5264.4891	24481.3446	0.2150	0.0000	4.6428	44
45	0.0002	4.6504	21.5972	6396.3542	29745.8336	0.2150	0.0000	4.6441	45
46	0.0001	4.6506	21.6030	7771.5704	36142.1879	0.2150	0.0000	4.6452	46
47	0.0001	4.6507	21.6079	9442.4580	43913.7583	0.2150	0.0000	4.6462	47
48	0.0001	4.6508	21.6120	11472.5865	53356.2163	0.2150	0.0000	4.6470	48
49	0.0001	4.6508	21.6154	13939.1926	64828.8028	0.2150	0.0000	4.6476	49
50	0.0001	4.6509	21.6183	16936.1190	78767.9954	0.2150	0.0000	4.6482	50
51	0.0000	4.6509	21.6207	20577.3846	95704.1144	0.2150	0.0000	4.6487	51
52	0.0000	4.6510	21.6228	25001.5223	116281.4990	0.2150	0.0000	4.6491	52
53	0.0000	4.6510	21.6245	30376.8496	141283.0213	0.2150	0.0000	4.6494	53
54	0.0000	4.6510	21.6259	36907.8722	171659.8708	0.2150	0.0000	4.6497	54
55	0.0000	4.6511	21.6271	44843.0648	208567.7431	0.2150	0.0000	4.6499	55
60	0.0000	4.6511	21.6308	118734.4312	552248.5172	0.2150	0.0000	4.6507	60
65	0.0000	4.6511	21.6323	314382.2847	1462238.5334	0.2150	0.0000	4.6510	65
70	0.0000	4.6512	21.6329	832414.1526	3871689.0817	0.2150	0.0000	4.6511	70
75	0.0000	4.6512	21.6331	2204046.9681	************	0.2150	0.0000	4.6511	75
80	0.0000	4.6512	21.6332	5835824.6587	************	0.2150	0.0000	4.6511	80
85	0.0000	4.6512	21.6333	************	************	0.2150	0.0000	4.6512	85
90	0.0000	4.6512	21.6333	************	************	0.2150	0.0000	4.6512	90
95	0.0000	4.6512	21.6333	************	************	0.2150	0.0000	4.6512	95
100	0.0000	4.6512	21.6333	************	************	0.2150	0.0000	4.6512	100

$I = 21.75\%$

n	(P/F)	(P/A)	(P/G)	(F/P)	(F/A)	(A/P)	(A/F)	(A/G)	n
1	0.8214	0.8214	0.0000	1.2175	1.0000	1.2175	1.0000	0.0000	1
2	0.6746	1.4960	0.6746	1.4823	2.2175	0.6685	0.4510	0.4510	2
3	0.5541	2.0501	1.7828	1.8047	3.6998	0.4878	0.2703	0.8696	3
4	0.4551	2.5052	3.1482	2.1972	5.5045	0.3992	0.1817	1.2567	4
5	0.3738	2.8790	4.6434	2.6751	7.7017	0.3473	0.1298	1.6129	5
6	0.3070	3.1861	6.1786	3.2570	10.3769	0.3139	0.0964	1.9393	6
7	0.2522	3.4382	7.6917	3.9654	13.6338	0.2908	0.0733	2.2371	7
8	0.2071	3.6454	9.1416	4.8278	17.5992	0.2743	0.0568	2.5077	8
9	0.1701	3.8155	10.5027	5.8779	22.4270	0.2621	0.0446	2.7526	9
10	0.1397	3.9552	11.7603	7.1563	28.3049	0.2528	0.0353	2.9734	10
11	0.1148	4.0700	12.9080	8.7128	35.4612	0.2457	0.0282	3.1715	11
12	0.0943	4.1643	13.9450	10.6079	44.1741	0.2401	0.0226	3.3487	12
13	0.0774	4.2417	14.8742	12.9151	54.7819	0.2358	0.0183	3.5066	13
14	0.0636	4.3053	15.7009	15.7241	67.6970	0.2323	0.0148	3.6469	14
15	0.0522	4.3575	16.4322	19.1441	83.4211	0.2295	0.0120	3.7710	15
16	0.0429	4.4004	17.0758	23.3079	102.5651	0.2272	0.0097	3.8805	16
17	0.0352	4.4357	17.6396	28.3774	125.8731	0.2254	0.0079	3.9768	17
18	0.0289	4.4646	18.1316	34.5495	154.2505	0.2240	0.0065	4.0612	18
19	0.0238	4.4884	18.5596	42.0640	188.7999	0.2228	0.0053	4.1350	19
20	0.0195	4.5079	18.9306	51.2129	230.8639	0.2218	0.0043	4.1994	20
21	0.0160	4.5240	19.2513	62.3517	282.0768	0.2210	0.0035	4.2554	21
22	0.0132	4.5371	19.5280	75.9132	344.4285	0.2204	0.0029	4.3040	22
23	0.0108	4.5480	19.7660	92.4243	420.3417	0.2199	0.0024	4.3461	23
24	0.0089	4.5568	19.9704	112.5266	512.7661	0.2195	0.0020	4.3825	24
25	0.0073	4.5641	20.1456	137.0012	625.2927	0.2191	0.0016	4.4139	25
26	0.0060	4.5701	20.2955	166.7989	762.2939	0.2188	0.0013	4.4409	26
27	0.0049	4.5751	20.4235	203.0777	929.0928	0.2186	0.0011	4.4641	27
28	0.0040	4.5791	20.5327	247.2471	1132.1704	0.2184	0.0009	4.4840	28
29	0.0033	4.5824	20.6257	301.0233	1379.4175	0.2182	0.0007	4.5010	29
30	0.0027	4.5852	20.7048	366.4959	1680.4408	0.2181	0.0006	4.5156	30
31	0.0022	4.5874	20.7721	446.2087	2046.9367	0.2180	0.0005	4.5281	31
32	0.0018	4.5892	20.8291	543.2591	2493.1454	0.2179	0.0004	4.5387	32
33	0.0015	4.5907	20.8775	661.4180	3036.4046	0.2178	0.0003	4.5477	33
34	0.0012	4.5920	20.9185	805.2764	3697.8226	0.2178	0.0003	4.5554	34
35	0.0010	4.5930	20.9532	980.4240	4503.0990	0.2177	0.0002	4.5620	35
36	0.0008	4.5938	20.9825	1193.6663	5483.5230	0.2177	0.0002	4.5675	36
37	0.0007	4.5945	21.0073	1453.2887	6677.1893	0.2176	0.0001	4.5722	37
38	0.0006	4.5951	21.0282	1769.3789	8130.4779	0.2176	0.0001	4.5762	38
39	0.0005	4.5956	21.0458	2154.2189	9899.8569	0.2176	0.0001	4.5796	39
40	0.0004	4.5959	21.0607	2622.7615	12054.0757	0.2176	0.0001	4.5824	40
41	0.0003	4.5963	21.0732	3193.2121	14676.8372	0.2176	0.0001	4.5849	41
42	0.0003	4.5965	21.0837	3887.7357	17870.0493	0.2176	0.0001	4.5869	42
43	0.0002	4.5967	21.0926	4733.3182	21757.7850	0.2175	0.0000	4.5886	43
44	0.0002	4.5969	21.1001	5762.8150	26491.1033	0.2175	0.0000	4.5901	44
45	0.0001	4.5970	21.1064	7016.2272	32253.9182	0.2175	0.0000	4.5913	45
46	0.0001	4.5972	21.1116	8542.2566	39270.1454	0.2175	0.0000	4.5923	46
47	0.0001	4.5973	21.1160	10400.1975	47812.4021	0.2175	0.0000	4.5932	47
48	0.0001	4.5973	21.1198	12662.2404	58212.5995	0.2175	0.0000	4.5939	48
49	0.0001	4.5974	21.1229	15416.2777	70874.8399	0.2175	0.0000	4.5945	49
50	0.0001	4.5975	21.1255	18769.3181	86291.1176	0.2175	0.0000	4.5950	50
51	0.0000	4.5975	21.1277	22851.6448	105060.4357	0.2175	0.0000	4.5955	51
52	0.0000	4.5975	21.1295	27821.8775	127912.0805	0.2175	0.0000	4.5958	52
53	0.0000	4.5976	21.1310	33873.1359	155733.9580	0.2175	0.0000	4.5961	53
54	0.0000	4.5976	21.1323	41240.5429	189607.0938	0.2175	0.0000	4.5964	54
55	0.0000	4.5976	21.1334	50210.3610	230847.6367	0.2175	0.0000	4.5966	55
60	0.0000	4.5977	21.1366	134319.2299	617555.0800	0.2175	0.0000	4.5973	60
65	0.0000	4.5977	21.1380	359321.3665	1652047.6621	0.2175	0.0000	4.5975	65
70	0.0000	4.5977	21.1385	961231.2735	4419449.5334	0.2175	0.0000	4.5976	70
75	0.0000	4.5977	21.1387	2571418.3662	************	0.2175	0.0000	4.5977	75
80	0.0000	4.5977	21.1388	6878877.7438	************	0.2175	0.0000	4.5977	80
85	0.0000	4.5977	21.1388	************	************	0.2175	0.0000	4.5977	85
90	0.0000	4.5977	21.1388	************	************	0.2175	0.0000	4.5977	90
95	0.0000	4.5977	21.1389	************	************	0.2175	0.0000	4.5977	95
100	0.0000	4.5977	21.1389	************	************	0.2175	0.0000	4.5977	100

ENGINEERING ECONOMIC ANALYSIS

$I = 22.00\%$

n	(P/F)	(P/A)	(P/G)	(F/P)	(F/A)	(A/P)	(A/F)	(A/G)	n
1	0.8197	0.8197	0.0000	1.2200	1.0000	1.2200	1.0000	0.0000	1
2	0.6719	1.4915	0.6719	1.4884	2.2200	0.6705	0.4505	0.4505	2
3	0.5507	2.0422	1.7733	1.8158	3.7084	0.4897	0.2697	0.8683	3
4	0.4514	2.4936	3.1275	2.2153	5.5242	0.4010	0.1810	1.2542	4
5	0.3700	2.8636	4.6075	2.7027	7.7396	0.3492	0.1292	1.6090	5
6	0.3033	3.1669	6.1239	3.2973	10.4423	0.3158	0.0958	1.9337	6
7	0.2486	3.4155	7.6154	4.0227	13.7396	0.2928	0.0728	2.2297	7
8	0.2038	3.6193	9.0417	4.9077	17.7623	0.2763	0.0563	2.4982	8
9	0.1670	3.7863	10.3779	5.9874	22.6700	0.2641	0.0441	2.7409	9
10	0.1369	3.9232	11.6100	7.3046	28.6574	0.2549	0.0349	2.9593	10
11	0.1122	4.0354	12.7321	8.9117	35.9620	0.2478	0.0278	3.1551	11
12	0.0920	4.1274	13.7438	10.8722	44.8737	0.2423	0.0223	3.3299	12
13	0.0754	4.2028	14.6485	13.2641	55.7459	0.2379	0.0179	3.4855	13
14	0.0618	4.2646	15.4519	16.1822	69.0100	0.2345	0.0145	3.6233	14
15	0.0507	4.3152	16.1610	19.7423	85.1922	0.2317	0.0117	3.7451	15
16	0.0415	4.3567	16.7838	24.0856	104.9345	0.2295	0.0095	3.8524	16
17	0.0340	4.3908	17.3283	29.3844	129.0201	0.2278	0.0078	3.9465	17
18	0.0279	4.4187	17.8025	35.8490	158.4045	0.2263	0.0063	4.0289	18
19	0.0229	4.4415	18.2141	43.7358	194.2535	0.2251	0.0051	4.1009	19
20	0.0187	4.4603	18.5702	53.3576	237.9893	0.2242	0.0042	4.1635	20
21	0.0154	4.4756	18.8774	65.0963	291.3469	0.2234	0.0034	4.2178	21
22	0.0126	4.4882	19.1418	79.4175	356.4432	0.2228	0.0028	4.2649	22
23	0.0103	4.4985	19.3689	96.8894	435.8607	0.2223	0.0023	4.3056	23
24	0.0085	4.5070	19.5635	118.2050	532.7501	0.2219	0.0019	4.3407	24
25	0.0069	4.5139	19.7299	144.2101	650.9551	0.2215	0.0015	4.3709	25
26	0.0057	4.5196	19.8720	175.9364	795.1653	0.2213	0.0013	4.3968	26
27	0.0047	4.5243	19.9931	214.6424	971.1016	0.2210	0.0010	4.4191	27
28	0.0038	4.5281	20.0962	261.8637	1185.7440	0.2208	0.0008	4.4381	28
29	0.0031	4.5312	20.1839	319.4737	1447.6077	0.2207	0.0007	4.4544	29
30	0.0026	4.5338	20.2583	389.7579	1767.0813	0.2206	0.0006	4.4683	30
31	0.0021	4.5359	20.3214	475.5046	2156.8392	0.2205	0.0005	4.4801	31
32	0.0017	4.5376	20.3748	580.1156	2632.3439	0.2204	0.0004	4.4902	32
33	0.0014	4.5390	20.4200	707.7411	3212.4595	0.2203	0.0003	4.4988	33
34	0.0012	4.5402	20.4582	863.4441	3920.2006	0.2203	0.0003	4.5060	34
35	0.0009	4.5411	20.4905	1053.4018	4783.6447	0.2202	0.0002	4.5122	35
36	0.0008	4.5419	20.5178	1285.1502	5837.0466	0.2202	0.0002	4.5174	36
37	0.0006	4.5426	20.5407	1567.8833	7122.1968	0.2201	0.0001	4.5218	37
38	0.0005	4.5431	20.5601	1912.8176	8690.0801	0.2201	0.0001	4.5256	38
39	0.0004	4.5435	20.5763	2333.6375	10602.8978	0.2201	0.0001	4.5287	39
40	0.0004	4.5439	20.5900	2847.0378	12936.5353	0.2201	0.0001	4.5314	40
41	0.0003	4.5441	20.6016	3473.3861	15783.5730	0.2201	0.0001	4.5336	41
42	0.0002	4.5444	20.6112	4237.5310	19256.5931	0.2201	0.0001	4.5355	42
43	0.0002	4.5446	20.6194	5169.7878	23494.4901	0.2200	0.0000	4.5371	43
44	0.0002	4.5447	20.6262	6307.1411	28664.2779	0.2200	0.0000	4.5385	44
45	0.0001	4.5449	20.6319	7694.7122	34971.4191	0.2200	0.0000	4.5396	45
46	0.0001	4.5450	20.6367	9387.5489	42666.1312	0.2200	0.0000	4.5406	46
47	0.0001	4.5451	20.6407	11452.8096	52053.6801	0.2200	0.0000	4.5414	47
48	0.0001	4.5451	20.6441	13972.4277	63506.4897	0.2200	0.0000	4.5420	48
49	0.0001	4.5452	20.6469	17046.3618	77478.9175	0.2200	0.0000	4.5426	49
50	0.0000	4.5452	20.6492	20796.5615	94525.2793	0.2200	0.0000	4.5431	50
51	0.0000	4.5453	20.6512	25371.8050	115321.8408	0.2200	0.0000	4.5434	51
52	0.0000	4.5453	20.6529	30953.6021	140693.6458	0.2200	0.0000	4.5438	52
53	0.0000	4.5453	20.6542	37763.3945	171647.2478	0.2200	0.0000	4.5441	53
54	0.0000	4.5454	20.6554	46071.3413	209410.6423	0.2200	0.0000	4.5443	54
55	0.0000	4.5454	20.6563	56207.0364	255481.9837	0.2200	0.0000	4.5445	55
60	0.0000	4.5454	20.6592	151911.2161	690500.9824	0.2200	0.0000	4.5451	60
65	0.0000	4.5454	20.6604	410571.6839	1866230.3813	0.2200	0.0000	4.5453	65
70	0.0000	4.5455	20.6609	1109655.4416	5043883.8256	0.2200	0.0000	4.5454	70
75	0.0000	4.5455	20.6610	2999074.8205	************	0.2200	0.0000	4.5454	75
80	0.0000	4.5455	20.6611	8105623.9993	************	0.2200	0.0000	4.5454	80
85	0.0000	4.5455	20.6611	************	************	0.2200	0.0000	4.5455	85
90	0.0000	4.5455	20.6611	************	************	0.2200	0.0000	4.5455	90
95	0.0000	4.5455	20.6612	************	************	0.2200	0.0000	4.5455	95
100	0.0000	4.5455	20.6612	************	************	0.2200	0.0000	4.5455	100

$I = 22.25\%$

n	(P/F)	(P/A)	(P/G)	(F/P)	(F/A)	(A/P)	(A/F)	(A/G)	n
1	0.8180	0.8180	0.0000	1.2225	1.0000	1.2225	1.0000	0.0000	1
2	0.6691	1.4871	0.6691	1.4945	2.2225	0.6724	0.4499	0.4499	2
3	0.5473	2.0344	1.7638	1.8270	3.7170	0.4915	0.2690	0.8670	3
4	0.4477	2.4822	3.1069	2.2335	5.5440	0.4029	0.1804	1.2517	4
5	0.3662	2.8484	4.5719	2.7305	7.7776	0.3511	0.1286	1.6051	5
6	0.2996	3.1480	6.0697	3.3381	10.5081	0.3177	0.0952	1.9281	6
7	0.2451	3.3930	7.5401	4.0808	13.8462	0.2947	0.0722	2.2222	7
8	0.2005	3.5935	8.9432	4.9887	17.9269	0.2783	0.0558	2.4887	8
9	0.1640	3.7574	10.2550	6.0987	22.9157	0.2661	0.0436	2.7292	9
10	0.1341	3.8916	11.4621	7.4557	29.0144	0.2570	0.0345	2.9454	10
11	0.1097	4.0013	12.5592	9.1146	36.4701	0.2499	0.0274	3.1388	11
12	0.0897	4.0910	13.5464	11.1426	45.5847	0.2444	0.0219	3.3113	12
13	0.0734	4.1644	14.4274	13.6218	56.7273	0.2401	0.0176	3.4644	13
14	0.0601	4.2245	15.2080	16.6527	70.3491	0.2367	0.0142	3.6000	14
15	0.0491	4.2736	15.8957	20.3579	87.0018	0.2340	0.0115	3.7195	15
16	0.0402	4.3138	16.4984	24.8875	107.3597	0.2318	0.0093	3.8246	16
17	0.0329	4.3467	17.0243	30.4250	132.2472	0.2301	0.0076	3.9166	17
18	0.0269	4.3735	17.4814	37.1946	162.6723	0.2286	0.0061	3.9971	18
19	0.0220	4.3955	17.8772	45.4704	199.8668	0.2275	0.0050	4.0671	19
20	0.0180	4.4135	18.2190	55.5875	245.3372	0.2266	0.0041	4.1280	20
21	0.0147	4.4282	18.5134	67.9558	300.9247	0.2258	0.0033	4.1807	21
22	0.0120	4.4403	18.7661	83.0759	368.8805	0.2252	0.0027	4.2263	22
23	0.0098	4.4501	18.9828	101.5603	451.9564	0.2247	0.0022	4.2657	23
24	0.0081	4.4582	19.1680	124.1575	553.5167	0.2243	0.0018	4.2995	24
25	0.0066	4.4648	19.3261	151.7825	677.6741	0.2240	0.0015	4.3286	25
26	0.0054	4.4702	19.4609	185.5541	829.4566	0.2237	0.0012	4.3535	26
27	0.0044	4.4746	19.5755	226.8399	1015.0107	0.2235	0.0010	4.3748	27
28	0.0036	4.4782	19.6728	277.3118	1241.8506	0.2233	0.0008	4.3930	28
29	0.0029	4.4811	19.7554	339.0136	1519.1624	0.2232	0.0007	4.4086	29
30	0.0024	4.4835	19.8254	414.4442	1858.1760	0.2230	0.0005	4.4218	30
31	0.0020	4.4855	19.8846	506.6580	2272.6202	0.2229	0.0004	4.4331	31
32	0.0016	4.4871	19.9347	619.3894	2779.2782	0.2229	0.0004	4.4426	32
33	0.0013	4.4884	19.9769	757.2035	3398.6676	0.2228	0.0003	4.4507	33
34	0.0011	4.4895	20.0126	925.6813	4155.8712	0.2227	0.0002	4.4576	34
35	0.0009	4.4904	20.0426	1131.6454	5081.5525	0.2227	0.0002	4.4634	35
36	0.0007	4.4911	20.0679	1383.4365	6213.1979	0.2227	0.0002	4.4683	36
37	0.0006	4.4917	20.0892	1691.2512	7596.6344	0.2226	0.0001	4.4725	37
38	0.0005	4.4922	20.1071	2067.5545	9287.8856	0.2226	0.0001	4.4760	38
39	0.0004	4.4926	20.1221	2527.5854	11355.4402	0.2226	0.0001	4.4789	39
40	0.0003	4.4929	20.1348	3089.9732	13883.0256	0.2226	0.0001	4.4814	40
41	0.0003	4.4932	20.1453	3777.4922	16972.9988	0.2226	0.0001	4.4835	41
42	0.0002	4.4934	20.1542	4617.9843	20750.4910	0.2225	0.0000	4.4853	42
43	0.0002	4.4936	20.1617	5645.4857	25368.4753	0.2225	0.0000	4.4868	43
44	0.0001	4.4937	20.1679	6901.6063	31013.9610	0.2225	0.0000	4.4880	44
45	0.0001	4.4938	20.1731	8437.2137	37915.5673	0.2225	0.0000	4.4890	45
46	0.0001	4.4939	20.1775	10314.4938	46352.7811	0.2225	0.0000	4.4899	46
47	0.0001	4.4940	20.1811	12609.4687	56667.2749	0.2225	0.0000	4.4907	47
48	0.0001	4.4941	20.1842	15415.0754	69276.7435	0.2225	0.0000	4.4913	48
49	0.0001	4.4941	20.1867	18844.9297	84691.8189	0.2225	0.0000	4.4918	49
50	0.0000	4.4942	20.1888	23037.9266	103536.7487	0.2225	0.0000	4.4922	50
51	0.0000	4.4942	20.1906	28163.8652	126574.6752	0.2225	0.0000	4.4926	51
52	0.0000	4.4943	20.1921	34430.3253	154738.5405	0.2225	0.0000	4.4929	52
53	0.0000	4.4943	20.1933	42091.0726	189168.8657	0.2225	0.0000	4.4931	53
54	0.0000	4.4943	20.1944	51456.3363	231259.9383	0.2225	0.0000	4.4933	54
55	0.0000	4.4943	20.1952	62905.3711	282716.2746	0.2225	0.0000	4.4935	55
60	0.0000	4.4944	20.1978	171763.9693	771968.4015	0.2225	0.0000	4.4940	60
65	0.0000	4.4944	20.1988	469003.8488	2107877.9723	0.2225	0.0000	4.4942	65
70	0.0000	4.4944	20.1992	1280621.3729	5755597.1816	0.2225	0.0000	4.4943	70
75	0.0000	4.4944	20.1994	3496754.0348	************	0.2225	0.0000	4.4944	75
80	0.0000	4.4944	20.1994	9547934.3378	************	0.2225	0.0000	4.4944	80
85	0.0000	4.4944	20.1995	************	************	0.2225	0.0000	4.4944	85
90	0.0000	4.4944	20.1995	************	************	0.2225	0.0000	4.4944	90
95	0.0000	4.4944	20.1995	************	************	0.2225	0.0000	4.4944	95
100	0.0000	4.4944	20.1995	************	************	0.2225	0.0000	4.4944	100

ENGINEERING ECONOMIC ANALYSIS

$I = 22.50\%$

n	(P/F)	(P/A)	(P/G)	(F/P)	(F/A)	(A/P)	(A/F)	(A/G)	n
1	0.8163	0.8163	0.0000	1.2250	1.0000	1.2250	1.0000	0.0000	1
2	0.6664	1.4827	0.6664	1.5006	2.2250	0.6744	0.4494	0.4494	2
3	0.5440	2.0267	1.7544	1.8383	3.7256	0.4934	0.2684	0.8656	3
4	0.4441	2.4708	3.0866	2.2519	5.5639	0.4047	0.1797	1.2492	4
5	0.3625	2.8333	4.5366	2.7585	7.8158	0.3529	0.1279	1.6012	5
6	0.2959	3.1292	6.0163	3.3792	10.5743	0.3196	0.0946	1.9226	6
7	0.2416	3.3708	7.4657	4.1395	13.9535	0.2967	0.0717	2.2148	7
8	0.1972	3.5680	8.8461	5.0709	18.0931	0.2803	0.0553	2.4793	8
9	0.1610	3.7290	10.1340	6.2119	23.1640	0.2682	0.0432	2.7176	9
10	0.1314	3.8604	11.3167	7.6096	29.3759	0.2590	0.0340	2.9315	10
11	0.1073	3.9677	12.3894	9.3217	36.9855	0.2520	0.0270	3.1226	11
12	0.0876	4.0552	13.3527	11.4191	46.3072	0.2466	0.0216	3.2927	12
13	0.0715	4.1267	14.2106	13.9884	57.7264	0.2423	0.0173	3.4436	13
14	0.0584	4.1851	14.9692	17.1358	71.7148	0.2389	0.0139	3.5768	14
15	0.0476	4.2327	15.6362	20.9914	88.8507	0.2363	0.0113	3.6941	15
16	0.0389	4.2716	16.2195	25.7145	109.8420	0.2341	0.0091	3.7971	16
17	0.0317	4.3034	16.7274	31.5002	135.5565	0.2324	0.0074	3.8871	17
18	0.0259	4.3293	17.1680	38.5878	167.0567	0.2310	0.0060	3.9656	18
19	0.0212	4.3504	17.5488	47.2700	205.6445	0.2299	0.0049	4.0338	19
20	0.0173	4.3677	17.8769	57.9058	252.9145	0.2290	0.0040	4.0930	20
21	0.0141	4.3818	18.1589	70.9346	310.8203	0.2282	0.0032	4.1442	21
22	0.0115	4.3933	18.4005	86.8948	381.7548	0.2276	0.0026	4.1883	22
23	0.0094	4.4027	18.6072	106.4462	468.6496	0.2271	0.0021	4.2263	23
24	0.0077	4.4104	18.7836	130.3966	575.0958	0.2267	0.0017	4.2590	24
25	0.0063	4.4166	18.9338	159.7358	705.4924	0.2264	0.0014	4.2870	25
26	0.0051	4.4217	19.0616	195.6763	865.2282	0.2262	0.0012	4.3109	26
27	0.0042	4.4259	19.1701	239.7035	1060.9045	0.2259	0.0009	4.3313	27
28	0.0034	4.4293	19.2620	293.6368	1300.6080	0.2258	0.0008	4.3488	28
29	0.0028	4.4321	19.3399	359.7051	1594.2448	0.2256	0.0006	4.3636	29
30	0.0023	4.4344	19.4057	440.6387	1953.9499	0.2255	0.0005	4.3762	30
31	0.0019	4.4362	19.4612	539.7824	2394.5886	0.2254	0.0004	4.3869	31
32	0.0015	4.4377	19.5081	661.2335	2934.3710	0.2253	0.0003	4.3960	32
33	0.0012	4.4390	19.5476	810.0110	3595.6045	0.2253	0.0003	4.4037	33
34	0.0010	4.4400	19.5809	992.2635	4405.6156	0.2252	0.0002	4.4101	34
35	0.0008	4.4408	19.6089	1215.5228	5397.8790	0.2252	0.0002	4.4156	35
36	0.0007	4.4415	19.6324	1489.0154	6613.4018	0.2252	0.0002	4.4203	36
37	0.0005	4.4420	19.6521	1824.0439	8102.4172	0.2251	0.0001	4.4241	37
38	0.0004	4.4425	19.6687	2234.4538	9926.4611	0.2251	0.0001	4.4274	38
39	0.0004	4.4428	19.6825	2737.2058	12160.9149	0.2251	0.0001	4.4302	39
40	0.0003	4.4431	19.6942	3353.0772	14898.1207	0.2251	0.0001	4.4325	40
41	0.0002	4.4434	19.7039	4107.5195	18251.1979	0.2251	0.0001	4.4345	41
42	0.0002	4.4436	19.7121	5031.7114	22358.7174	0.2250	0.0000	4.4361	42
43	0.0002	4.4437	19.7189	6163.8465	27390.4288	0.2250	0.0000	4.4375	43
44	0.0001	4.4439	19.7246	7550.7120	33554.2753	0.2250	0.0000	4.4386	44
45	0.0001	4.4440	19.7293	9249.6221	41104.9873	0.2250	0.0000	4.4396	45
46	0.0001	4.4441	19.7333	11330.7871	50354.6094	0.2250	0.0000	4.4404	46
47	0.0001	4.4441	19.7366	13880.2142	61685.3965	0.2250	0.0000	4.4411	47
48	0.0001	4.4442	19.7394	17003.2624	75565.6108	0.2250	0.0000	4.4416	48
49	0.0000	4.4442	19.7417	20828.9965	92568.8732	0.2250	0.0000	4.4421	49
50	0.0000	4.4443	19.7436	25515.5207	113397.8697	0.2250	0.0000	4.4425	50
51	0.0000	4.4443	19.7452	31256.5128	138913.3903	0.2250	0.0000	4.4428	51
52	0.0000	4.4443	19.7465	38289.2282	170169.9032	0.2250	0.0000	4.4431	52
53	0.0000	4.4443	19.7476	46904.3046	208459.1314	0.2250	0.0000	4.4433	53
54	0.0000	4.4444	19.7486	57457.7731	255363.4359	0.2250	0.0000	4.4435	54
55	0.0000	4.4444	19.7493	70385.7720	312821.2090	0.2250	0.0000	4.4437	55
60	0.0000	4.4444	19.7516	194162.4851	862939.9340	0.2250	0.0000	4.4441	60
65	0.0000	4.4444	19.7525	535606.4096	2380468.4869	0.2250	0.0000	4.4443	65
70	0.0000	4.4444	19.7529	1477495.6436	6566642.8605	0.2250	0.0000	4.4444	70
75	0.0000	4.4444	19.7530	4075741.6975	************	0.2250	0.0000	4.4444	75
80	0.0000	4.4444	19.7531	************	************	0.2250	0.0000	4.4444	80
85	0.0000	4.4444	19.7531	************	************	0.2250	0.0000	4.4444	85
90	0.0000	4.4444	19.7531	************	************	0.2250	0.0000	4.4444	90
95	0.0000	4.4444	19.7531	************	************	0.2250	0.0000	4.4444	95
100	0.0000	4.4444	19.7531	************	************	0.2250	0.0000	4.4444	100

$I = 22.75\%$

n	(P/F)	(P/A)	(P/G)	(F/P)	(F/A)	(A/P)	(A/F)	(A/G)	n
1	0.8147	0.8147	0.0000	1.2275	1.0000	1.2275	1.0000	0.0000	1
2	0.6637	1.4783	0.6637	1.5068	2.2275	0.6764	0.4489	0.4489	2
3	0.5407	2.0190	1.7450	1.8495	3.7343	0.4953	0.2678	0.8643	3
4	0.4405	2.4595	3.0664	2.2703	5.5838	0.4066	0.1791	1.2468	4
5	0.3588	2.8183	4.5018	2.7868	7.8541	0.3548	0.1273	1.5973	5
6	0.2923	3.1106	5.9634	3.4208	10.6409	0.3215	0.0940	1.9171	6
7	0.2381	3.3488	7.3923	4.1990	14.0617	0.2986	0.0711	2.2075	7
8	0.1940	3.5428	8.7504	5.1543	18.2608	0.2823	0.0548	2.4699	8
9	0.1581	3.7009	10.0148	6.3269	23.4151	0.2702	0.0427	2.7061	9
10	0.1288	3.8296	11.1737	7.7663	29.7420	0.2611	0.0336	2.9177	10
11	0.1049	3.9345	12.2226	9.5332	37.5084	0.2542	0.0267	3.1065	11
12	0.0855	4.0200	13.1627	11.7019	47.0415	0.2488	0.0213	3.2743	12
13	0.0696	4.0896	13.9981	14.3641	58.7435	0.2445	0.0170	3.4229	13
14	0.0567	4.1463	14.7354	17.6320	73.1076	0.2412	0.0137	3.5539	14
15	0.0462	4.1925	15.3822	21.6433	90.7396	0.2385	0.0110	3.6690	15
16	0.0376	4.2302	15.9468	26.5671	112.3828	0.2364	0.0089	3.7698	16
17	0.0307	4.2608	16.4375	32.6111	138.9499	0.2347	0.0072	3.8578	17
18	0.0250	4.2858	16.8621	40.0301	171.5610	0.2333	0.0058	3.9344	18
19	0.0204	4.3061	17.2285	49.1370	211.5911	0.2322	0.0047	4.0009	19
20	0.0166	4.3227	17.5435	60.3157	260.7281	0.2313	0.0038	4.0584	20
21	0.0135	4.3362	17.8136	74.0375	321.0438	0.2306	0.0031	4.1081	21
22	0.0110	4.3472	18.0447	90.8810	395.0812	0.2300	0.0025	4.1508	22
23	0.0090	4.3562	18.2419	111.5564	485.9622	0.2296	0.0021	4.1876	23
24	0.0073	4.3635	18.4098	136.9355	597.5186	0.2292	0.0017	4.2191	24
25	0.0059	4.3695	18.5526	168.0883	734.4541	0.2289	0.0014	4.2460	25
26	0.0048	4.3743	18.6738	206.3284	902.5424	0.2286	0.0011	4.2690	26
27	0.0039	4.3782	18.7765	253.2681	1108.8708	0.2284	0.0009	4.2886	27
28	0.0032	4.3815	18.8633	310.8866	1362.1390	0.2282	0.0007	4.3052	28
29	0.0026	4.3841	18.9367	381.6133	1673.0256	0.2281	0.0006	4.3194	29
30	0.0021	4.3862	18.9986	468.4303	2054.6389	0.2280	0.0005	4.3314	30
31	0.0017	4.3880	19.0508	574.9983	2523.0692	0.2279	0.0004	4.3416	31
32	0.0014	4.3894	19.0947	705.8104	3098.0675	0.2278	0.0003	4.3502	32
33	0.0012	4.3905	19.1316	866.3822	3803.8778	0.2278	0.0003	4.3575	33
34	0.0009	4.3915	19.1626	1063.4842	4670.2601	0.2277	0.0002	4.3636	34
35	0.0008	4.3922	19.1887	1305.4268	5733.7442	0.2277	0.0002	4.3688	35
36	0.0006	4.3929	19.2105	1602.4114	7039.1710	0.2276	0.0001	4.3731	36
37	0.0005	4.3934	19.2288	1966.9600	8641.5824	0.2276	0.0001	4.3768	37
38	0.0004	4.3938	19.2442	2414.4434	10608.5424	0.2276	0.0001	4.3799	38
39	0.0003	4.3941	19.2570	2963.7293	13022.9859	0.2276	0.0001	4.3824	39
40	0.0003	4.3944	19.2677	3637.9777	15986.7151	0.2276	0.0001	4.3846	40
41	0.0002	4.3946	19.2767	4465.6176	19624.6928	0.2276	0.0001	4.3864	41
42	0.0002	4.3948	19.2841	5481.5456	24090.3104	0.2275	0.0000	4.3879	42
43	0.0001	4.3950	19.2904	6728.5973	29571.8561	0.2275	0.0000	4.3892	43
44	0.0001	4.3951	19.2956	8259.3531	36300.4533	0.2275	0.0000	4.3903	44
45	0.0001	4.3952	19.2999	10138.3560	44559.8065	0.2275	0.0000	4.3912	45
46	0.0001	4.3953	19.3035	12444.8320	54698.1624	0.2275	0.0000	4.3919	46
47	0.0001	4.3953	19.3065	15276.0312	67142.9944	0.2275	0.0000	4.3925	47
48	0.0001	4.3954	19.3091	18751.3283	82419.0256	0.2275	0.0000	4.3930	48
49	0.0000	4.3954	19.3111	23017.2555	101170.3539	0.2275	0.0000	4.3935	49
50	0.0000	4.3954	19.3129	28253.6812	124187.6095	0.2275	0.0000	4.3938	50
51	0.0000	4.3955	19.3143	34681.3936	152441.2906	0.2275	0.0000	4.3941	51
52	0.0000	4.3955	19.3155	42571.4107	187122.6842	0.2275	0.0000	4.3944	52
53	0.0000	4.3955	19.3165	52256.4066	229694.0949	0.2275	0.0000	4.3946	53
54	0.0000	4.3955	19.3173	64144.7391	281950.5015	0.2275	0.0000	4.3948	54
55	0.0000	4.3955	19.3180	78737.6672	346095.2406	0.2275	0.0000	4.3949	55
60	0.0000	4.3956	19.3200	219426.9910	964509.8506	0.2275	0.0000	4.3953	60
65	0.0000	4.3956	19.3208	611501.5352	2687914.4404	0.2275	0.0000	4.3955	65
70	0.0000	4.3956	19.3211	1704139.1572	7490717.1745	0.2275	0.0000	4.3956	70
75	0.0000	4.3956	19.3213	4749113.6162	************	0.2275	0.0000	4.3956	75
80	0.0000	4.3956	19.3213	************	************	0.2275	0.0000	4.3956	80
85	0.0000	4.3956	19.3213	************	************	0.2275	0.0000	4.3956	85
90	0.0000	4.3956	19.3213	************	************	0.2275	0.0000	4.3956	90
95	0.0000	4.3956	19.3213	************	************	0.2275	0.0000	4.3956	95
100	0.0000	4.3956	19.3213	************	************	0.2275	0.0000	4.3956	100

$I = 23.00\%$

n	(P/F)	(P/A)	(P/G)	(F/P)	(F/A)	(A/P)	(A/F)	(A/G)	n
1	0.8130	0.8130	0.0000	1.2300	1.0000	1.2300	1.0000	0.0000	1
2	0.6610	1.4740	0.6610	1.5129	2.2300	0.6784	0.4484	0.4484	2
3	0.5374	2.0114	1.7358	1.8609	3.7429	0.4972	0.2672	0.8630	3
4	0.4369	2.4483	3.0464	2.2889	5.6038	0.4085	0.1785	1.2443	4
5	0.3552	2.8035	4.4672	2.8153	7.8926	0.3567	0.1267	1.5935	5
6	0.2888	3.0923	5.9112	3.4628	10.7079	0.3234	0.0934	1.9116	6
7	0.2348	3.3270	7.3198	4.2593	14.1708	0.3006	0.0706	2.2001	7
8	0.1909	3.5179	8.6560	5.2389	18.4300	0.2843	0.0543	2.4605	8
9	0.1552	3.6731	9.8975	6.4439	23.6690	0.2722	0.0422	2.6946	9
10	0.1262	3.7993	11.0330	7.9259	30.1128	0.2632	0.0332	2.9040	1
11	0.1026	3.9018	12.0588	9.7489	38.0388	0.2563	0.0263	3.0905	11
12	0.0834	3.9852	12.9761	11.9912	47.7877	0.2509	0.0209	3.2560	12
13	0.0678	4.0530	13.7897	14.7491	59.7788	0.2467	0.0167	3.4023	13
14	0.0551	4.1082	14.5063	18.1414	74.5280	0.2434	0.0134	3.5311	14
15	0.0448	4.1530	15.1337	22.3140	92.6694	0.2408	0.0108	3.6441	15
16	0.0364	4.1894	15.6802	27.4462	114.9834	0.2387	0.0087	3.7428	16
17	0.0296	4.2190	16.1542	33.7588	142.4295	0.2370	0.0070	3.8289	17
18	0.0241	4.2431	16.5636	41.5233	176.1883	0.2357	0.0057	3.9036	18
19	0.0196	4.2627	16.9160	51.0737	217.7116	0.2346	0.0046	3.9684	19
20	0.0159	4.2786	17.2185	62.8206	268.7853	0.2337	0.0037	4.0243	20
21	0.0129	4.2916	17.4773	77.2694	331.6059	0.2330	0.0030	4.0725	21
22	0.0105	4.3021	17.6983	95.0413	408.8753	0.2324	0.0024	4.1139	22
23	0.0086	4.3106	17.8865	116.9008	503.9166	0.2320	0.0020	4.1494	23
24	0.0070	4.3176	18.0464	143.7880	620.8174	0.2316	0.0016	4.1797	24
25	0.0057	4.3232	18.1821	176.8593	764.6054	0.2313	0.0013	4.2057	25
26	0.0046	4.3278	18.2970	217.5369	941.4647	0.2311	0.0011	4.2278	26
27	0.0037	4.3316	18.3942	267.5704	1159.0016	0.2309	0.0009	4.2465	27
28	0.0030	4.3346	18.4763	329.1115	1426.5719	0.2307	0.0007	4.2625	28
29	0.0025	4.3371	18.5454	404.8072	1755.6835	0.2306	0.0006	4.2760	29
30	0.0020	4.3391	18.6037	497.9129	2160.4907	0.2305	0.0005	4.2875	30
31	0.0016	4.3407	18.6526	612.4328	2658.4036	0.2304	0.0004	4.2971	31
32	0.0013	4.3421	18.6938	753.2924	3270.8364	0.2303	0.0003	4.3053	32
33	0.0011	4.3431	18.7283	926.5496	4024.1287	0.2302	0.0002	4.3122	33
34	0.0009	4.3440	18.7573	1139.6560	4950.6783	0.2302	0.0002	4.3180	34
35	0.0007	4.3447	18.7815	1401.7769	6090.3344	0.2302	0.0002	4.3228	35
36	0.0006	4.3453	18.8018	1724.1856	7492.1113	0.2301	0.0001	4.3269	36
37	0.0005	4.3458	18.8188	2120.7483	9216.2969	0.2301	0.0001	4.3304	37
38	0.0004	4.3462	18.8330	2608.5204	11337.0451	0.2301	0.0001	4.3333	38
39	0.0003	4.3465	18.8449	3208.4801	13945.5655	0.2301	0.0001	4.3357	39
40	0.0003	4.3467	18.8547	3946.4305	17154.0456	0.2301	0.0001	4.3377	40
41	0.0002	4.3469	18.8630	4854.1095	21100.4761	0.2300	0.0000	4.3394	41
42	0.0002	4.3471	18.8698	5970.5547	25954.5856	0.2300	0.0000	4.3408	42
43	0.0001	4.3472	18.8756	7343.7823	31925.1403	0.2300	0.0000	4.3420	43
44	0.0001	4.3473	18.8803	9032.8522	39268.9225	0.2300	0.0000	4.3430	44
45	0.0001	4.3474	18.8843	11110.4082	48301.7747	0.2300	0.0000	4.3438	45
46	0.0001	4.3475	18.8876	13665.8021	59412.1829	0.2300	0.0000	4.3445	46
47	0.0001	4.3476	18.8903	16808.9365	73077.9850	0.2300	0.0000	4.3450	47
48	0.0000	4.3476	18.8926	20674.9919	89886.9215	0.2300	0.0000	4.3455	48
49	0.0000	4.3477	18.8945	25430.2401	110561.9135	0.2300	0.0000	4.3459	49
50	0.0000	4.3477	18.8960	31279.1953	135992.1536	0.2300	0.0000	4.3462	50
51	0.0000	4.3477	18.8973	38473.4102	167271.3489	0.2300	0.0000	4.3465	51
52	0.0000	4.3477	18.8984	47322.2946	205744.7591	0.2300	0.0000	4.3467	52
53	0.0000	4.3478	18.8993	58206.4224	253067.0537	0.2300	0.0000	4.3469	53
54	0.0000	4.3478	18.9000	71593.8995	311273.4761	0.2300	0.0000	4.3471	54
55	0.0000	4.3478	18.9007	88060.4964	382867.3756	0.2300	0.0000	4.3472	55
60	0.0000	4.3478	18.9025	247917.2160	1077896.5914	0.2300	0.0000	4.3476	60
65	0.0000	4.3478	18.9032	697962.7475	3034616.2935	0.2300	0.0000	4.3477	65
70	0.0000	4.3478	18.9034	1964978.4905	8543380.3934	0.2300	0.0000	4.3478	70
75	0.0000	4.3478	18.9035	5532015.1138	***********	0.2300	0.0000	4.3478	75
80	0.0000	4.3478	18.9036	***********	***********	0.2300	0.0000	4.3478	80
85	0.0000	4.3478	18.9036	***********	***********	0.2300	0.0000	4.3478	85
90	0.0000	4.3478	18.9036	***********	***********	0.2300	0.0000	4.3478	90
95	0.0000	4.3478	18.9036	***********	***********	0.2300	0.0000	4.3478	95
100	0.0000	4.3478	18.9036	***********	***********	0.2300	0.0000	4.3478	100

$I = 23.25\%$

n	(P/F)	(P/A)	(P/G)	(F/P)	(F/A)	(A/P)	(A/F)	(A/G)	n
1	0.8114	0.8114	0.0000	1.2325	1.0000	1.2325	1.0000	0.0000	1
2	0.6583	1.4697	0.6583	1.5191	2.2325	0.6804	0.4479	0.4479	2
3	0.5341	2.0038	1.7265	1.8722	3.7516	0.4991	0.2666	0.8616	3
4	0.4334	2.4371	3.0266	2.3075	5.6238	0.4103	0.1778	1.2419	4
5	0.3516	2.7888	4.4331	2.8440	7.9313	0.3586	0.1261	1.5896	5
6	0.2853	3.0740	5.8595	3.5053	10.7754	0.3253	0.0928	1.9061	6
7	0.2315	3.3055	7.2483	4.3202	14.2806	0.3025	0.0700	2.1928	7
8	0.1878	3.4933	8.5629	5.3247	18.6009	0.2863	0.0538	2.4512	8
9	0.1524	3.6457	9.7820	6.5627	23.9256	0.2743	0.0418	2.6832	9
10	0.1236	3.7693	10.8946	8.0885	30.4883	0.2653	0.0328	2.8903	1
11	0.1003	3.8696	11.8977	9.9691	38.5768	0.2584	0.0259	3.0746	11
12	0.0814	3.9510	12.7930	12.2869	48.5459	0.2531	0.0206	3.2379	12
13	0.0660	4.0171	13.5854	15.1436	60.8328	0.2489	0.0164	3.3819	13
14	0.0536	4.0706	14.2819	18.6645	75.9765	0.2457	0.0132	3.5085	14
15	0.0435	4.1141	14.8905	23.0040	94.6410	0.2431	0.0106	3.6194	15
16	0.0353	4.1494	15.4196	28.3525	117.6450	0.2410	0.0085	3.7161	16
17	0.0286	4.1780	15.8774	34.9444	145.9975	0.2393	0.0068	3.8003	17
18	0.0232	4.2012	16.2722	43.0690	180.9419	0.2380	0.0055	3.8732	18
19	0.0188	4.2200	16.6113	53.0825	224.0109	0.2370	0.0045	3.9363	19
20	0.0153	4.2353	16.9017	65.4242	277.0934	0.2361	0.0036	3.9906	20
21	0.0124	4.2477	17.1497	80.6354	342.5176	0.2354	0.0029	4.0374	21
22	0.0101	4.2578	17.3610	99.3831	423.1530	0.2349	0.0024	4.0775	22
23	0.0082	4.2660	17.5406	122.4896	522.5361	0.2344	0.0019	4.1118	23
24	0.0066	4.2726	17.6930	150.9685	645.0257	0.2341	0.0016	4.1410	24
25	0.0054	4.2780	17.8219	186.0686	795.9942	0.2338	0.0013	4.1660	25
26	0.0044	4.2823	17.9310	229.3296	982.0628	0.2335	0.0010	4.1872	26
27	0.0035	4.2859	18.0229	282.6487	1211.3924	0.2333	0.0008	4.2052	27
28	0.0029	4.2887	18.1004	348.3646	1494.0412	0.2332	0.0007	4.2205	28
29	0.0023	4.2911	18.1657	429.3593	1842.4057	0.2330	0.0005	4.2334	29
30	0.0019	4.2929	18.2205	529.1854	2271.7651	0.2329	0.0004	4.2443	30
31	0.0015	4.2945	18.2665	652.2210	2800.9505	0.2329	0.0004	4.2535	31
32	0.0012	4.2957	18.3050	803.8624	3453.1714	0.2328	0.0003	4.2612	32
33	0.0010	4.2967	18.3373	990.7604	4257.0338	0.2327	0.0002	4.2677	33
34	0.0008	4.2976	18.3643	1221.1121	5247.7942	0.2327	0.0002	4.2732	34
35	0.0007	4.2982	18.3869	1505.0207	6468.9063	0.2327	0.0002	4.2778	35
36	0.0005	4.2988	18.4058	1854.9380	7973.9270	0.2326	0.0001	4.2817	36
37	0.0004	4.2992	18.4215	2286.2111	9828.8651	0.2326	0.0001	4.2849	37
38	0.0004	4.2995	18.4347	2817.7552	12115.0762	0.2326	0.0001	4.2876	38
39	0.0003	4.2998	18.4456	3472.8833	14932.8314	0.2326	0.0001	4.2898	39
40	0.0002	4.3001	18.4547	4280.3287	18405.7147	0.2326	0.0001	4.2917	40
41	0.0002	4.3003	18.4623	5275.5051	22686.0434	0.2325	0.0000	4.2933	41
42	0.0002	4.3004	18.4686	6502.0600	27961.5485	0.2325	0.0000	4.2946	42
43	0.0001	4.3005	18.4739	8013.7890	34463.6085	0.2325	0.0000	4.2957	43
44	0.0001	4.3006	18.4782	9876.9949	42477.3974	0.2325	0.0000	4.2966	44
45	0.0001	4.3007	18.4818	12173.3962	52354.3923	0.2325	0.0000	4.2974	45
46	0.0001	4.3008	18.4848	15003.7108	64527.7886	0.2325	0.0000	4.2980	46
47	0.0001	4.3008	18.4873	18492.0736	79531.4994	0.2325	0.0000	4.2985	47
48	0.0000	4.3009	18.4894	22791.4807	98023.5730	0.2325	0.0000	4.2990	48
49	0.0000	4.3009	18.4911	28090.5000	120815.0538	0.2325	0.0000	4.2993	49
50	0.0000	4.3010	18.4925	34621.5412	148905.5538	0.2325	0.0000	4.2996	50
51	0.0000	4.3010	18.4937	42671.0496	183527.0950	0.2325	0.0000	4.2999	51
52	0.0000	4.3010	18.4946	52592.0686	226198.1446	0.2325	0.0000	4.3001	52
53	0.0000	4.3010	18.4954	64819.7246	278790.2132	0.2325	0.0000	4.3003	53
54	0.0000	4.3010	18.4961	79890.3105	343609.9378	0.2325	0.0000	4.3004	54
55	0.0000	4.3010	18.4967	98464.8077	423500.2483	0.2325	0.0000	4.3005	55
60	0.0000	4.3011	18.4983	280037.1680	1204456.6364	0.2325	0.0000	4.3009	60
65	0.0000	4.3011	18.4989	796434.9624	3425522.4189	0.2325	0.0000	4.3010	65
70	0.0000	4.3011	18.4991	2265087.3594	9742306.9222	0.2325	0.0000	4.3010	70
75	0.0000	4.3011	18.4992	6441983.3232	************	0.2325	0.0000	4.3011	75
80	0.0000	4.3011	18.4992	************	************	0.2325	0.0000	4.3011	80
85	0.0000	4.3011	18.4992	************	************	0.2325	0.0000	4.3011	85
90	0.0000	4.3011	18.4992	************	************	0.2325	0.0000	4.3011	90
95	0.0000	4.3011	18.4992	************	************	0.2325	0.0000	4.3011	95
100	0.0000	4.3011	18.4992	************	************	0.2325	0.0000	4.3011	100

$I = 23.50\%$

n	(P/F)	(P/A)	(P/G)	(F/P)	(F/A)	(A/P)	(A/F)	(A/G)	n
1	0.8097	0.8097	0.0000	1.2350	1.0000	1.2350	1.0000	0.0000	1
2	0.6556	1.4654	0.6556	1.5252	2.2350	0.6824	0.4474	0.4474	2
3	0.5309	1.9962	1.7174	1.8837	3.7602	0.5009	0.2659	0.8603	3
4	0.4299	2.4261	3.0070	2.3263	5.6439	0.4122	0.1772	1.2394	4
5	0.3481	2.7742	4.3993	2.8730	7.9702	0.3605	0.1255	1.5858	5
6	0.2818	3.0560	5.8085	3.5481	10.8432	0.3272	0.0922	1.9007	6
7	0.2282	3.2842	7.1777	4.3820	14.3913	0.3045	0.0695	2.1855	7
8	0.1848	3.4690	8.4712	5.4117	18.7733	0.2883	0.0533	2.4420	8
9	0.1496	3.6186	9.6682	6.6835	24.1850	0.2763	0.0413	2.6718	9
10	0.1212	3.7398	10.7586	8.2541	30.8685	0.2674	0.0324	2.8768	1
11	0.0981	3.8379	11.7395	10.1938	39.1226	0.2606	0.0256	3.0589	11
12	0.0794	3.9173	12.6133	12.5894	49.3164	0.2553	0.0203	3.2199	12
13	0.0643	3.9816	13.3851	15.5479	61.9058	0.2512	0.0162	3.3617	13
14	0.0521	4.0337	14.0621	19.2016	77.4536	0.2479	0.0129	3.4862	14
15	0.0422	4.0759	14.6525	23.7140	96.6552	0.2453	0.0103	3.5949	15
16	0.0341	4.1100	15.1647	29.2868	120.3692	0.2433	0.0083	3.6897	16
17	0.0276	4.1377	15.6070	36.1691	149.6559	0.2417	0.0067	3.7719	17
18	0.0224	4.1601	15.9876	44.6689	185.8251	0.2404	0.0054	3.8431	18
19	0.0181	4.1782	16.3139	55.1661	230.4940	0.2393	0.0043	3.9045	19
20	0.0147	4.1929	16.5928	68.1301	285.6601	0.2385	0.0035	3.9574	20
21	0.0119	4.2047	16.8305	84.1407	353.7902	0.2378	0.0028	4.0027	21
22	0.0096	4.2144	17.0326	103.9138	437.9309	0.2373	0.0023	4.0415	22
23	0.0078	4.2222	17.2040	128.3335	541.8447	0.2368	0.0018	4.0747	23
24	0.0063	4.2285	17.3491	158.4919	670.1782	0.2365	0.0015	4.1029	24
25	0.0051	4.2336	17.4717	195.7375	828.6700	0.2362	0.0012	4.1269	25
26	0.0041	4.2377	17.5752	241.7358	1024.4075	0.2360	0.0010	4.1473	26
27	0.0033	4.2411	17.6622	298.5437	1266.1432	0.2358	0.0008	4.1646	27
28	0.0027	4.2438	17.7355	368.7014	1564.6869	0.2356	0.0006	4.1792	28
29	0.0022	4.2460	17.7970	455.3463	1933.3883	0.2355	0.0005	4.1915	29
30	0.0018	4.2478	17.8485	562.3526	2388.7346	0.2354	0.0004	4.2019	30
31	0.0014	4.2492	17.8917	694.5055	2951.0872	0.2353	0.0003	4.2106	31
32	0.0012	4.2504	17.9279	857.7143	3645.5927	0.2353	0.0003	4.2180	32
33	0.0009	4.2513	17.9581	1059.2771	4503.3070	0.2352	0.0002	4.2241	33
34	0.0008	4.2521	17.9833	1308.2073	5562.5841	0.2352	0.0002	4.2293	34
35	0.0006	4.2527	18.0043	1615.6360	6870.7914	0.2351	0.0001	4.2336	35
36	0.0005	4.2532	18.0219	1995.3104	8486.4274	0.2351	0.0001	4.2373	36
37	0.0004	4.2536	18.0365	2464.2084	10481.7378	0.2351	0.0001	4.2403	37
38	0.0003	4.2539	18.0487	3043.2974	12945.9462	0.2351	0.0001	4.2428	38
39	0.0003	4.2542	18.0588	3758.4722	15989.2435	0.2351	0.0001	4.2449	39
40	0.0002	4.2544	18.0672	4641.7132	19747.7158	0.2351	0.0001	4.2467	40
41	0.0002	4.2546	18.0741	5732.5158	24389.4290	0.2350	0.0000	4.2482	41
42	0.0001	4.2547	18.0799	7079.6570	30121.9448	0.2350	0.0000	4.2494	42
43	0.0001	4.2548	18.0847	8743.3764	37201.6018	0.2350	0.0000	4.2504	43
44	0.0001	4.2549	18.0887	10798.0699	45944.9783	0.2350	0.0000	4.2512	44
45	0.0001	4.2550	18.0920	13335.6163	56743.0482	0.2350	0.0000	4.2519	45
46	0.0001	4.2551	18.0948	16469.4862	70078.6645	0.2350	0.0000	4.2525	46
47	0.0000	4.2551	18.0970	20339.8154	86548.1506	0.2350	0.0000	4.2530	47
48	0.0000	4.2551	18.0989	25119.6720	106887.9660	0.2350	0.0000	4.2534	48
49	0.0000	4.2552	18.1004	31022.7949	132007.6380	0.2350	0.0000	4.2537	49
50	0.0000	4.2552	18.1017	38313.1518	163030.4330	0.2350	0.0000	4.2540	50
51	0.0000	4.2552	18.1028	47316.7424	201343.5847	0.2350	0.0000	4.2542	51
52	0.0000	4.2552	18.1036	58436.1769	248660.3272	0.2350	0.0000	4.2544	52
53	0.0000	4.2553	18.1044	72168.6784	307096.5040	0.2350	0.0000	4.2546	53
54	0.0000	4.2553	18.1050	89128.3179	379265.1825	0.2350	0.0000	4.2547	54
55	0.0000	4.2553	18.1055	110073.4726	468393.5004	0.2350	0.0000	4.2548	55
60	0.0000	4.2553	18.1069	316240.4765	1345699.9001	0.2350	0.0000	4.2551	60
65	0.0000	4.2553	18.1074	908557.1359	3866196.3229	0.2350	0.0000	4.2552	65
70	0.0000	4.2553	18.1076	2610279.6146	************	0.2350	0.0000	4.2553	70
75	0.0000	4.2553	18.1077	7499318.8621	************	0.2350	0.0000	4.2553	75
80	0.0000	4.2553	18.1077	************	************	0.2350	0.0000	4.2553	80
85	0.0000	4.2553	18.1077	************	************	0.2350	0.0000	4.2553	85
90	0.0000	4.2553	18.1077	************	************	0.2350	0.0000	4.2553	90
95	0.0000	4.2553	18.1077	************	************	0.2350	0.0000	4.2553	95
100	0.0000	4.2553	18.1077	************	************	0.2350	0.0000	4.2553	100

$I = 23.75\%$

n	(P/F)	(P/A)	(P/G)	(F/P)	(F/A)	(A/P)	(A/F)	(A/G)	n
1	0.8081	0.8081	0.0000	1.2375	1.0000	1.2375	1.0000	0.0000	1
2	0.6530	1.4611	0.6530	1.5314	2.2375	0.6844	0.4469	0.4469	2
3	0.5277	1.9887	1.7083	1.8951	3.7689	0.5028	0.2653	0.8590	3
4	0.4264	2.4151	2.9875	2.3452	5.6640	0.4141	0.1766	1.2370	4
5	0.3446	2.7597	4.3658	2.9022	8.0092	0.3624	0.1249	1.5820	5
6	0.2784	3.0382	5.7580	3.5915	10.9114	0.3291	0.0916	1.8952	6
7	0.2250	3.2632	7.1080	4.4444	14.5029	0.3065	0.0690	2.1783	7
8	0.1818	3.4450	8.3807	5.5000	18.9473	0.2903	0.0528	2.4327	8
9	0.1469	3.5919	9.5561	6.8062	24.4473	0.2784	0.0409	2.6605	9
10	0.1187	3.7106	10.6247	8.4227	31.2535	0.2695	0.0320	2.8633	1
11	0.0959	3.8066	11.5841	10.4231	39.6762	0.2627	0.0252	3.0432	11
12	0.0775	3.8841	12.4369	12.8986	50.0994	0.2575	0.0200	3.2020	12
13	0.0626	3.9467	13.1887	15.9620	62.9980	0.2534	0.0159	3.3417	13
14	0.0506	3.9974	13.8468	19.7530	78.9600	0.2502	0.0127	3.4640	14
15	0.0409	4.0383	14.4195	24.4443	98.7130	0.2476	0.0101	3.5707	15
16	0.0331	4.0713	14.9154	30.2499	123.1573	0.2456	0.0081	3.6635	16
17	0.0267	4.0980	15.3428	37.4342	153.4071	0.2440	0.0065	3.7439	17
18	0.0216	4.1196	15.7098	46.3248	190.8413	0.2427	0.0052	3.8134	18
19	0.0174	4.1371	16.0238	57.3270	237.1662	0.2417	0.0042	3.8732	19
20	0.0141	4.1512	16.2916	70.9421	294.4931	0.2409	0.0034	3.9246	20
21	0.0114	4.1626	16.5194	87.7909	365.4352	0.2402	0.0027	3.9686	21
22	0.0092	4.1718	16.7127	108.6412	453.2261	0.2397	0.0022	4.0061	22
23	0.0074	4.1792	16.8763	134.4435	561.8673	0.2393	0.0018	4.0382	23
24	0.0060	4.1852	17.0146	166.3738	696.3108	0.2389	0.0014	4.0654	24
25	0.0049	4.1901	17.1312	205.8876	862.6846	0.2387	0.0012	4.0885	25
26	0.0039	4.1940	17.2293	254.7859	1068.5722	0.2384	0.0009	4.1081	26
27	0.0032	4.1972	17.3117	315.2976	1323.3581	0.2383	0.0008	4.1246	27
28	0.0026	4.1997	17.3809	390.1807	1638.6557	0.2381	0.0006	4.1386	28
29	0.0021	4.2018	17.4389	482.8486	2028.8364	0.2380	0.0005	4.1503	29
30	0.0017	4.2035	17.4875	597.5252	2511.6851	0.2379	0.0004	4.1602	30
31	0.0014	4.2048	17.5280	739.4374	3109.2103	0.2378	0.0003	4.1685	31
32	0.0011	4.2059	17.5619	915.0538	3848.6477	0.2378	0.0003	4.1755	32
33	0.0009	4.2068	17.5902	1132.3791	4763.7015	0.2377	0.0002	4.1814	33
34	0.0007	4.2075	17.6137	1401.3192	5896.0807	0.2377	0.0002	4.1862	34
35	0.0006	4.2081	17.6333	1734.1325	7297.3998	0.2376	0.0001	4.1903	35
36	0.0005	4.2086	17.6496	2145.9889	9031.5323	0.2376	0.0001	4.1937	36
37	0.0004	4.2089	17.6632	2655.6613	11177.5212	0.2376	0.0001	4.1966	37
38	0.0003	4.2092	17.6745	3286.3808	13833.1825	0.2376	0.0001	4.1990	38
39	0.0002	4.2095	17.6838	4066.8963	17119.5633	0.2376	0.0001	4.2009	39
40	0.0002	4.2097	17.6915	5032.7841	21186.4596	0.2375	0.0000	4.2026	40
41	0.0002	4.2099	17.6980	6228.0704	26219.2437	0.2375	0.0000	4.2039	41
42	0.0001	4.2100	17.7033	7707.2371	32447.3141	0.2375	0.0000	4.2051	42
43	0.0001	4.2101	17.7077	9537.7059	40154.5512	0.2375	0.0000	4.2060	43
44	0.0001	4.2102	17.7113	11802.9111	49692.2571	0.2375	0.0000	4.2068	44
45	0.0001	4.2102	17.7143	14606.1024	61495.1682	0.2375	0.0000	4.2074	45
46	0.0001	4.2103	17.7168	18075.0518	76101.2706	0.2375	0.0000	4.2080	46
47	0.0000	4.2103	17.7189	22367.8766	94176.3224	0.2375	0.0000	4.2084	47
48	0.0000	4.2104	17.7206	27680.2472	116544.1989	0.2375	0.0000	4.2088	48
49	0.0000	4.2104	17.7220	34254.3060	144224.4462	0.2375	0.0000	4.2091	49
50	0.0000	4.2104	17.7231	42389.7036	178478.7522	0.2375	0.0000	4.2093	50
51	0.0000	4.2104	17.7241	52457.2583	220868.4558	0.2375	0.0000	4.2096	51
52	0.0000	4.2105	17.7249	64915.8571	273325.7141	0.2375	0.0000	4.2097	52
53	0.0000	4.2105	17.7255	80333.3731	338241.5711	0.2375	0.0000	4.2099	53
54	0.0000	4.2105	17.7261	99412.5493	418574.9443	0.2375	0.0000	4.2100	54
55	0.0000	4.2105	17.7265	123023.0297	517987.4936	0.2375	0.0000	4.2101	55
60	0.0000	4.2105	17.7278	357036.3683	1503306.8140	0.2375	0.0000	4.2104	60
65	0.0000	4.2105	17.7283	1036187.8471	4362891.9877	0.2375	0.0000	4.2105	65
70	0.0000	4.2105	17.7284	3007215.3699	***********	0.2375	0.0000	4.2105	70
75	0.0000	4.2105	17.7285	8727514.3270	***********	0.2375	0.0000	4.2105	75
80	0.0000	4.2105	17.7285	***********	***********	0.2375	0.0000	4.2105	80
85	0.0000	4.2105	17.7285	***********	***********	0.2375	0.0000	4.2105	85
90	0.0000	4.2105	17.7285	***********	***********	0.2375	0.0000	4.2105	90
95	0.0000	4.2105	17.7285	***********	***********	0.2375	0.0000	4.2105	95
100	0.0000	4.2105	17.7285	***********	***********	0.2375	0.0000	4.2105	100

$I = 24.00\%$

n	(P/F)	(P/A)	(P/G)	(F/P)	(F/A)	(A/P)	(A/F)	(A/G)	n
1	0.8065	0.8065	0.0000	1.2400	1.0000	1.2400	1.0000	0.0000	1
2	0.6504	1.4568	0.6504	1.5376	2.2400	0.6864	0.4464	0.4464	2
3	0.5245	1.9813	1.6993	1.9066	3.7776	0.5047	0.2647	0.8577	3
4	0.4230	2.4043	2.9683	2.3642	5.6842	0.4159	0.1759	1.2346	4
5	0.3411	2.7454	4.3327	2.9316	8.0484	0.3642	0.1242	1.5782	5
6	0.2751	3.0205	5.7081	3.6352	10.9801	0.3311	0.0911	1.8898	6
7	0.2218	3.2423	7.0392	4.5077	14.6153	0.3084	0.0684	2.1710	7
8	0.1789	3.4212	8.2915	5.5895	19.1229	0.2923	0.0523	2.4236	8
9	0.1443	3.5655	9.4458	6.9310	24.7125	0.2805	0.0405	2.6492	9
10	0.1164	3.6819	10.4930	8.5944	31.6434	0.2716	0.0316	2.8499	1
11	0.0938	3.7757	11.4313	10.6571	40.2379	0.2649	0.0249	3.0276	11
12	0.0757	3.8514	12.2637	13.2148	50.8950	0.2596	0.0196	3.1843	12
13	0.0610	3.9124	12.9960	16.3863	64.1097	0.2556	0.0156	3.3218	13
14	0.0492	3.9616	13.6358	20.3191	80.4961	0.2524	0.0124	3.4420	14
15	0.0397	4.0013	14.1915	25.1956	100.8151	0.2499	0.0099	3.5467	15
16	0.0320	4.0333	14.6716	31.2426	126.0108	0.2479	0.0079	3.6376	16
17	0.0258	4.0591	15.0846	38.7408	157.2534	0.2464	0.0064	3.7162	17
18	0.0208	4.0799	15.4385	48.0386	195.9942	0.2451	0.0051	3.7840	18
19	0.0168	4.0967	15.7406	59.5679	244.0328	0.2441	0.0041	3.8423	19
20	0.0135	4.1103	15.9979	73.8641	303.6006	0.2433	0.0033	3.8922	20
21	0.0109	4.1212	16.2162	91.5915	377.4648	0.2426	0.0026	3.9349	21
22	0.0088	4.1300	16.4011	113.5735	469.0563	0.2421	0.0021	3.9712	22
23	0.0071	4.1371	16.5574	140.8312	582.6298	0.2417	0.0017	4.0022	23
24	0.0057	4.1428	16.6891	174.6306	723.4610	0.2414	0.0014	4.0284	24
25	0.0046	4.1474	16.7999	216.5420	898.0916	0.2411	0.0011	4.0507	25
26	0.0037	4.1511	16.8930	268.5121	1114.6336	0.2409	0.0009	4.0695	26
27	0.0030	4.1542	16.9711	332.9550	1383.1457	0.2407	0.0007	4.0853	27
28	0.0024	4.1566	17.0365	412.8642	1716.1007	0.2406	0.0006	4.0987	28
29	0.0020	4.1585	17.0912	511.9516	2128.9648	0.2405	0.0005	4.1099	29
30	0.0016	4.1601	17.1369	634.8199	2640.9164	0.2404	0.0004	4.1193	30
31	0.0013	4.1614	17.1750	787.1767	3275.7363	0.2403	0.0003	4.1272	31
32	0.0010	4.1624	17.2067	976.0991	4062.9130	0.2402	0.0002	4.1338	32
33	0.0008	4.1632	17.2332	1210.3629	5039.0122	0.2402	0.0002	4.1394	33
34	0.0007	4.1639	17.2552	1500.8500	6249.3751	0.2402	0.0002	4.1440	34
35	0.0005	4.1644	17.2734	1861.0540	7750.2251	0.2401	0.0001	4.1479	35
36	0.0004	4.1649	17.2886	2307.7070	9611.2791	0.2401	0.0001	4.1511	36
37	0.0003	4.1652	17.3012	2861.5567	11918.9861	0.2401	0.0001	4.1537	37
38	0.0003	4.1655	17.3116	3548.3303	14780.5428	0.2401	0.0001	4.1560	38
39	0.0002	4.1657	17.3202	4399.9295	18328.8731	0.2401	0.0001	4.1578	39
40	0.0002	4.1659	17.3274	5455.9126	22728.8026	0.2400	0.0000	4.1593	40
41	0.0001	4.1661	17.3333	6765.3317	28184.7152	0.2400	0.0000	4.1606	41
42	0.0001	4.1662	17.3382	8389.0113	34950.0469	0.2400	0.0000	4.1617	42
43	0.0001	4.1663	17.3422	10402.3740	43339.0581	0.2400	0.0000	4.1625	43
44	0.0001	4.1663	17.3456	12898.9437	53741.4321	0.2400	0.0000	4.1633	44
45	0.0001	4.1664	17.3483	15994.6902	66640.3758	0.2400	0.0000	4.1639	45
46	0.0001	4.1665	17.3506	19833.4158	82635.0660	0.2400	0.0000	4.1643	46
47	0.0000	4.1665	17.3524	24593.4356	102468.4818	0.2400	0.0000	4.1648	47
48	0.0000	4.1665	17.3540	30495.8602	127061.9174	0.2400	0.0000	4.1651	48
49	0.0000	4.1666	17.3553	37814.8666	157557.7776	0.2400	0.0000	4.1654	49
50	0.0000	4.1666	17.3563	46890.4346	195372.6442	0.2400	0.0000	4.1656	50
51	0.0000	4.1666	17.3572	58144.1389	242263.0788	0.2400	0.0000	4.1658	51
52	0.0000	4.1666	17.3579	72098.7323	300407.2178	0.2400	0.0000	4.1659	52
53	0.0000	4.1666	17.3584	89402.4280	372505.9500	0.2400	0.0000	4.1661	53
54	0.0000	4.1666	17.3589	110859.0107	461908.3780	0.2400	0.0000	4.1662	54
55	0.0000	4.1666	17.3593	137465.1733	572767.3888	0.2400	0.0000	4.1663	55
60	0.0000	4.1667	17.3604	402996.3473	1679147.2802	0.2400	0.0000	4.1665	60
65	0.0000	4.1667	17.3609	1181434.1917	4922638.2987	0.2400	0.0000	4.1666	65
70	0.0000	4.1667	17.3610	3463522.0859	************	0.2400	0.0000	4.1666	70
75	0.0000	4.1667	17.3611	************	************	0.2400	0.0000	4.1667	75
80	0.0000	4.1667	17.3611	************	************	0.2400	0.0000	4.1667	80
85	0.0000	4.1667	17.3611	************	************	0.2400	0.0000	4.1667	85
90	0.0000	4.1667	17.3611	************	************	0.2400	0.0000	4.1667	90
95	0.0000	4.1667	17.3611	************	************	0.2400	0.0000	4.1667	95
100	0.0000	4.1667	17.3611	************	************	0.2400	0.0000	4.1667	100

$I = 24.25\%$

n	(P/F)	(P/A)	(P/G)	(F/P)	(F/A)	(A/P)	(A/F)	(A/G)	n
1	0.8048	0.8048	0.0000	1.2425	1.0000	1.2425	1.0000	0.0000	1
2	0.6477	1.4526	0.6477	1.5438	2.2425	0.6884	0.4459	0.4459	2
3	0.5213	1.9739	1.6904	1.9182	3.7863	0.5066	0.2641	0.8564	3
4	0.4196	2.3935	2.9491	2.3833	5.7045	0.4178	0.1753	1.2322	4
5	0.3377	2.7312	4.2999	2.9613	8.0878	0.3661	0.1236	1.5744	5
6	0.2718	3.0030	5.6588	3.6794	11.0491	0.3330	0.0905	1.8844	6
7	0.2187	3.2217	6.9712	4.5717	14.7285	0.3104	0.0679	2.1638	7
8	0.1760	3.3977	8.2036	5.6803	19.3002	0.2943	0.0518	2.4144	8
9	0.1417	3.5394	9.3371	7.0578	24.9805	0.2825	0.0400	2.6380	9
10	0.1140	3.6535	10.3634	8.7693	32.0383	0.2737	0.0312	2.8366	1
11	0.0918	3.7452	11.2812	10.8958	40.8076	0.2670	0.0245	3.0121	11
12	0.0739	3.8191	12.0937	13.5381	51.7034	0.2618	0.0193	3.1666	12
13	0.0594	3.8786	12.8071	16.8211	65.2415	0.2578	0.0153	3.3020	13
14	0.0478	3.9264	13.4291	20.9002	82.0625	0.2547	0.0122	3.4202	14
15	0.0385	3.9649	13.9682	25.9684	102.9627	0.2522	0.0097	3.5230	15
16	0.0310	3.9959	14.4331	32.2658	128.9311	0.2503	0.0078	3.6120	16
17	0.0249	4.0209	14.8322	40.0902	161.1969	0.2487	0.0062	3.6888	17
18	0.0201	4.0409	15.1735	49.8121	201.2872	0.2475	0.0050	3.7550	18
19	0.0162	4.0571	15.4643	61.8916	251.0993	0.2465	0.0040	3.8117	19
20	0.0130	4.0701	15.7114	76.9003	312.9909	0.2457	0.0032	3.8602	20
21	0.0105	4.0806	15.9207	95.5486	389.8911	0.2451	0.0026	3.9016	21
22	0.0084	4.0890	16.0976	118.7191	485.4398	0.2446	0.0021	3.9368	22
23	0.0068	4.0958	16.2467	147.5085	604.1589	0.2442	0.0017	3.9667	23
24	0.0055	4.1012	16.3722	183.2794	751.6674	0.2438	0.0013	3.9920	24
25	0.0044	4.1056	16.4776	227.7246	934.9468	0.2436	0.0011	4.0134	25
26	0.0035	4.1091	16.5660	282.9478	1162.6714	0.2434	0.0009	4.0315	26
27	0.0028	4.1120	16.6399	351.5627	1445.6192	0.2432	0.0007	4.0467	27
28	0.0023	4.1143	16.7017	436.8166	1797.1818	0.2431	0.0006	4.0595	28
29	0.0018	4.1161	16.7533	542.7446	2233.9984	0.2429	0.0004	4.0702	29
30	0.0015	4.1176	16.7963	674.3602	2776.7430	0.2429	0.0004	4.0792	30
31	0.0012	4.1188	16.8321	837.8925	3451.1032	0.2428	0.0003	4.0867	31
32	0.0010	4.1198	16.8619	1041.0815	4288.9958	0.2427	0.0002	4.0929	32
33	0.0008	4.1205	16.8866	1293.5437	5330.0772	0.2427	0.0002	4.0982	33
34	0.0006	4.1211	16.9072	1607.2281	6623.6210	0.2427	0.0002	4.1025	34
35	0.0005	4.1216	16.9242	1996.9809	8230.8490	0.2426	0.0001	4.1062	35
36	0.0004	4.1220	16.9383	2481.2488	10227.8299	0.2426	0.0001	4.1092	36
37	0.0003	4.1224	16.9500	3082.9516	12709.0787	0.2426	0.0001	4.1117	37
38	0.0003	4.1226	16.9596	3830.5673	15792.0303	0.2426	0.0001	4.1138	38
39	0.0002	4.1228	16.9676	4759.4799	19622.5976	0.2426	0.0001	4.1155	39
40	0.0002	4.1230	16.9742	5913.6538	24382.0775	0.2425	0.0000	4.1169	40
41	0.0001	4.1232	16.9797	7347.7148	30295.7313	0.2425	0.0000	4.1181	41
42	0.0001	4.1233	16.9842	9129.5357	37643.4462	0.2425	0.0000	4.1191	42
43	0.0001	4.1233	16.9879	11343.4481	46772.9819	0.2425	0.0000	4.1199	43
44	0.0001	4.1234	16.9909	14094.2343	58116.4300	0.2425	0.0000	4.1206	44
45	0.0001	4.1235	16.9934	17512.0861	72210.6642	0.2425	0.0000	4.1211	45
46	0.0000	4.1235	16.9955	21758.7670	89722.7503	0.2425	0.0000	4.1216	46
47	0.0000	4.1236	16.9972	27035.2679	111481.5173	0.2425	0.0000	4.1220	47
48	0.0000	4.1236	16.9986	33591.3204	138516.7852	0.2425	0.0000	4.1223	48
49	0.0000	4.1236	16.9997	41737.2156	172108.1056	0.2425	0.0000	4.1225	49
50	0.0000	4.1236	17.0007	51858.4904	213845.3212	0.2425	0.0000	4.1227	50
51	0.0000	4.1236	17.0015	64434.1743	265703.8116	0.2425	0.0000	4.1229	51
52	0.0000	4.1237	17.0021	80059.4616	330137.9859	0.2425	0.0000	4.1231	52
53	0.0000	4.1237	17.0026	99473.8810	410197.4475	0.2425	0.0000	4.1232	53
54	0.0000	4.1237	17.0031	123596.2972	509671.3285	0.2425	0.0000	4.1233	54
55	0.0000	4.1237	17.0034	153568.3992	633267.6257	0.2425	0.0000	4.1234	55
60	0.0000	4.1237	17.0044	454761.6613	1875301.6960	0.2425	0.0000	4.1236	60
65	0.0000	4.1237	17.0048	1346684.4064	5553333.6347	0.2425	0.0000	4.1237	65
70	0.0000	4.1237	17.0049	3987932.6797	************	0.2425	0.0000	4.1237	70
75	0.0000	4.1237	17.0050	************	************	0.2425	0.0000	4.1237	75
80	0.0000	4.1237	17.0050	************	************	0.2425	0.0000	4.1237	80
85	0.0000	4.1237	17.0050	************	************	0.2425	0.0000	4.1237	85
90	0.0000	4.1237	17.0050	************	************	0.2425	0.0000	4.1237	90
95	0.0000	4.1237	17.0050	************	************	0.2425	0.0000	4.1237	95
100	0.0000	4.1237	17.0050	************	************	0.2425	0.0000	4.1237	100

ENGINEERING ECONOMIC ANALYSIS

$I = 24.50\%$

n	(P/F)	(P/A)	(P/G)	(F/P)	(F/A)	(A/P)	(A/F)	(A/G)	n
1	0.8032	0.8032	0.0000	1.2450	1.0000	1.2450	1.0000	0.0000	1
2	0.6452	1.4484	0.6452	1.5500	2.2450	0.6904	0.4454	0.4454	2
3	0.5182	1.9666	1.6815	1.9298	3.7950	0.5085	0.2635	0.8551	3
4	0.4162	2.3828	2.9302	2.4026	5.7248	0.4197	0.1747	1.2297	4
5	0.3343	2.7171	4.2674	2.9912	8.1274	0.3680	0.1230	1.5706	5
6	0.2685	2.9856	5.6101	3.7241	11.1186	0.3349	0.0899	1.8790	6
7	0.2157	3.2013	6.9042	4.6364	14.8426	0.3124	0.0674	2.1567	7
8	0.1732	3.3745	8.1168	5.7724	19.4791	0.2963	0.0513	2.4053	8
9	0.1391	3.5137	9.2300	7.1866	25.2515	0.2846	0.0396	2.6269	9
10	0.1118	3.6254	10.2359	8.9473	32.4381	0.2758	0.0308	2.8233	1
11	0.0898	3.7152	11.1336	11.1394	41.3854	0.2692	0.0242	2.9968	11
12	0.0721	3.7873	11.9268	13.8686	52.5248	0.2640	0.0190	3.1491	12
13	0.0579	3.8452	12.6218	17.2664	66.3934	0.2601	0.0151	3.2824	13
14	0.0465	3.8918	13.2265	21.4967	83.6598	0.2570	0.0120	3.3986	14
15	0.0374	3.9291	13.7496	26.7633	105.1565	0.2545	0.0095	3.4994	15
16	0.0300	3.9591	14.1998	33.3204	131.9198	0.2526	0.0076	3.5866	16
17	0.0241	3.9832	14.5855	41.4838	165.2402	0.2511	0.0061	3.6617	17
18	0.0194	4.0026	14.9146	51.6474	206.7240	0.2498	0.0048	3.7262	18
19	0.0156	4.0182	15.1946	64.3010	258.3714	0.2489	0.0039	3.7815	19
20	0.0125	4.0306	15.4319	80.0547	322.6724	0.2481	0.0031	3.8286	20
21	0.0100	4.0407	15.6326	99.6681	402.7271	0.2475	0.0025	3.8688	21
22	0.0081	4.0487	15.8018	124.0868	502.3953	0.2470	0.0020	3.9029	22
23	0.0065	4.0552	15.9442	154.4881	626.4821	0.2466	0.0016	3.9318	23
24	0.0052	4.0604	16.0638	192.3377	780.9702	0.2463	0.0013	3.9562	24
25	0.0042	4.0646	16.1640	239.4604	973.3079	0.2460	0.0010	3.9768	25
26	0.0034	4.0679	16.2479	298.1283	1212.7684	0.2458	0.0008	3.9941	26
27	0.0027	4.0706	16.3179	371.1697	1510.8966	0.2457	0.0007	4.0087	27
28	0.0022	4.0728	16.3764	462.1062	1882.0663	0.2455	0.0005	4.0209	28
29	0.0017	4.0745	16.4250	575.3223	2344.1726	0.2454	0.0004	4.0311	29
30	0.0014	4.0759	16.4655	716.2762	2919.4948	0.2453	0.0003	4.0397	30
31	0.0011	4.0771	16.4992	891.7639	3635.7711	0.2453	0.0003	4.0468	31
32	0.0009	4.0780	16.5271	1110.2461	4527.5350	0.2452	0.0002	4.0528	32
33	0.0007	4.0787	16.5502	1382.2564	5637.7811	0.2452	0.0002	4.0577	33
34	0.0006	4.0793	16.5694	1720.9092	7020.0374	0.2451	0.0001	4.0619	34
35	0.0005	4.0797	16.5853	2142.5319	8740.9466	0.2451	0.0001	4.0653	35
36	0.0004	4.0801	16.5984	2667.4522	10883.4785	0.2451	0.0001	4.0681	36
37	0.0003	4.0804	16.6092	3320.9780	13550.9307	0.2451	0.0001	4.0705	37
38	0.0002	4.0806	16.6182	4134.6177	16871.9088	0.2451	0.0001	4.0724	38
39	0.0002	4.0808	16.6256	5147.5990	21006.5264	0.2450	0.0000	4.0741	39
40	0.0002	4.0810	16.6317	6408.7607	26154.1254	0.2450	0.0000	4.0754	40
41	0.0001	4.0811	16.6367	7978.9071	32562.8861	0.2450	0.0000	4.0765	41
42	0.0001	4.0812	16.6408	9933.7393	40541.7932	0.2450	0.0000	4.0774	42
43	0.0001	4.0813	16.6442	12367.5055	50475.5326	0.2450	0.0000	4.0782	43
44	0.0001	4.0814	16.6470	15397.5443	62843.0381	0.2450	0.0000	4.0788	44
45	0.0001	4.0814	16.6493	19169.9427	78240.5824	0.2450	0.0000	4.0793	45
46	0.0000	4.0815	16.6512	23866.5786	97410.5251	0.2450	0.0000	4.0797	46
47	0.0000	4.0815	16.6527	29713.8904	121277.1037	0.2450	0.0000	4.0801	47
48	0.0000	4.0815	16.6540	36993.7936	150990.9941	0.2450	0.0000	4.0803	48
49	0.0000	4.0815	16.6550	46057.2730	187984.7877	0.2450	0.0000	4.0806	49
50	0.0000	4.0816	16.6559	57341.3049	234042.0607	0.2450	0.0000	4.0808	50
51	0.0000	4.0816	16.6566	71389.9246	291383.3655	0.2450	0.0000	4.0809	51
52	0.0000	4.0816	16.6571	88880.4561	362773.2901	0.2450	0.0000	4.0810	52
53	0.0000	4.0816	16.6576	110656.1678	451653.7462	0.2450	0.0000	4.0812	53
54	0.0000	4.0816	16.6580	137766.9289	562309.9140	0.2450	0.0000	4.0812	54
55	0.0000	4.0816	16.6583	171519.8265	700076.8429	0.2450	0.0000	4.0813	55
60	0.0000	4.0816	16.6592	513051.6467	2094084.2724	0.2450	0.0000	4.0815	60
65	0.0000	4.0816	16.6595	1534644.6972	6263851.8253	0.2450	0.0000	4.0816	65
70	0.0000	4.0816	16.6597	4590443.0121	************	0.2450	0.0000	4.0816	70
75	0.0000	4.0816	16.6597	************	************	0.2450	0.0000	4.0816	75
80	0.0000	4.0816	16.6597	************	************	0.2450	0.0000	4.0816	80
85	0.0000	4.0816	16.6597	************	************	0.2450	0.0000	4.0816	85
90	0.0000	4.0816	16.6597	************	************	0.2450	0.0000	4.0816	90
95	0.0000	4.0816	16.6597	************	************	0.2450	0.0000	4.0816	95
100	0.0000	4.0816	16.6597	************	************	0.2450	0.0000	4.0816	100

$I = 24.75\%$

n	(P/F)	(P/A)	(P/G)	(F/P)	(F/A)	(A/P)	(A/F)	(A/G)	n
1	0.8016	0.8016	0.0000	1.2475	1.0000	1.2475	1.0000	0.0000	1
2	0.6426	1.4442	0.6426	1.5563	2.2475	0.6924	0.4449	0.4449	2
3	0.5151	1.9593	1.6727	1.9414	3.8038	0.5104	0.2629	0.8538	3
4	0.4129	2.3721	2.9114	2.4219	5.7452	0.4216	0.1741	1.2273	4
5	0.3310	2.7031	4.2353	3.0214	8.1671	0.3699	0.1224	1.5668	5
6	0.2653	2.9684	5.5619	3.7691	11.1885	0.3369	0.0894	1.8737	6
7	0.2127	3.1811	6.8379	4.7020	14.9576	0.3144	0.0669	2.1495	7
8	0.1705	3.3516	8.0313	5.8658	19.6596	0.2984	0.0509	2.3963	8
9	0.1367	3.4883	9.1246	7.3175	25.5254	0.2867	0.0392	2.6158	9
10	0.1095	3.5978	10.1105	9.1286	32.8429	0.2779	0.0304	2.8102	1
11	0.0878	3.6856	10.9886	11.3880	41.9716	0.2713	0.0238	2.9815	11
12	0.0704	3.7560	11.7629	14.2065	53.3595	0.2662	0.0187	3.1318	12
13	0.0564	3.8124	12.4400	17.7226	67.5660	0.2623	0.0148	3.2630	13
14	0.0452	3.8577	13.0280	22.1089	85.2886	0.2592	0.0117	3.3772	14
15	0.0363	3.8939	13.5356	27.5809	107.3975	0.2568	0.0093	3.4761	15
16	0.0291	3.9230	13.9715	34.4072	134.9784	0.2549	0.0074	3.5615	16
17	0.0233	3.9463	14.3443	42.9229	169.3856	0.2534	0.0059	3.6349	17
18	0.0187	3.9649	14.6618	53.5464	212.3085	0.2522	0.0047	3.6978	18
19	0.0150	3.9799	14.9312	66.7991	265.8549	0.2513	0.0038	3.7516	19
20	0.0120	3.9919	15.1592	83.3319	332.6540	0.2505	0.0030	3.7975	20
21	0.0096	4.0015	15.3516	103.9565	415.9858	0.2499	0.0024	3.8364	21
22	0.0077	4.0092	15.5136	129.6857	519.9423	0.2494	0.0019	3.8694	22
23	0.0062	4.0154	15.6496	161.7829	649.6281	0.2490	0.0015	3.8974	23
24	0.0050	4.0204	15.7635	201.8242	811.4110	0.2487	0.0012	3.9209	24
25	0.0040	4.0244	15.8588	251.7757	1013.2353	0.2485	0.0010	3.9407	25
26	0.0032	4.0275	15.9384	314.0902	1265.0110	0.2483	0.0008	3.9574	26
27	0.0026	4.0301	16.0048	391.8275	1579.1012	0.2481	0.0006	3.9713	27
28	0.0020	4.0321	16.0600	488.8049	1970.9287	0.2480	0.0005	3.9830	28
29	0.0016	4.0338	16.1059	609.7841	2459.7336	0.2479	0.0004	3.9928	29
30	0.0013	4.0351	16.1441	760.7056	3069.5177	0.2478	0.0003	4.0009	30
31	0.0011	4.0361	16.1757	948.9803	3830.2233	0.2478	0.0003	4.0077	31
32	0.0008	4.0370	16.2019	1183.8529	4779.2036	0.2477	0.0002	4.0134	32
33	0.0007	4.0377	16.2235	1476.8565	5963.0565	0.2477	0.0002	4.0180	33
34	0.0005	4.0382	16.2414	1842.3785	7439.9129	0.2476	0.0001	4.0219	34
35	0.0004	4.0386	16.2562	2298.3671	9282.2914	0.2476	0.0001	4.0252	35
36	0.0003	4.0390	16.2684	2867.2130	11580.6585	0.2476	0.0001	4.0278	36
37	0.0003	4.0393	16.2785	3576.8482	14447.8715	0.2476	0.0001	4.0301	37
38	0.0002	4.0395	16.2868	4462.1181	18024.7197	0.2476	0.0001	4.0319	38
39	0.0002	4.0397	16.2936	5566.4924	22486.8378	0.2475	0.0000	4.0334	39
40	0.0001	4.0398	16.2992	6944.1992	28053.3302	0.2475	0.0000	4.0346	40
41	0.0001	4.0399	16.3039	8662.8885	34997.5294	0.2475	0.0000	4.0357	41
42	0.0001	4.0400	16.3077	10806.9534	43660.4179	0.2475	0.0000	4.0365	42
43	0.0001	4.0401	16.3108	13481.6744	54467.3714	0.2475	0.0000	4.0372	43
44	0.0001	4.0402	16.3133	16818.3888	67949.0458	0.2475	0.0000	4.0378	44
45	0.0000	4.0402	16.3154	20980.9401	84767.4346	0.2475	0.0000	4.0383	45
46	0.0000	4.0402	16.3171	26173.7227	105748.3747	0.2475	0.0000	4.0386	46
47	0.0000	4.0403	16.3185	32651.7191	131922.0974	0.2475	0.0000	4.0390	47
48	0.0000	4.0403	16.3197	40733.0196	164573.8166	0.2475	0.0000	4.0392	48
49	0.0000	4.0403	16.3206	50814.4420	205306.8362	0.2475	0.0000	4.0394	49
50	0.0000	4.0403	16.3214	63391.0163	256121.2781	0.2475	0.0000	4.0396	50
51	0.0000	4.0404	16.3221	79080.2929	319512.2944	0.2475	0.0000	4.0398	51
52	0.0000	4.0404	16.3226	98652.6654	398592.5873	0.2475	0.0000	4.0399	52
53	0.0000	4.0404	16.3230	123069.2000	497245.2527	0.2475	0.0000	4.0400	53
54	0.0000	4.0404	16.3233	153528.8271	620314.4527	0.2475	0.0000	4.0401	54
55	0.0000	4.0404	16.3236	191527.2117	773843.2798	0.2475	0.0000	4.0401	55
60	0.0000	4.0404	16.3244	578673.0512	2338068.8936	0.2475	0.0000	4.0403	60
65	0.0000	4.0404	16.3247	1748380.8024	7064160.8179	0.2475	0.0000	4.0404	65
70	0.0000	4.0404	16.3248	5282491.4243	************	0.2475	0.0000	4.0404	70
75	0.0000	4.0404	16.3248	************	************	0.2475	0.0000	4.0404	75
80	0.0000	4.0404	16.3249	************	************	0.2475	0.0000	4.0404	80
85	0.0000	4.0404	16.3249	************	************	0.2475	0.0000	4.0404	85
90	0.0000	4.0404	16.3249	************	************	0.2475	0.0000	4.0404	90
95	0.0000	4.0404	16.3249	************	************	0.2475	0.0000	4.0404	95
100	0.0000	4.0404	16.3249	************	************	0.2475	0.0000	4.0404	100

$I = 25.00\%$

n	(P/F)	(P/A)	(P/G)	(F/P)	(F/A)	(A/P)	(A/F)	(A/G)	n
1	0.8000	0.8000	0.0000	1.2500	1.0000	1.2500	1.0000	0.0000	1
2	0.6400	1.4400	0.6400	1.5625	2.2500	0.6944	0.4444	0.4444	2
3	0.5120	1.9520	1.6640	1.9531	3.8125	0.5123	0.2623	0.8525	3
4	0.4096	2.3616	2.8928	2.4414	5.7656	0.4234	0.1734	1.2249	4
5	0.3277	2.6893	4.2035	3.0518	8.2070	0.3718	0.1218	1.5631	5
6	0.2621	2.9514	5.5142	3.8147	11.2588	0.3388	0.0888	1.8683	6
7	0.2097	3.1611	6.7725	4.7684	15.0735	0.3163	0.0663	2.1424	7
8	0.1678	3.3289	7.9469	5.9605	19.8419	0.3004	0.0504	2.3872	8
9	0.1342	3.4631	9.0207	7.4506	25.8023	0.2888	0.0388	2.6048	9
10	0.1074	3.5705	9.9870	9.3132	33.2529	0.2801	0.0301	2.7971	10
11	0.0859	3.6564	10.8460	11.6415	42.5661	0.2735	0.0235	2.9663	11
12	0.0687	3.7251	11.6020	14.5519	54.2077	0.2684	0.0184	3.1145	12
13	0.0550	3.7801	12.2617	18.1899	68.7596	0.2645	0.0145	3.2437	13
14	0.0440	3.8241	12.8334	22.7374	86.9495	0.2615	0.0115	3.3559	14
15	0.0352	3.8593	13.3260	28.4217	109.6868	0.2591	0.0091	3.4530	15
16	0.0281	3.8874	13.7482	35.5271	138.1085	0.2572	0.0072	3.5366	16
17	0.0225	3.9099	14.1085	44.4089	173.6357	0.2558	0.0058	3.6084	17
18	0.0180	3.9279	14.4147	55.5112	218.0446	0.2546	0.0046	3.6698	18
19	0.0144	3.9424	14.6741	69.3889	273.5558	0.2537	0.0037	3.7222	19
20	0.0115	3.9539	14.8932	86.7362	342.9447	0.2529	0.0029	3.7667	20
21	0.0092	3.9631	15.0777	108.4202	429.6809	0.2523	0.0023	3.8045	21
22	0.0074	3.9705	15.2326	135.5253	538.1011	0.2519	0.0019	3.8365	22
23	0.0059	3.9764	15.3625	169.4066	673.6264	0.2515	0.0015	3.8634	23
24	0.0047	3.9811	15.4711	211.7582	843.0329	0.2512	0.0012	3.8861	24
25	0.0038	3.9849	15.5618	264.6978	1054.7912	0.2509	0.0009	3.9052	25
26	0.0030	3.9879	15.6373	330.8722	1319.4890	0.2508	0.0008	3.9212	26
27	0.0024	3.9903	15.7002	413.5903	1650.3612	0.2506	0.0006	3.9346	27
28	0.0019	3.9923	15.7524	516.9879	2063.9515	0.2505	0.0005	3.9457	28
29	0.0015	3.9938	15.7957	646.2349	2580.9394	0.2504	0.0004	3.9551	29
30	0.0012	3.9950	15.8316	807.7936	3227.1743	0.2503	0.0003	3.9628	30
31	0.0010	3.9960	15.8614	1009.7420	4034.9678	0.2502	0.0002	3.9693	31
32	0.0008	3.9968	15.8859	1262.1774	5044.7098	0.2502	0.0002	3.9746	32
33	0.0006	3.9975	15.9062	1577.7218	6306.8872	0.2502	0.0002	3.9791	33
34	0.0005	3.9980	15.9229	1972.1523	7884.6091	0.2501	0.0001	3.9828	34
35	0.0004	3.9984	15.9367	2465.1903	9856.7613	0.2501	0.0001	3.9858	35
36	0.0003	3.9987	15.9481	3081.4879	12321.9516	0.2501	0.0001	3.9883	36
37	0.0003	3.9990	15.9574	3851.8599	15403.4396	0.2501	0.0001	3.9904	37
38	0.0002	3.9992	15.9651	4814.8249	19255.2994	0.2501	0.0001	3.9921	38
39	0.0002	3.9993	15.9714	6018.5311	24070.1243	0.2500	0.0000	3.9935	39
40	0.0001	3.9995	15.9766	7523.1638	30088.6554	0.2500	0.0000	3.9947	40
41	0.0001	3.9996	15.9809	9403.9548	37611.8192	0.2500	0.0000	3.9956	41
42	0.0001	3.9997	15.9843	11754.9435	47015.7740	0.2500	0.0000	3.9964	42
43	0.0001	3.9997	15.9872	14693.6794	58770.7175	0.2500	0.0000	3.9971	43
44	0.0001	3.9998	15.9895	18367.0992	73464.3969	0.2500	0.0000	3.9976	44
45	0.0000	3.9998	15.9915	22958.8740	91831.4962	0.2500	0.0000	3.9980	45
46	0.0000	3.9999	15.9930	28698.5925	114790.3702	0.2500	0.0000	3.9984	46
47	0.0000	3.9999	15.9943	35873.2407	143488.9627	0.2500	0.0000	3.9987	47
48	0.0000	3.9999	15.9954	44841.5509	179362.2034	0.2500	0.0000	3.9989	48
49	0.0000	3.9999	15.9962	56051.9386	224203.7543	0.2500	0.0000	3.9991	49
50	0.0000	3.9999	15.9969	70064.9232	280255.6929	0.2500	0.0000	3.9993	50
51	0.0000	4.0000	15.9975	87581.1540	350320.6161	0.2500	0.0000	3.9994	51
52	0.0000	4.0000	15.9980	109476.4425	437901.7701	0.2500	0.0000	3.9995	52
53	0.0000	4.0000	15.9983	136845.5532	547378.2126	0.2500	0.0000	3.9996	53
54	0.0000	4.0000	15.9986	171056.9414	684223.7658	0.2500	0.0000	3.9997	54
55	0.0000	4.0000	15.9989	213821.1768	855280.7072	0.2500	0.0000	3.9997	55
60	0.0000	4.0000	15.9996	652530.4468	2610117.7872	0.2500	0.0000	3.9999	60
65	0.0000	4.0000	15.9999	1991364.8889	7965455.5557	0.2500	0.0000	4.0000	65
70	0.0000	4.0000	16.0000	6077163.3573	************	0.2500	0.0000	4.0000	70
75	0.0000	4.0000	16.0000	************	************	0.2500	0.0000	4.0000	75
80	0.0000	4.0000	16.0000	************	************	0.2500	0.0000	4.0000	80
85	0.0000	4.0000	16.0000	************	************	0.2500	0.0000	4.0000	85
90	0.0000	4.0000	16.0000	************	************	0.2500	0.0000	4.0000	90
95	0.0000	4.0000	16.0000	************	************	0.2500	0.0000	4.0000	95
100	0.0000	4.0000	16.0000	************	************	0.2500	0.0000	4.0000	100

$I = 30.00\%$

n	(P/F)	(P/A)	(P/G)	(F/P)	(F/A)	(A/P)	(A/F)	(A/G)	n
1	0.7692	0.7692	0.0000	1.3000	1.0000	1.3000	1.0000	0.0000	1
2	0.5917	1.3609	0.5917	1.6900	2.3000	0.7348	0.4348	0.4348	2
3	0.4552	1.8161	1.5020	2.1970	3.9900	0.5506	0.2506	0.8271	3
4	0.3501	2.1662	2.5524	2.8561	6.1870	0.4616	0.1616	1.1783	4
5	0.2693	2.4356	3.6297	3.7129	9.0431	0.4106	0.1106	1.4903	5
6	0.2072	2.6427	4.6656	4.8268	12.7560	0.3784	0.0784	1.7654	6
7	0.1594	2.8021	5.6218	6.2749	17.5828	0.3569	0.0569	2.0063	7
8	0.1226	2.9247	6.4800	8.1573	23.8577	0.3419	0.0419	2.2156	8
9	0.0943	3.0190	7.2343	10.6045	32.0150	0.3312	0.0312	2.3963	9
10	0.0725	3.0915	7.8872	13.7858	42.6195	0.3235	0.0235	2.5512	10
11	0.0558	3.1473	8.4452	17.9216	56.4053	0.3177	0.0177	2.6833	11
12	0.0429	3.1903	8.9173	23.2981	74.3270	0.3135	0.0135	2.7952	12
13	0.0330	3.2233	9.3135	30.2875	97.6250	0.3102	0.0102	2.8895	13
14	0.0254	3.2487	9.6437	39.3738	127.9125	0.3078	0.0078	2.9685	14
15	0.0195	3.2682	9.9172	51.1859	167.2863	0.3060	0.0060	3.0344	15
16	0.0150	3.2832	10.1426	66.5417	218.4722	0.3046	0.0046	3.0892	16
17	0.0116	3.2948	10.3276	86.5042	285.0139	0.3035	0.0035	3.1345	17
18	0.0089	3.3037	10.4788	112.4554	371.5180	0.3027	0.0027	3.1718	18
19	0.0068	3.3105	10.6019	146.1920	483.9734	0.3021	0.0021	3.2025	19
20	0.0053	3.3158	10.7019	190.0496	630.1655	0.3016	0.0016	3.2275	20
21	0.0040	3.3198	10.7828	247.0645	820.2151	0.3012	0.0012	3.2480	21
22	0.0031	3.3230	10.8482	321.1839	1067.2796	0.3009	0.0009	3.2646	22
23	0.0024	3.3254	10.9009	417.5391	1388.4635	0.3007	0.0007	3.2781	23
24	0.0018	3.3272	10.9433	542.8008	1806.0026	0.3006	0.0006	3.2890	24
25	0.0014	3.3286	10.9773	705.6410	2348.8033	0.3004	0.0004	3.2979	25
26	0.0011	3.3297	11.0045	917.3333	3054.4443	0.3003	0.0003	3.3050	26
27	0.0008	3.3305	11.0263	1192.5333	3971.7776	0.3003	0.0003	3.3107	27
28	0.0006	3.3312	11.0437	1550.2933	5164.3109	0.3002	0.0002	3.3153	28
29	0.0005	3.3317	11.0576	2015.3813	6714.6042	0.3001	0.0001	3.3189	29
30	0.0004	3.3321	11.0687	2619.9956	8729.9855	0.3001	0.0001	3.3219	30
31	0.0003	3.3324	11.0775	3405.9943	11349.9811	0.3001	0.0001	3.3242	31
32	0.0002	3.3326	11.0845	4427.7926	14755.9755	0.3001	0.0001	3.3261	32
33	0.0002	3.3328	11.0901	5756.1304	19183.7681	0.3001	0.0001	3.3276	33
34	0.0001	3.3329	11.0945	7482.9696	24939.8985	0.3000	0.0000	3.3288	34
35	0.0001	3.3330	11.0980	9727.8604	32422.8681	0.3000	0.0000	3.3297	35
36	0.0001	3.3331	11.1007	12646.2186	42150.7285	0.3000	0.0000	3.3305	36
37	0.0001	3.3331	11.1029	16440.0841	54796.9471	0.3000	0.0000	3.3311	37
38	0.0000	3.3332	11.1047	21372.1094	71237.0312	0.3000	0.0000	3.3316	38
39	0.0000	3.3332	11.1060	27783.7422	92609.1405	0.3000	0.0000	3.3319	39
40	0.0000	3.3332	11.1071	36118.8648	120392.8827	0.3000	0.0000	3.3322	40
41	0.0000	3.3333	11.1080	46954.5243	156511.7475	0.3000	0.0000	3.3325	41
42	0.0000	3.3333	11.1086	61040.8815	203466.2718	0.3000	0.0000	3.3326	42
43	0.0000	3.3333	11.1092	79353.1460	264507.1533	0.3000	0.0000	3.3328	43
44	0.0000	3.3333	11.1096	103159.0898	343860.2993	0.3000	0.0000	3.3329	44
45	0.0000	3.3333	11.1099	134106.8167	447019.3890	0.3000	0.0000	3.3330	45
46	0.0000	3.3333	11.1102	174338.8617	581126.2058	0.3000	0.0000	3.3331	46
47	0.0000	3.3333	11.1104	226640.5202	755465.0675	0.3000	0.0000	3.3331	47
48	0.0000	3.3333	11.1105	294632.6763	982105.5877	0.3000	0.0000	3.3332	48
49	0.0000	3.3333	11.1107	383022.4792	1276738.2640	0.3000	0.0000	3.3332	49
50	0.0000	3.3333	11.1108	497929.2230	1659760.7433	0.3000	0.0000	3.3332	50

$I = 35.00\%$

n	(P/F)	(P/A)	(P/G)	(F/P)	(F/A)	(A/P)	(A/F)	(A/G)	n
1	0.7407	0.7407	0.0000	1.3500	1.0000	1.3500	1.0000	0.0000	1
2	0.5487	1.2894	0.5487	1.8225	2.3500	0.7755	0.4255	0.4255	2
3	0.4064	1.6959	1.3616	2.4604	4.1725	0.5897	0.2397	0.8029	3
4	0.3011	1.9969	2.2648	3.3215	6.6329	0.5008	0.1508	1.1341	4
5	0.2230	2.2200	3.1568	4.4840	9.9544	0.4505	0.1005	1.4220	5
6	0.1652	2.3852	3.9828	6.0534	14.4384	0.4193	0.0693	1.6698	6
7	0.1224	2.5075	4.7170	8.1722	20.4919	0.3988	0.0488	1.8811	7
8	0.0906	2.5982	5.3515	11.0324	28.6640	0.3849	0.0349	2.0597	8
9	0.0671	2.6653	5.8886	14.8937	39.6964	0.3752	0.0252	2.2094	9
10	0.0497	2.7150	6.3363	20.1066	54.5902	0.3683	0.0183	2.3338	10
11	0.0368	2.7519	6.7047	27.1439	74.6967	0.3634	0.0134	2.4364	11
12	0.0273	2.7792	7.0049	36.6442	101.8406	0.3598	0.0098	2.5205	12
13	0.0202	2.7994	7.2474	49.4697	138.4848	0.3572	0.0072	2.5889	13
14	0.0150	2.8144	7.4421	66.7841	187.9544	0.3553	0.0053	2.6443	14
15	0.0111	2.8255	7.5974	90.1585	254.7385	0.3539	0.0039	2.6889	15
16	0.0082	2.8337	7.7206	121.7139	344.8970	0.3529	0.0029	2.7246	16
17	0.0061	2.8398	7.8180	164.3138	466.6109	0.3521	0.0021	2.7530	17
18	0.0045	2.8443	7.8946	221.8236	630.9247	0.3516	0.0016	2.7756	18
19	0.0033	2.8476	7.9547	299.4619	852.7483	0.3512	0.0012	2.7935	19
20	0.0025	2.8501	8.0017	404.2736	1152.2103	0.3509	0.0009	2.8075	20
21	0.0018	2.8519	8.0384	545.7693	1556.4838	0.3506	0.0006	2.8186	21
22	0.0014	2.8533	8.0669	736.7886	2102.2532	0.3505	0.0005	2.8272	22
23	0.0010	2.8543	8.0890	994.6646	2839.0418	0.3504	0.0004	2.8340	23
24	0.0007	2.8550	8.1061	1342.7973	3833.7064	0.3503	0.0003	2.8393	24
25	0.0006	2.8556	8.1194	1812.7763	5176.5037	0.3502	0.0002	2.8433	25
26	0.0004	2.8560	8.1296	2447.2480	6989.2800	0.3501	0.0001	2.8465	26
27	0.0003	2.8563	8.1374	3303.7848	9436.5280	0.3501	0.0001	2.8490	27
28	0.0002	2.8565	8.1435	4460.1095	12740.3128	0.3501	0.0001	2.8509	28
29	0.0002	2.8567	8.1481	6021.1478	17200.4222	0.3501	0.0001	2.8523	29
30	0.0001	2.8568	8.1517	8128.5495	23221.5700	0.3500	0.0000	2.8535	30
31	0.0001	2.8569	8.1545	10973.5418	31350.1195	0.3500	0.0000	2.8543	31
32	0.0001	2.8569	8.1565	14814.2815	42323.6613	0.3500	0.0000	2.8550	32
33	0.0001	2.8570	8.1581	19999.2800	57137.9428	0.3500	0.0000	2.8555	33
34	0.0000	2.8570	8.1594	26999.0280	77137.2228	0.3500	0.0000	2.8559	34
35	0.0000	2.8571	8.1603	36448.6878	104136.2508	0.3500	0.0000	2.8562	35
36	0.0000	2.8571	8.1610	49205.7285	140584.9385	0.3500	0.0000	2.8564	36
37	0.0000	2.8571	8.1616	66427.7334	189790.6670	0.3500	0.0000	2.8566	37
38	0.0000	2.8571	8.1620	89677.4402	256218.4004	0.3500	0.0000	2.8567	38
39	0.0000	2.8571	8.1623	121064.5442	345895.8406	0.3500	0.0000	2.8568	39
40	0.0000	2.8571	8.1625	163437.1347	466960.3848	0.3500	0.0000	2.8569	40
41	0.0000	2.8571	8.1627	220640.1318	630397.5195	0.3500	0.0000	2.8570	41
42	0.0000	2.8571	8.1628	297864.1780	851037.6513	0.3500	0.0000	2.8570	42
43	0.0000	2.8571	8.1629	402116.6402	1148901.8293	0.3500	0.0000	2.8570	43
44	0.0000	2.8571	8.1630	542857.4643	1551018.4695	0.3500	0.0000	2.8571	44
45	0.0000	2.8571	8.1631	732857.5768	2093875.9338	0.3500	0.0000	2.8571	45
46	0.0000	2.8571	8.1631	989357.7287	2826733.5107	0.3500	0.0000	2.8571	46
47	0.0000	2.8571	8.1632	1335632.9338	3816091.2394	0.3500	0.0000	2.8571	47
48	0.0000	2.8571	8.1632	1803104.4606	5151724.1732	0.3500	0.0000	2.8571	48
49	0.0000	2.8571	8.1632	2434191.0218	6954828.6338	0.3500	0.0000	2.8571	49
50	0.0000	2.8571	8.1632	3286157.8795	9389019.6556	0.3500	0.0000	2.8571	50

$I = 40.00\%$

n	(P/F)	(P/A)	(P/G)	(F/P)	(F/A)	(A/P)	(A/F)	(A/G)	n
1	0.7143	0.7143	0.0000	1.4000	1.0000	1.4000	1.0000	0.0000	1
2	0.5102	1.2245	0.5102	1.9600	2.4000	0.8167	0.4167	0.4167	2
3	0.3644	1.5889	1.2391	2.7440	4.3600	0.6294	0.2294	0.7798	3
4	0.2603	1.8492	2.0200	3.8416	7.1040	0.5408	0.1408	1.0923	4
5	0.1859	2.0352	2.7637	5.3782	10.9456	0.4914	0.0914	1.3580	5
6	0.1328	2.1680	3.4278	7.5295	16.3238	0.4613	0.0613	1.5811	6
7	0.0949	2.2628	3.9970	10.5414	23.8534	0.4419	0.0419	1.7664	7
8	0.0678	2.3306	4.4713	14.7579	34.3947	0.4291	0.0291	1.9185	8
9	0.0484	2.3790	4.8585	20.6610	49.1526	0.4203	0.0203	2.0422	9
10	0.0346	2.4136	5.1696	28.9255	69.8137	0.4143	0.0143	2.1419	10
11	0.0247	2.4383	5.4166	40.4957	98.7391	0.4101	0.0101	2.2215	11
12	0.0176	2.4559	5.6106	56.6939	139.2348	0.4072	0.0072	2.2845	12
13	0.0126	2.4685	5.7618	79.3715	195.9287	0.4051	0.0051	2.3341	13
14	0.0090	2.4775	5.8788	111.1201	275.3002	0.4036	0.0036	2.3729	14
15	0.0064	2.4839	5.9688	155.5681	386.4202	0.4026	0.0026	2.4030	15
16	0.0046	2.4885	6.0376	217.7953	541.9883	0.4018	0.0018	2.4262	16
17	0.0033	2.4918	6.0901	304.9135	759.7837	0.4013	0.0013	2.4441	17
18	0.0023	2.4941	6.1299	426.8789	1064.6971	0.4009	0.0009	2.4577	18
19	0.0017	2.4958	6.1601	597.6304	1491.5760	0.4007	0.0007	2.4682	19
20	0.0012	2.4970	6.1828	836.6826	2089.2064	0.4005	0.0005	2.4761	20
21	0.0009	2.4979	6.1998	1171.3556	2925.8889	0.4003	0.0003	2.4821	21
22	0.0006	2.4985	6.2127	1639.8978	4097.2445	0.4002	0.0002	2.4866	22
23	0.0004	2.4989	6.2222	2295.8569	5737.1423	0.4002	0.0002	2.4900	23
24	0.0003	2.4992	6.2294	3214.1997	8032.9993	0.4001	0.0001	2.4925	24
25	0.0002	2.4994	6.2347	4499.8796	11247.1990	0.4001	0.0001	2.4944	25
26	0.0002	2.4996	6.2387	6299.8314	15747.0785	0.4001	0.0001	2.4959	26
27	0.0001	2.4997	6.2416	8819.7640	22046.9099	0.4000	0.0000	2.4969	27
28	0.0001	2.4998	6.2438	12347.6696	30866.6739	0.4000	0.0000	2.4977	28
29	0.0001	2.4999	6.2454	17286.7374	43214.3435	0.4000	0.0000	2.4983	29
30	0.0000	2.4999	6.2466	24201.4324	60501.0809	0.4000	0.0000	2.4988	30
31	0.0000	2.4999	6.2475	33882.0053	84702.5132	0.4000	0.0000	2.4991	31
32	0.0000	2.4999	6.2482	47434.8074	118584.5185	0.4000	0.0000	2.4993	32
33	0.0000	2.5000	6.2487	66408.7304	166019.3260	0.4000	0.0000	2.4995	33
34	0.0000	2.5000	6.2490	92972.2225	232428.0563	0.4000	0.0000	2.4996	34
35	0.0000	2.5000	6.2493	130161.1116	325400.2789	0.4000	0.0000	2.4997	35
36	0.0000	2.5000	6.2495	182225.5562	455561.3904	0.4000	0.0000	2.4998	36
37	0.0000	2.5000	6.2496	255115.7786	637786.9466	0.4000	0.0000	2.4999	37
38	0.0000	2.5000	6.2497	357162.0901	892902.7252	0.4000	0.0000	2.4999	38
39	0.0000	2.5000	6.2498	500026.9261	1250064.8153	0.4000	0.0000	2.4999	39
40	0.0000	2.5000	6.2498	700037.6966	1750091.7415	0.4000	0.0000	2.4999	40
41	0.0000	2.5000	6.2499	980052.7752	2450129.4381	0.4000	0.0000	2.5000	41
42	0.0000	2.5000	6.2499	1372073.8853	3430182.2133	0.4000	0.0000	2.5000	42
43	0.0000	2.5000	6.2499	1920903.4394	4802256.0986	0.4000	0.0000	2.5000	43
44	0.0000	2.5000	6.2500	2689264.8152	6723159.5381	0.4000	0.0000	2.5000	44
45	0.0000	2.5000	6.2500	3764970.7413	9412424.3533	0.4000	0.0000	2.5000	45

I = 45.00%

n	(P/F)	(P/A)	(P/G)	(F/P)	(F/A)	(A/P)	(A/F)	(A/G)	n
1	0.6897	0.6897	0.0000	1.4500	1.0000	1.4500	1.0000	0.0000	1
2	0.4756	1.1653	0.4756	2.1025	2.4500	0.8582	0.4082	0.4082	2
3	0.3280	1.4933	1.1317	3.0486	4.5525	0.6697	0.2197	0.7578	3
4	0.2262	1.7195	1.8103	4.4205	7.6011	0.5816	0.1316	1.0528	4
5	0.1560	1.8755	2.4344	6.4097	12.0216	0.5332	0.0832	1.2980	5
6	0.1076	1.9831	2.9723	9.2941	18.4314	0.5043	0.0543	1.4988	6
7	0.0742	2.0573	3.4176	13.4765	27.7255	0.4861	0.0361	1.6612	7
8	0.0512	2.1085	3.7758	19.5409	41.2019	0.4743	0.0243	1.7907	8
9	0.0353	2.1438	4.0581	28.3343	60.7428	0.4665	0.0165	1.8930	9
10	0.0243	2.1681	4.2772	41.0847	89.0771	0.4612	0.0112	1.9728	10
11	0.0168	2.1849	4.4450	59.5728	130.1618	0.4577	0.0077	2.0344	11
12	0.0116	2.1965	4.5724	86.3806	189.7346	0.4553	0.0053	2.0817	12
13	0.0080	2.2045	4.6682	125.2518	276.1151	0.4536	0.0036	2.1176	13
14	0.0055	2.2100	4.7398	181.6151	401.3670	0.4525	0.0025	2.1447	14
15	0.0038	2.2138	4.7929	263.3419	582.9821	0.4517	0.0017	2.1650	15
16	0.0026	2.2164	4.8322	381.8458	846.3240	0.4512	0.0012	2.1802	16
17	0.0018	2.2182	4.8611	553.6764	1228.1699	0.4508	0.0008	2.1915	17
18	0.0012	2.2195	4.8823	802.8308	1781.8463	0.4506	0.0006	2.1998	18
19	0.0009	2.2203	4.8978	1164.1047	2584.6771	0.4504	0.0004	2.2059	19
20	0.0006	2.2209	4.9090	1687.9518	3748.7818	0.4503	0.0003	2.2104	20
21	0.0004	2.2213	4.9172	2447.5301	5436.7336	0.4502	0.0002	2.2136	21
22	0.0003	2.2216	4.9231	3548.9187	7884.2638	0.4501	0.0001	2.2160	22
23	0.0002	2.2218	4.9274	5145.9321	11433.1824	0.4501	0.0001	2.2178	23
24	0.0001	2.2219	4.9305	7461.6015	16579.1145	0.4501	0.0001	2.2190	24
25	0.0001	2.2220	4.9327	10819.3222	24040.7161	0.4500	0.0000	2.2199	25
26	0.0001	2.2221	4.9343	15688.0172	34860.0383	0.4500	0.0000	2.2206	26
27	0.0000	2.2221	4.9354	22747.6250	50548.0556	0.4500	0.0000	2.2210	27
28	0.0000	2.2222	4.9362	32984.0563	73295.6806	0.4500	0.0000	2.2214	28
29	0.0000	2.2222	4.9368	47826.8816	106279.7368	0.4500	0.0000	2.2216	29
30	0.0000	2.2222	4.9372	69348.9783	154106.6184	0.4500	0.0000	2.2218	30
31	0.0000	2.2222	4.9375	100556.0185	223455.5967	0.4500	0.0000	2.2219	31
32	0.0000	2.2222	4.9378	145806.2269	324011.6152	0.4500	0.0000	2.2220	32
33	0.0000	2.2222	4.9379	211419.0289	469817.8421	0.4500	0.0000	2.2221	33
34	0.0000	2.2222	4.9380	306557.5920	681236.8710	0.4500	0.0000	2.2221	34
35	0.0000	2.2222	4.9381	444508.5083	987794.4630	0.4500	0.0000	2.2221	35

$I = 50.00\%$

n	(P/F)	(P/A)	(P/G)	(F/P)	(F/A)	(A/P)	(A/F)	(A/G)	n
1	0.6667	0.6667	0.0000	1.5000	1.0000	1.5000	1.0000	0.0000	1
2	0.4444	1.1111	0.4444	2.2500	2.5000	0.9000	0.4000	0.4000	2
3	0.2963	1.4074	1.0370	3.3750	4.7500	0.7105	0.2105	0.7368	3
4	0.1975	1.6049	1.6296	5.0625	8.1250	0.6231	0.1231	1.0154	4
5	0.1317	1.7366	2.1564	7.5937	13.1875	0.5758	0.0758	1.2417	5
6	0.0878	1.8244	2.5953	11.3906	20.7812	0.5481	0.0481	1.4226	6
7	0.0585	1.8829	2.9465	17.0859	32.1719	0.5311	0.0311	1.5648	7
8	0.0390	1.9220	3.2196	25.6289	49.2578	0.5203	0.0203	1.6752	8
9	0.0260	1.9480	3.4277	38.4434	74.8867	0.5134	0.0134	1.7596	9
10	0.0173	1.9653	3.5838	57.6650	113.3301	0.5088	0.0088	1.8235	10
11	0.0116	1.9769	3.6994	86.4976	170.9951	0.5058	0.0058	1.8713	11
12	0.0077	1.9846	3.7842	129.7463	257.4927	0.5039	0.0039	1.9068	12
13	0.0051	1.9897	3.8459	194.6195	387.2390	0.5026	0.0026	1.9329	13
14	0.0034	1.9931	3.8904	291.9293	581.8585	0.5017	0.0017	1.9519	14
15	0.0023	1.9954	3.9224	437.8939	873.7878	0.5011	0.0011	1.9657	15
16	0.0015	1.9970	3.9452	656.8408	1311.6817	0.5008	0.0008	1.9756	16
17	0.0010	1.9980	3.9614	985.2613	1968.5225	0.5005	0.0005	1.9827	17
18	0.0007	1.9986	3.9729	1477.8919	2953.7838	0.5003	0.0003	1.9878	18
19	0.0005	1.9991	3.9811	2216.8378	4431.6756	0.5002	0.0002	1.9914	19
20	0.0003	1.9994	3.9868	3325.2567	6648.5135	0.5002	0.0002	1.9940	20
21	0.0002	1.9996	3.9908	4987.8851	9973.7702	0.5001	0.0001	1.9958	21
22	0.0001	1.9997	3.9936	7481.8276	14961.6553	0.5001	0.0001	1.9971	22
23	0.0001	1.9998	3.9955	11222.7415	22443.4829	0.5000	0.0000	1.9980	23
24	0.0001	1.9999	3.9969	16834.1122	33666.2244	0.5000	0.0000	1.9986	24
25	0.0000	1.9999	3.9979	25251.1683	50500.3366	0.5000	0.0000	1.9990	25
26	0.0000	1.9999	3.9985	37876.7524	75751.5049	0.5000	0.0000	1.9993	26
27	0.0000	2.0000	3.9990	56815.1287	113628.2573	0.5000	0.0000	1.9995	27
28	0.0000	2.0000	3.9993	85222.6930	170443.3860	0.5000	0.0000	1.9997	28
29	0.0000	2.0000	3.9995	127834.0395	255666.0790	0.5000	0.0000	1.9998	29
30	0.0000	2.0000	3.9997	191751.0592	383500.1185	0.5000	0.0000	1.9998	30
31	0.0000	2.0000	3.9998	287626.5888	575251.1777	0.5000	0.0000	1.9999	31
32	0.0000	2.0000	3.9998	431439.8833	862877.7665	0.5000	0.0000	1.9999	32
33	0.0000	2.0000	3.9999	647159.8249	1294317.6498	0.5000	0.0000	1.9999	33
34	0.0000	2.0000	3.9999	970739.7374	1941477.4747	0.5000	0.0000	2.0000	34
35	0.0000	2.0000	3.9999	1456109.6060	2912217.2121	0.5000	0.0000	2.0000	35

APPENDIX A
Standard Cash Flow Factors

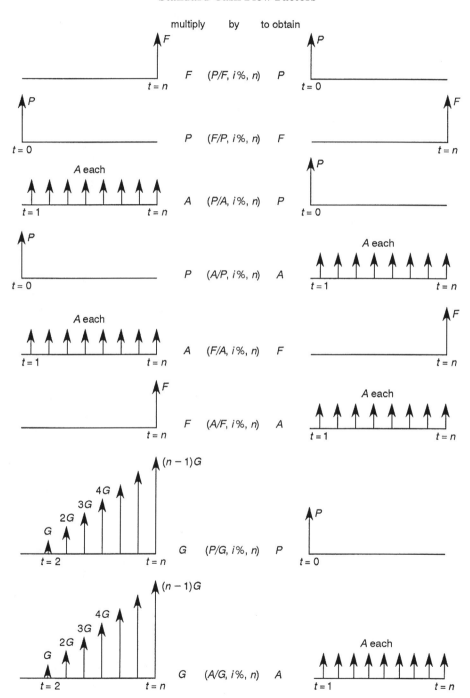

APPENDIX B
Consumer Price Index
All Urban Consumers–(CPI-U)
U.S. city averages
All items
(1982-84 = 100)

Year	CPI	Year	CPI	Year	CPI
1913	9.9	1947	22.3	1981	90.9
1914	10.0	1948	24.1	1982	96.5
1915	10.1	1949	23.8	1983	99.6
1916	10.9	1950	24.1	1984	103.9
1917	12.8	1951	26.0	1985	107.6
1918	15.1	1952	26.5	1986	109.6
1919	17.3	1953	26.7	1987	113.6
1920	20.0	1954	26.9	1988	118.3
1921	17.9	1955	26.8	1989	124.0
1922	16.8	1956	27.2	1990	130.7
1923	17.1	1957	28.1	1991	136.2
1924	17.1	1958	28.9	1992	140.3
1925	17.5	1959	29.1	1993	144.5
1926	17.7	1960	29.6	1994	148.2
1927	17.4	1961	29.9	1995	152.4
1928	17.1	1962	30.2	1996	156.9
1929	17.1	1963	30.6	1997	160.5
1930	16.7	1964	31.0	1998	163.0
1931	15.2	1965	31.5	1999	166.6
1932	13.7	1966	32.4	2000	172.2
1933	13.0	1967	33.4	2001	177.1
1934	13.4	1968	34.8	2002	179.9
1935	13.7	1969	36.7	2003	184.0
1936	13.9	1970	38.8	2004	188.9
1937	14.4	1971	40.5	2005	195.3
1938	14.1	1972	41.8	2006	201.6
1939	13.9	1973	44.4	2007	207.3
1940	14.0	1974	49.3	2008	215.3
1941	14.7	1975	53.8	2009	214.5
1942	16.3	1976	56.9	2010	218.1
1943	17.3	1977	60.6	2011	224.9
1944	17.6	1978	65.2	2012	229.6
1945	18.0	1979	72.6		
1946	19.5	1980	82.4		

Source: U.S. Department of Labor, Bureau of Labor Statistics, Washington, DC 20212.

APPENDIX C
Practice Problems

Basic Equivalence

1. If $250 is invested at 6% on January 1, year 1, how much will be accumulated by January 1, year 10?

2. How much must be invested on January 1, year 1, in order to accumulate $2000 on January 1, year 6 at 6%?

3. What is the present worth on January 1, year 1, of $2000 available at January 1, year 8, if interest is at 6%?

4. If $50 was invested at 6% on January 1, year 1, what equal year-end withdrawals could be made each year for 10 years, leaving nothing in the fund after the 10th withdrawal?

5. How much could be accumulated in a fund earning 6% at the end of 10 years if $20,000 is deposited at the end of each year for 10 years?

6. How much must be deposited at 6% each year beginning on January 1, year 1, in order to accumulate $5000 on the date of the last deposit, January 1, year 6?

7. What amount should be deposited at 6% in order to draw out $400 at the end of each year for seven years, and leave nothing in the fund at the end?

8. If $500 is invested now, $700 two years from now, and $900 four years from now (all at 4%), what will be the total amount in 10 years?

9. What is the compounded amount of $550 left for eight years with interest at nominal 6% compounded semiannually?

10. A savings certificate that costs $50 now will pay $75 in five years. What is the interest rate?

11. How much must be invested at the end of each month for 30 years in a sinking fund that is to amount to $50,000 at the end of 30 years if interest is nominal 4% compounded monthly?

12. How much would be accumulated in the sinking fund for problem 11 at the end of 18 years?

13. In 18 years, $20,000 is required for a child's college expenses. How much should be deposited each year starting on the day of birth so that this goal will be met? Assume that the first payment is made at birth, the last payment is on the child's 18th birthday, and that 5% interest is paid.

14. Starting on January 1, year 1, $50 is deposited in an account paying 6% annually. Each January 1 thereafter, up to and including January 1, year 10, another $50 will be deposited. Starting January 1, year 15 (the date of the first withdrawal), five uniform annual withdrawals are made. The last withdrawal will exhaust the fund. How much will be withdrawn each year?

Practical Loans

15. A $2000 loan is taken out at a bank. Monthly payments are $400 plus interest (10% nominal annual rate) on the unpaid balance. Round all values to the nearest whole dollar. (a) What will be the payments for the loan duration? (b) What principal remains to be paid off after the third payment? (c) What is the interest on the fourth payment?

Present Worth

16. Equipment is purchased for $12,000 and is expected to be sold after 10 years for $2000. The estimated maintenance is $1000 the first year, but is expected to increase by $200 each year thereafter. Using 10%, find the present worth of the project.

17. A fast-acting brake on a fast-turning lathe is estimated to save seven seconds per piece produced since the operator (paid at the rate of $15.00 per hour) does not have to wait as long for the lathe to stop. 40,000 pieces are produced annually. Assuming a three-year life, no salvage value, and an 8% interest rate, what should be the maximum purchase price of the brake?

Capitalized Cost

18. An item is purchased for $100,000. Annual costs are $18,000. Using 8%, what is the capitalized cost of perpetual service?

19. The heat loss from bare steam pipes costs a motel manager $400 annually. Two brands of insulation are available. Brand A will cost $120 and will reduce losses by 93%. Brand B will cost $70 and will reduce losses by 87%. (a) What are the capitalized costs of perpetual service for both brands? (b) Which insulation is economically superior? Use 10%.

Annual Cost

20. A new machine will cost $17,000 and will have an estimated salvage value of $14,000 in five years. Special tools for the new machine will cost $5000 and will have a resale value of $2500 at the end of five years. Maintenance costs are estimated at $200 per year. What will be the average annual cost of ownership during the next five years if interest is 6%?

21. A small building can be reroofed for $6000 (aluminum) or $3500 (shingles). The estimated lives are 50 years and 15 years, respectively. If the interest rate is 10%, which alternative is superior?

22. Designers need to decide how to condition the air in a new building with 40-year life. Two alternatives are available, both of which have 20-year lives. Costs are given below. Use 12% interest to find the best alternative.

	alternative A	alternative B
first cost	$90,000	$60,000
salvage in 20 years	$10,000	$6000
annual fuel costs	$3000	$5000
annual maintenance costs	$2200	$3000
extra annual income tax	$400	0

Rate of Return

23. A $14,000 plot of land can be purchased for $4000 down and $1200 per year for 12 years. What is the annual interest rate being charged?

24. A speculator in land and property pays $14,000 for a house that he expects to hold for 10 years. $1000 is spent in renovation and a monthly rent of $75 is collected from the tenants who live in the house. (Assume all rent is paid at the end of the year.) Taxes are $150 per year and maintenance costs are $250 per year. What must the sales price be in 10 years to realize a 10% rate of return?

25. An investor wishes to invest $40,000. Venture A, requiring $40,000, will return 8%. Venture B, requiring $10,000, will return 15%. What return on the remaining $30,000 is required to equal the overall profitability of venture A?

Benefit-Cost Ratio

26. A large sewer system will cost $175,000 annually. There will be favorable consequences to the general public of $500,000 annually, and adverse consequences to a small segment of the public of $50,000 annually. (a) What is the excess of benefits over costs? (b) What is the benefit-cost ratio?

27. A public works project has an initial cost of $1,000,000, benefits with a present worth of $1,500,000, and disbenefits with a present worth of $300,000. (a) What is the Winfrey benefit-cost ratio? (b) What is the excess of benefits over costs?

Useful Life

28. An asset costing $10,000 has the salvage values and operating costs shown below. Using 8%, when should the asset be replaced with an identical asset?

end of year	salvage	operating cost
1	$6000	$2300
2	$4000	$2500
3	$3200	$3300
4	$2500	$4800
5	$2000	$6800
6	$1800	$9500
7	$1700	$12,000

29. A car costs $6000. Annual costs are $400 the first year, $500 the second year, $600 the third year, and so on, increasing $100 each year. Use 10% to find the year in which the car should be replaced.

Depreciation

30. An asset is purchased for $500,000. The salvage value in 25 years is $100,000. What are the depreciations in the first three years using straight line, double declining balance, and sum-of-the-years' digits depreciation methods?

31. Determine the depreciation charge in the fifth year using sum-of-the-years' digits, double declining balance, and single line methods if an asset is purchased new for $12,000 and has a six-year expected life with $2000 salvage value.

32. Find the book value at the beginning and end of each year by single line, double declining balance, sum-of-the-years' digits, and 6% sinking fund methods for an asset with initial cost of $2500 and with salvage of $1100 in six years.

Income Taxes

33. If the corporate tax rate is 53%, what is the present worth of the six-year after-tax depreciation recovery for the asset in problem 32?

34. An agricultural corporation paying 53% in income taxes wants to build a grain elevator designed to last 25 years at a cost of $80,000, with no salvage value. Annual income generated will be $22,500 and annual expenditures will be $12,000. Using single line depreciation and a 10% interest rate, what is the 25-year after-tax present worth of the project?

35. Repeat problem 34 using sum-of-the-years' digits depreciation.

36. A small company pays 30% in income taxes. A $2000 machine with a four-year life and zero salvage value will generate revenues of $1200 annually. Using 8%, compare the present worth of the after-tax profits using single line and sum-of-the-years' digits depreciation.

Rate and Period Changes

37. What is the true effective annual rate of a credit card plan that charges 1.5% per month on an unpaid balance?

38. A loan company advertises that $100 borrowed for one year may be repaid by 12 monthly installments of $9.46. Assuming the difference between the amount repaid and the amount borrowed is interest only, what is the effective annual interest rate being charged?

39. What is the effective interest rate for a payment plan of 30 equal payments of $89.30 per month, when a lump sum of $2000 would have been an outright purchase?

Cost Accounting

40. The total estimated costs of producing 240 and 360 units are $3400 and $4000, respectively. (a) What is the average unit cost over the first 240 units? (b) What is the incremental cost? (c) What is the fixed cost? (d) If only 249 units are ultimately produced, what is the total profit or loss from the sale of units 240 through 249 at $10.47 each?

41. A machine with an eight-year life and $5000 salvage is purchased for $40,000. Annual costs of operation are $800. The machine operator is paid $12.90 per hour. Power is consumed by the machine at the rate of $2.15 per hour. 2000 units are produced each year on the machine, which requires 48 minutes per unit. Use a 10% interest rate to find the unit cost.

42. A factory with capacity of 700,000 units per year operates at 62% capacity. The annual income is $430,000, annual fixed costs are $190,000, and the variable costs are $0.348 per unit. (a) What is the current profit or loss? (b) What is the break-even point?

Break-Even Analysis

43. A fixture that costs $700 will save $0.06 per item produced. Maintenance will be $40 annually. 3500 units are produced annually. What is the pay-back period at 10%?

44. A hand tool costs $200 and requires $1.21 labor cost per unit. A machine tool costs $3600 and reduces labor to $0.75. What is the break-even point (in years) at 5% for an annual production of 4000 units?

Supplementary Practice Problems

1. A city engineer must find the capitalized cost of perpetual service from a water plant. A storage tank that costs $40,000 is needed now. In 10 years, a second $60,000 tank will be needed. At that time, a $7000 pumping system with life of eight years will be installed. Pump maintenance and operation costs are $800 the first year, $900 the second, and increases by $100 each year for the eight-year life. There is no salvage value. Use 10% to find the capitalized cost.

2. A construction company operates a bulldozer that initially costs $28,000. The first year maintenance is $600, the second year is $900, and $1200 the third year. The salvage value at the end of the third year is $15,000. If the company decides to keep the bulldozer beyond the third year, the maintenance costs are $900 the fourth year, $1150 the fifth year, and $1300 the sixth year. There is a $7000 cost at the end of the third year if the bulldozer is kept. The salvage value at the end of six years is $10,000. If the bulldozer is kept for two more years, another overhaul at the end of the sixth year will cost $14,000. Maintenance costs will be $9900 at the end of the seventh and eighth years. Compare the annual costs if the bulldozer is kept for three years, six years, or eight years. Assume the eight-year salvage value is zero. Use 10%.

3. During a slack period, a manufacturer can sell 3000 articles per month at a price of $1.20 each. A $200,000 investment is used which is depreciated over a 20-year life. An annual fixed cost of $20,000 must be considered, as well as an annual maintenance cost of $10,000. The production cost is $1.10 per item. The manufacturer's tax rate is 45%. Should the factory shut down?

4. A production order for 200 units is received. The item is made on a machine that has a $20 setup, an output of six items per hour, and a daily fixed cost of $100. The variable item cost is $1.00. What is the highest price the company could afford to buy the item for, instead of producing the item itself?

5. A car is purchased for $8000 and is run the same distance each year. Annual costs are $1000. The trade-in value is $3900 at the end of the first year, but decreases $550 each year. When should the car be traded in to minimize the annual cost?

6. A corporation needs to raise money and sells an issue of 4%, 20-year bonds worth $1,000,000 to an investment company for $970,000. An initial cost of $50,000 is incurred for legal fees, and accounting fees will run $10,000 annually. What is the cost of the bond issue to the corporation?

7. An investor purchased stock for $1000. His dividends in the first five years were $140, $150, $170, $50, and $30. He sold the stock for $700 at the end of the fifth year. What was the rate of return on his investment?

8. A manufacturer uses 1000 items uniformly throughout the year. The cost to produce the items in-house is $40 for setup and $5.20 per item. Insurance and taxes are estimated as 12% of average inventory, and it costs $0.80 to store each unit for one complete year. How many of these items should the manufacturer produce at a time?

9. An individual expects his return on an investment to be $10,000 the first year and to increase exponentially at 15% each year for the next five years. Using a 5% annual interest rate, compute the present worth of the investment using both the year-end and month-end conventions.

10. The demand for a product during the next 10 years will be such that revenues will be $100,000 the first year and will increase by $10,000 each year thereafter. If inflation is assumed to be 8% per year and the company uses 10% in its economic studies, what is the present worth of the revenues expressed in present dollars?

11. A company can increase its share of the market if it invests $40,000. The increased income would be $30,000 annually, but increased expenses would be $23,000. The equipment salvage at the end of 12 years would be $6000. The income tax rate is 52% and double declining balance depreciation is used. The company's minimum attractive rate of return is 6%. What is the present worth of the investment?

12. A company must invest in a precipitator to meet EPA regulations. Four alternatives are available, all of which have the same capacity. The company's effective annual interest rate for investments is 8%. Which precipitator should be recommended?

	A	B	C	D
cost	$4500	$5700	$6750	$7120
life	10	10	10	10
O&M costs	$2700	$2500	$2200	$2030
salvage	0	0	0	0

APPENDIX D
Solutions to Practice Problems

1. $F = P(F/P, 6\%, 9) = (\$250)(1.6895)$

$= \boxed{\$422}$

2. $P = F(P/F, 6\%, 5) = (\$2000)(0.7473)$

$= \boxed{\$1495}$

3. $P = F(P/F, 6\%, 7) = (\$2000)(0.6651)$

$= \boxed{\$1330}$

4. $A = P(A/P, 6\%, 10) = (\$50)(0.1359)$

$= \boxed{\$6.80}$

5. $F = A(F/A, 6\%, 10) = (\$20{,}000)(13.1808)$

$= \boxed{\$263{,}600}$

6. $A = F(A/F, 6\%, 6) = (\$5000)(0.1434)$

$= \boxed{\$717}$

7. $P = A(P/A, 6\%, 7) = (\$400)(5.5824)$

$= \boxed{\$2233}$

8. $F = (\$500)(F/P, 4\%, 10) + (\$700)(F/P, 4\%, 8) + (\$900)(F/P, 4\%, 6)$
     $= (\$500)(1.4802) + (\$700)(1.3686) + (\$900)(1.2653)$

$= \boxed{\$2837}$

9.
$$i = \frac{r}{k} = \frac{6\%}{2} = 3\%$$
$$n = (2)(8) = 16$$
$$F = P(F/P, 3\%, 16) = (\$550)(1.6047)$$
$$= \boxed{\$883}$$

10.
$$P = F(P/F, i\%, 5)$$
$$(P/F, i\%, 5) = \frac{P}{F} = \frac{50}{75} = 0.6667$$

From the table,
$$(P/F, 8\%, 5) = 0.6806$$
$$(P/F, 9\%, 5) = 0.6499$$
$$i = 8\% + (9\% - 8\%)\left(\frac{0.6667 - 0.6806}{0.6499 - 0.6806}\right)$$
$$= \boxed{8.45\%}$$

$(F/P, i, 5)$ can also be solved directly for i.
$$(1 + i)^5 = 1.5$$
$$i = 0.08447 \ (8.447\%)$$

11.
$$\phi = \frac{r}{k} = \frac{4\%}{12} = 0.3333\%$$
$$n = (30)(12) = 360$$
$$A = F(A/F, 0.3333\%, 360) = F\left[\frac{i}{(1 + i)^n - 1}\right]$$
$$= (\$50,000)\left[\frac{0.003333}{(1 + 0.003333)^{360} - 1}\right] = \boxed{\$72.05}$$

12.
$$n = (18)(12) = 216$$
$$F = A(F/A, 0.3333\%, 216) = A\left[\frac{(1 + i)^n - 1}{i}\right]$$
$$= (\$72.05)\left[\frac{(1 + 0.003333)^{216} - 1}{0.003333}\right]$$
$$= (\$72.05)(315.58)$$
$$= \boxed{\$22,738}$$

13. $A = F(A/F, 5\%, 19) = (\$20{,}000)(0.0327)$

$$= \boxed{\$654}$$

14. It helps to draw a cash flow diagram. Present worth as of January 1, year 0 of the deposits is

$$P_{\text{deposits}} = (\$50)(P/A, 6\%, 10)$$
$$= (\$50)(7.3601) = \$368.00$$

Present worth as of January 1, year 0 of the withdrawal is

$$P_{\text{withdrawal}} = (A_w)\big[(P/A, 6\%, 19) - (P/A, 6\%, 14)\big]$$
$$= (A_w)(11.1581 - 9.2950) = 1.8631 A_w$$

Since the last withdrawal exhausts the fund,

$$P_{\text{deposits}} = P_{\text{withdrawals}}$$
$$\$368.00 = 1.8631 A_w$$

$$A_w = \boxed{\$197.52}$$

15. (a) Number of compounding periods:

$$n = \frac{\$2000}{\$400} = 5$$

Effective interest rate:

$$i = \frac{r}{k} = \frac{0.10}{12} = 0.0083$$

payment = monthly payment + interest on unpaid balance

First month:

payment $= \$400 + (\$2000)(0.0083) = \boxed{\$417}$

Second month:

payment $= \$400 + (\$2000 - \$400)(0.0083) = \boxed{\$413}$

Third month:

$$\text{payment} = \$400 + (\$2000 - \$800)(0.0083) = \boxed{\$410}$$

Fourth month:

$$\text{payment} = \$400 + (\$2000 - \$1200)(0.0083) = \boxed{\$407}$$

Fifth month:

$$\text{payment} = \$400 + (\$2000 - \$1600)(0.0083) = \boxed{\$403}$$

(b) Principal remaining after the third payment is

$$\$2000 - (3)(\$400) = \boxed{\$800}$$

(c) Interest on the fourth payment is

$$\$407 - \$400 = \boxed{\$7}$$

16. $P = -\$12,000 + (\$2000)(P/F, 10\%, 10) - (\$1000)(P/A, 10\%, 10)$
$\qquad - (\$200)(P/G, 10\%, 10)$
$\quad = -\$12,000 + (\$2000)(0.3855) - (\$1000)(6.1446) - (\$200)(22.8913)$
$\quad = \boxed{-\$21,950}$

17. Amount of money saved per year:

$$A = \left(40,000\ \frac{\text{pieces}}{\text{yr}}\right)\left(7\ \frac{\text{sec}}{\text{piece}}\right)\left(\frac{1\ \text{hr}}{3600\ \text{sec}}\right)\left(\frac{\$15.00}{\text{hr}}\right)$$
$$= \$1166.67/\text{yr}$$

For a 3-year life, $n = 3$, the maximum purchase price, P, is

$$P = A(P/A, 8\%, 3) = (\$1166.67)(2.5771)$$
$$= \boxed{\$3006.63}$$

PROFESSIONAL PUBLICATIONS, INC.

18. The service has an infinite life; the capitalized cost, CC, is

$$CC = \text{initial cost} + \frac{\text{actual cost}}{i}$$

$$= \$100,000 + \frac{\$18,000}{0.08} = \boxed{\$325,000}$$

19. (a) Brand A:

$$CC = \$120 + \frac{(\$400)(1 - 0.93)}{0.10} = \boxed{\$400}$$

Brand B:

$$CC = \$70 + \frac{(\$400)(1 - 0.87)}{0.10} = \boxed{\$590}$$

(b) $\boxed{\text{Brand A is superior}}$

20. $\text{EUAC} = C_1(A/P, 6\%, 5) - S_1(A/F, 6\%, 5) + C_2(A/P, 6\%, 5)$
                 $- S_2(A/F, 6\%, 5) + \text{maintenance}$
          $= (\$17,000)(0.2374) - (\$14,000)(0.1774) + (\$5000)(0.2374)$
             $- (\$2500)(0.1774) + \$200$

$$= \boxed{\$2495.70}$$

21. Aluminum:

$$\text{EUAC} = (\$6000)(A/P, 10\%, 50)$$
$$= (\$6000)(0.1009) = \$605$$

Shingles:

$$\text{EUAC} = (\$3500)(A/P, 10\%, 15)$$
$$= (\$3500)(0.1315) = \$460$$

$$\boxed{\text{shingles are superior}}$$

22. Alternative A:

$$\begin{aligned}
\text{EUAC} &= C(A/P, 12\%, 20) + \text{other annual costs} - S(A/F, 12\%, 20) \\
&= (\$90{,}000)(0.1339) + (\$3000 + \$2200 + \$400) - (\$10{,}000)(0.0139) \\
&= \$17{,}512
\end{aligned}$$

Alternative B:

$$\begin{aligned}
\text{EUAC} &= (\$60{,}000)(0.1339) + (\$5000 + \$3000) - (\$6000)(0.0139) \\
&= \$15{,}951
\end{aligned}$$

$$\boxed{\text{alternative B is best}}$$

23.
$$\begin{aligned}
A &= (\text{amount of loan})(A/P, i\%, 12) \\
&= (\$14{,}000 - \$4000)(A/P, i\%, 12)
\end{aligned}$$
$$(A/P, i\%, 12) = \frac{A}{\$10{,}000} = \frac{\$1200}{\$10{,}000} = 0.1200$$

From the table,
$$(A/P, 6\%, 12) = 0.1193$$
$$(A/P, 7\%, 12) = 0.1259$$
$$i = 6\% + (7\% - 6\%)\left(\frac{0.1200 - 0.1193}{0.1259 - 0.1193}\right) = \boxed{6.1\%}$$

24.
$$\begin{aligned}
F &= (\$14{,}000 + \$1000)(F/P, 10\%, 10) \\
&\quad + \big[\$150 + \$250 - (\$75)(12)\big](F/A, 10\%, 10) \\
&= (\$15{,}000)(2.5937) - (\$500)(15.9374) \\
&= \boxed{\$30{,}937}
\end{aligned}$$

25.
$$\begin{aligned}
P_A(0.08) &= P_B(0.15) + (P_A - P_B)i \\
i &= \frac{P_A(0.08) - P_B(0.15)}{P_A - P_B} \\
&= \frac{(\$40{,}000)(0.08) - (\$10{,}000)(0.15)}{\$40{,}000 - \$10{,}000} \\
&= 0.0567 = \boxed{5.67\%}
\end{aligned}$$

26. (a) $B - C = \$500{,}000 - \$175{,}000 - \$50{,}000$

$$= \boxed{\$275{,}000}$$

(b) $\dfrac{B - D}{C} = \dfrac{\$500{,}000 - \$50{,}000}{\$175{,}000} = \boxed{2.57}$

27. (a) $\dfrac{B - D}{C} = \dfrac{\$1{,}500{,}000 - \$300{,}000}{\$1{,}000{,}000} = \boxed{1.2}$

(b) $B - C = \$1{,}500{,}000 - \$300{,}000 - \$1{,}000{,}000$

$$= \boxed{\$200{,}000}$$

28.   EUAC of capital $= (P - \text{salvage of defender})(A/P, 8\%, n) + Si$
EUAC of maintenance $=$ EUAC of operating costs

Year 1:

$$\text{total EUAC} = (\$10{,}000 - \$6000)(1.0800) + (\$6000)(0.08) + \$2300$$
$$= \$7100$$

Year 2:

$$\text{total EUAC} = (\$10{,}000 - \$4000)(0.5608) + (\$4000)(0.08)$$
$$+ [(\$2300)(P/F, 8\%, 1) + (\$2500)(P/F, 8\%, 2)](A/P, 8\%, 2)$$
$$= \$3685 + [(\$2300)(0.9259) + (\$2500)(0.8573)](0.5608)$$
$$= \$6081$$

Year 3:

$$\text{total EUAC} = (\$10{,}000 - \$3200)(0.3880) + (\$3200)(0.08) + [(\$2300)(0.9259)$$
$$+ (\$2500)(0.8573) + (\$3300)(P/F, 8\%, 3)](A/P, 8\%, 3)$$
$$= \$2894 + [\$2130 + \$2143 + (\$3300)(0.7938)](0.3880)$$
$$= \$5568$$

Year 4:

$$\text{total EUAC} = (\$10{,}000 - \$2500)(0.3019) + (\$2500)(0.08)$$
$$+ [\$2130 + \$2143 + \$2620 + (\$4800)(P/F, 8\%, 4)](A/P, 8\%, 4)$$
$$= \$2464 + [\$6893 + (\$4800)(0.7350)](0.3019)$$
$$= \$5610$$

Replace the asset at the end of year 3.

29.   EUAC of capital $= P(A/P, 10\%, n)$

EUAC of maintenance $= A + (\$100)(A/G, 10\%, n)$

year	EUAC of capital	EUAC of maintenance	total EUAC
1	($6000)(1.1000)	$400 + ($100)(0)	$7000
2	($6000)(0.5762)	$400 + ($100)(0.4762)	$3905
3	($6000)(0.4021)	$400 + ($100)(0.9366)	$2906
4	($6000)(0.3155)	$400 + ($100)(1.3812)	$2431
5	($6000)(0.2638)	$400 + ($100)(1.8101)	$2164
6	($6000)(0.2296)	$400 + ($100)(2.2236)	$2000
7	($6000)(0.2054)	$400 + ($100)(2.6216)	$1895
8	($6000)(0.1874)	$400 + ($100)(3.0045)	$1825
9	($6000)(0.1736)	$400 + ($100)(3.3724)	$1779
10	($6000)(0.1627)	$400 + ($100)(3.7255)	$1749
11	($6000)(0.1540)	$400 + ($100)(4.0641)	$1730
12	($6000)(0.1468)	$400 + ($100)(4.3884)	$1720
13	($6000)(0.1408)	$400 + ($100)(4.6988)	$1715
14	($6000)(0.1357)	$400 + ($100)(4.9955)	$1714
15	($6000)(0.1315)	$400 + ($100)(5.2789)	$1717

The car should be replaced at the end of year 14.

30. SL:

$$\text{annual depreciation} = \frac{\$500{,}000 - \$100{,}000}{25 \text{ yr}}$$
$$= \$16{,}000/\text{yr}$$

$$D_1 = D_2 = D_3 = \boxed{\$16{,}000}$$

DDB:

$$D_j = \left(\frac{2C}{n}\right)\left(1 - \frac{2}{n}\right)^{j-1}$$

$$D_1 = \left[\frac{(2)(\$500,000)}{25}\right]\left(1 - \frac{2}{25}\right)^{1-1} = \boxed{\$40,000}$$

$$D_2 = \left[\frac{(2)(\$500,000)}{25}\right]\left(1 - \frac{2}{25}\right)^{2-1} = \boxed{\$36,800}$$

$$D_3 = \left[\frac{(2)(\$500,000)}{25}\right]\left(1 - \frac{2}{25}\right)^{3-1} = \boxed{\$33,860}$$

SOYD: $\qquad T = \frac{1}{2}n(n+1) = \left(\frac{1}{2}\right)(25)(25+1) = 325$

Depreciation in year j:

$$D_j = \frac{(c - S_n)(n - J + 1)}{T}$$

$$D_1 = \frac{(\$500,000 - \$100,000)(25 - 1 + 1)}{325} = \boxed{\$30,770}$$

$$D_2 = \frac{(\$500,000 - \$100,000)(25 - 2 + 1)}{325} = \boxed{\$29,540}$$

$$D_3 = \frac{(\$500,000 - \$100,000)(25 - 3 + 1)}{325} = \boxed{\$28,310}$$

31. SOYD: $\qquad T = \frac{1}{2}n(n+1) = \left(\frac{1}{2}\right)(6)(6+1) = 21$

Year 5:

$$D_5 = \frac{(C - S_n)(n - j + 1)}{T}$$

$$= \frac{(\$12,000 - \$2000)(6 - 5 + 1)}{21}$$

$$= \boxed{\$952}$$

DDB:

$$D_j = \left(\frac{2C}{n}\right)\left(1 - \frac{2}{n}\right)^{j-1}$$

$$D_5 = \left[\frac{(2)(\$12,000)}{6}\right]\left(1 - \frac{2}{6}\right)^{5-1} = \$790$$

Check to see if the book value has dropped below the salvage value (which is not permitted).

$$BV_5 = (\$12{,}000)\left(1 - \frac{2}{6}\right)^5 = 1580$$

Since the book value would be less than $2000, the maximum depreciation allowed is the difference between the previous book value and $2000.

$$BV_4 = (\$12{,}000)\left(1 - \frac{2}{6}\right)^4 = \$2370$$

$$D_5 = BV_4 - S_5 = \$2370 - \$2000 = \boxed{\$370}$$

SL:
$$D_5 = \frac{C - S_n}{n} = \frac{\$12{,}000 - \$2000}{6} = \boxed{\$1667}$$

32. SL:
$$D = \frac{C - S}{n} = \frac{\$2500 - \$1100}{6} = \$233$$

$$BV_1 = \$2500 - \$233 = \boxed{\$2267}$$

$$BV_2 = \$2267 - \$233 = \boxed{\$2034}$$

$$BV_3 = \$2034 - \$233 = \boxed{\$1801}$$

$$BV_4 = \$1801 - \$233 = \boxed{\$1568}$$

$$BV_5 = \$1568 - \$233 = \boxed{\$1335}$$

$$BV_6 = \$1335 - \$233 = \boxed{\$1102}$$

DDB:

$$D_j = \frac{2C}{n}\left(1 - \frac{2}{n}\right)^{j-1}$$

$$D_1 = \left[\frac{(2)(\$2500)}{6}\right]\left(1 - \frac{2}{6}\right)^{1-1} = \$833$$

$$BV_1 = \$2500 - \$833 = \boxed{\$1667}$$

$$D_2 = \left[\frac{(2)(\$2500)}{6}\right]\left(1 - \frac{2}{6}\right)^{2-1} = \$556$$

$$BV_2 = \$1667 - \$556 = \boxed{\$1111}$$

$$D_3 = \left(\frac{\$2500}{3}\right)\left(1 - \frac{2}{6}\right)^{3-1} = \$370$$

$$BV_3 = \$1111 - \$370 = \$741$$

Since book value cannot be less than salvage value,

$$BV_3 = \boxed{\$1100}$$

$$D_3 = \$1111 - \$1100 = \$11$$

SOYD:

$$T = \tfrac{1}{2}n(n + 1) = \left(\tfrac{1}{2}\right)(6)(6 + 1) = 21$$

$$D_j = \frac{(C - S_n)(n - j + 1)}{T}$$

$$C - S_n = \$2500 - \$1100 = \$1400$$

$$D_1 = \frac{(\$1400)(6 - 1 + 1)}{21} = \$400$$

$$BV_1 = \$2500 - \$400 = \boxed{\$2100}$$

$$D_2 = \frac{(\$1400)(6 - 2 + 1)}{21} = \$333$$

$$BV_2 = \$2100 - \$333 = \boxed{\$1767}$$

$$D_3 = \frac{(\$1400)(6 - 3 + 1)}{21} = \$267$$

$$BV_3 = \$1767 - \$267 = \boxed{\$1500}$$

$$D_4 = \frac{(\$1400)(6 - 4 + 1)}{21} = \$200$$

$$BV_4 = \$1500 - \$200 = \boxed{\$1300}$$

$$D_5 = \frac{(\$1400)(6 - 5 + 1)}{21} = \$133$$

$$BV_5 = \$1300 - \$133 = \boxed{\$1167}$$

$$D_6 = \frac{(\$1400)(6 - 6 + 1)}{21} = \$67$$

$$BV_6 = \$1167 - \$67 = \boxed{\$1100}$$

Sinking fund:

$$D_j = (C - S_n)(A/F, i\%, n)(F/P, i\%, j - 1)$$
$$C - S_n = \$2500 - \$1100 = \$1400$$

$$D_1 = (\$1400)(A/F, 6\%, 6)(F/P, 6\%, 0)$$
$$= (\$1400)(0.1434)(1) = \$201$$

$$BV_1 = \$2500 - \$201 = \boxed{\$2299}$$

$$D_2 = (\$201)(F/P, 6\%, 1) = (\$201)(1.0600) = \$213$$

$$BV_2 = \$2299 - \$213 = \boxed{\$2086}$$

$$D_3 = (\$201)(F/P, 6\%, 2) = (\$201)(1.1236) = \$226$$

$$BV_3 = \$2086 - \$226 = \boxed{\$1860}$$

$$D_4 = (\$201)(F/P, 6\%, 3) = (\$201)(1.1910) = \$239$$

$$BV_4 = \$1860 - \$239 = \boxed{\$1621}$$

$$D_5 = (\$201)(F/P, 6\%, 4) = (\$201)(1.2625) = \$254$$

$$BV_5 = \$1621 - \$254 = \boxed{\$1367}$$

$$D_6 = (\$201)(F/P, 6\%, 5) = (\$201)(1.3382) = \$269$$

$$BV_6 = \$1367 - \$269 = \boxed{\$1098}$$

33. SL: annual depreciation, $D = \$233$

$$DR = (0.53)D(P/A, 6\%, 6)$$

$$= (0.53)(\$233)(4.9173) = \boxed{\$607}$$

DDB: $DR = (0.53)[D_1(P/F, 6\%, 1) + D_2(P/F, 6\%, 2) + D_3(P/F, 6\%, 3)]$

$$= (0.53)[(\$833)(0.9434) + (\$556)(0.8900) + (\$11)(0.8396)]$$

$$= \boxed{\$684}$$

SOYD: The depreciation begins at \$400 per year and decreases \$67 per year after the first year.

$$A = \$400$$

$$G = \$67$$

$$DR = (0.53)[A(P/A, 6\%, 6) - G(P/G, 6\%, 6)]$$

$$= (0.53)[(\$400)(4.9173) - (\$67)(11.4594)]$$

$$= \boxed{\$635}$$

Sinking fund: The present worth at the end of each year of depreciation is equal to the depreciation at the end of the first year (\$201).

$$DR = (0.53)[(\$201)(6)(P/F, 6\%, 1)]$$

$$= (0.53)[(\$201)(6)(0.9434)] = \boxed{\$603}$$

34.   annual depreciation $= \dfrac{\$80,000}{25} = \$3200$

annual taxable income $= \$22,500 - \$12,000 - \$3200 = \$7300$

annual income (net) $= \$22,500 - \$12,000 - (0.53)(\$7300) = \$6631$

$$P_t = -\$80,000 + A(P/A, 10\%, 25)$$

$$= -\$80,000 + (\$6631)(9.0770) = \boxed{-\$19,810}$$

35. $T = \frac{1}{2}n(n + 1) = \left(\frac{1}{2}\right)(25)(25 + 1) = 325$

End of the first year depreciation:

$$D_1 = \frac{Cn}{T} = \frac{(\$80,000)(25)}{325} = \$6154$$

Decreasing depreciation gradient:

$$G = \frac{C}{T} = \frac{\$80{,}000}{325} = \$246$$

$$
\begin{aligned}
P_t &= -\$80{,}000 + (\$22{,}500 - \$12{,}000)(P/A, 10\%, 25)(0.47) \\
&\quad + (\$6154)(P/A, 10\%, 25)(0.53) - (\$246)(P/G, 10\%, 25)(0.53) \\
&= -\$80{,}000 + (\$10{,}500)(9.0770)(0.47) + (\$6154)(9.0770)(0.53) \\
&\quad - (\$246)(67.6964)(0.53)
\end{aligned}
$$

$$= \boxed{-\$14{,}430}$$

36. SL: $\qquad$ annual depreciation $= \dfrac{C}{n} = \dfrac{\$2000}{4} = \$500$

$$
\begin{aligned}
P_t &= -\$2000 + (\$1200)(1 - 0.30)(P/A, 8\%, 4) + D(0.30)(P/A, 8\%, 4) \\
&= -\$2000 + (\$1200)(0.70)(3.3121) + (\$500)(0.30)(3.3121)
\end{aligned}
$$

$$= \boxed{\$1279}$$

SOYD: $\qquad T = \tfrac{1}{2}n(n+1) = \left(\tfrac{1}{2}\right)(4)(4+1) = 10$

End of the first year depreciation:

$$D_1 = \frac{Cn}{T} = \frac{(\$2000)(4)}{10} = \$800$$

Decreasing depreciation gradient:

$$G = \frac{C}{T} = \frac{\$2000}{10} = \$200$$

$$
\begin{aligned}
P_t &= -\$2000 + (\$1200)(0.70)(P/A, 8\%, 4) + (\$800)(0.30)(P/A, 8\%, 4) \\
&\quad - (\$200)(0.30)(P/G, 8\%, 4) \\
&= -\$2000 + (\$1200)(0.70)(3.3121) + (\$800)(0.30)(3.3121) \\
&\quad - (\$200)(0.30)(4.6501)
\end{aligned}
$$

$$= \boxed{\$1298}$$

37. Effective annual rate:

$$i = (1 + \phi)^k - 1 = (1 + 0.015)^{12} - 1$$

$$= 0.1956 = \boxed{19.56\%}$$

38. $$P = A(P/A, i\%, 12)$$

$$(P/A, i\%, 12) = \frac{P}{A} = \frac{\$100}{\$9.46} = 10.5708$$

At $n = 12$ in the (P/A) column, $\phi = 2\%$.

Effective annual rate:

$$i = (1 + \phi)^k - 1 = (1 + 0.02)^{12} - 1$$

$$= 0.2682 = \boxed{26.82\%}$$

39. $$P = A(P/A, i\%, n)$$

$$(P/A, i\%, 30) = \frac{\$2000}{\$89.30} = 22.3964$$

From the table, monthly effective interest rate is 2%.

effective annual interest rate $= (1 + 0.02)^{12} - 1 = 0.2682 = \boxed{26.82\%}$

40. (a) average unit cost of first 240 units $= \dfrac{\$3400}{240 \text{ units}} = \boxed{\$14.17/\text{unit}}$

(b) incremental cost $= \dfrac{\$4000 - \$3400}{360 \text{ units} - 240 \text{ units}} = \dfrac{\$600}{120 \text{ units}} = \boxed{\$5.00/\text{unit}}$

(c) fixed cost $= \$3400 - \left(\dfrac{\$5.00}{\text{unit}}\right)(240 \text{ units}) = \boxed{\$2200}$

(d) Without the fixed cost:

$$\text{profit} = R - C = (10)(\$10.47 - \$5) = \boxed{\$54.70}$$

With the fixed cost:

$$\text{profit} = (10)(\$10.47) - (10)\left(\$5.00 + \frac{\$2200}{249}\right)$$

$$= -\$33.65 = \boxed{\text{loss of } \$33.65}$$

41. Annual power cost:

$$A_1 = \left[\frac{(2000 \text{ units})\left(48 \dfrac{\text{min}}{\text{unit}}\right)}{60 \dfrac{\text{min}}{\text{hr}}}\right]\left(\frac{\$2.15}{\text{hr}}\right) = \$3440$$

Annual labor cost:

$$A_2 = (1600 \text{ hr}) \left(\frac{\$12.90}{\text{hr}} \right) = \$20,640$$

$$\text{unit cost} = \frac{C(A/P, 10\%, 8) - S_8(A/F, 10\%, 8) + A_1 + A_2 + \text{operation cost}}{2000 \text{ units}}$$

$$= \frac{(\$40,000)(0.1874) - (\$5000)(0.0874) + \$3440 + \$20,640 + \$800}{2000 \text{ units}}$$

$$= \boxed{\$15.97/\text{unit}}$$

42. (a)
$$\text{total cost} = (700,000)(0.62)(\$0.348) + \$190,000$$
$$= \$341,032$$
$$\text{profit} = R - C = \$430,000 - \$341,032$$
$$= \boxed{\$88,968}$$

(b)
$$p = \frac{\$430,000}{(700,000 \text{ units})(0.62)} = \$0.991/\text{unit}$$

Break-even point:

$$Q^* = \frac{f}{p - a} = \frac{\$190,000}{\dfrac{\$0.991}{\text{unit}} - \dfrac{\$0.348}{\text{unit}}} = \boxed{295,500 \text{ units}}$$

43.
$$\text{annual amount saved} = (3500)(\$0.06) = \$210$$

Pay-back period (traditional definition):

$$\frac{\text{initial investment}}{\text{net annual profit}} = \frac{\$700}{\dfrac{\$210}{\text{yr}} - \dfrac{\$40}{\text{yr}}}$$
$$= \boxed{4.12 \text{ yr}}$$

Pay-back period at 10% interest:

$$P = A(P/A, 10\%, n)$$
$$\$700 = (\$210 - \$40)(P/A, 10\%, n)$$

$$(P/A, 10\%, n) = \frac{\$700}{\$170} = 4.1176 = \frac{(1 + i)^n - 1}{i(1 + i)^n}$$

$$(4.1176)(0.10)(1.10)^n - (1.10)^n = -1$$

$$n = \frac{\ln(1.6999)}{\ln(1.10)} = \boxed{5.57 \text{ yr}}$$

44.

$$\text{EUAC}_{\text{hand tool}} = (\$200)(A/P, 5\%, n) + (4000)(\$1.21)$$
$$\text{EUAC}_{\text{machine tool}} = (\$3600)(A/P, 5\%, n) + (4000)(\$0.75)$$

$$(\$3400)(A/P, 5\%, n) = \$1840$$

$$(A/P, 5\%, n) = \frac{\$1840}{\$3400} = 0.5412$$

$$0.5412 = \frac{i(1 + i)^n}{(1 + i)^n - 1} = \frac{(0.05)(1 + 0.05)^n}{(1 + 0.05)^n - 1}$$

$$(0.05)(1.05)^n = (0.5412)(1.05)^n - 0.5412$$

$$(1.05)^n = \frac{0.5412}{0.4912} = 1.1018$$

Take the log of both sides of the equation.

$$n = \frac{\log(1.1018)}{\log(1.05)} = \boxed{1.99 \text{ yr}}$$

Solutions to Supplementary Practice Problems

1. This is a capitalized cost problem within another capitalized cost problem. At the 10-year point, the annualized cost of all expenses encountered from that point on is

$$A = (\$60,000)(A/P, 10\%, \infty) + (\$7000)(A/P, 10\%, 8) + \$800$$
$$+ (\$100)(A/G, 10\%, 8)$$
$$= (\$60,000)(0.10) + (\$7000)(0.1874) + \$800 + (\$100)(3.0045)$$
$$= \$8412.25$$

The present worth at $t = 10$ for perpetual service is

$$P_{10} = (\$8412.25)(P/A, 10\%, \infty)$$
$$= \frac{\$8412.25}{0.10}$$
$$= \$84{,}122.50$$

The capitalized cost of the entire project is

$$CC = \$40{,}000 + P_{10}(P/F, 10\%, 10)$$
$$= \$40{,}000 + (\$84{,}122.50)(0.3855)$$

$$= \boxed{\$72{,}429}$$

2. Kept for 3 years:

$$\text{EUAC}(3) = (\$28{,}000)(A/P, 10\%, 3) + \$600 + (\$300)(A/G, 10\%, 3)$$
$$- (\$15{,}000)(A/F, 10\%, 3)$$
$$= (\$28{,}000)(0.4021) + \$600 + (\$300)(0.9366) - (\$15{,}000)(0.3021)$$

$$= \boxed{\$7608}$$

Kept for 6 years:

$$\text{EUAC}(6) = [(\$28{,}000)(A/P, 10\%, 6) - (\$10{,}000)(A/F, 10\%, 6)]$$
$$+ [(\$600)(P/A, 10\%, 3) + (\$300)(P/G, 10\%, 3)$$
$$+ (\$7000)(P/F, 10\%, 3) + (\$900)(P/F, 10\%, 4)$$
$$+ (\$1150)(P/F, 10\%, 5) + (\$1300)(P/F, 10\%, 6)](A/P, 10\%, 6)$$
$$= [(\$28{,}000)(0.2296) - (\$10{,}000)(0.1296)] + [(\$600)(2.4869)$$
$$+ (\$300)(2.3291) + (\$7000)(0.7513) + (\$900)(0.6830)$$
$$+ (\$1150)(0.6209) + (\$1300)(0.5645)](0.2296)$$
$$= \$5132 + (\$9512)(0.2296)$$

$$= \boxed{\$7317}$$

Kept for 8 years:

$$
\begin{aligned}
\text{EUAC(8)} &= (\$28{,}000)(A/P, 10\%, 8) + [(\$600)(P/A, 10\%, 3) \\
&\quad + (\$300)(P/G, 10\%, 3) + (\$7000)(P/F, 10\%, 3) \\
&\quad + (\$900)(P/F, 10\%, 4) + (\$1150)(P/F, 10\%, 5) \\
&\quad + (\$1300 + \$14{,}000)(P/F, 10\%, 6) + (\$9900)(P/F, 10\%, 7) \\
&\quad + (\$9900)(P/F, 10\%, 8)](A/P, 10\%, 8) \\
&= (\$28{,}000)(0.1874) + [(\$600)(2.4869) + (\$300)(2.3291) \\
&\quad + (\$7000)(0.7513) + (\$900)(0.6830) + (\$1150)(0.6209) \\
&\quad + (\$15{,}300)(0.5645) + (\$9900)(0.5132) + (\$9900)(0.4665)](0.1874) \\
&= \$5247 + (\$27{,}115)(0.1874)
\end{aligned}
$$

$$
= \boxed{\$10{,}328}
$$

3. $\text{after-tax revenue per year} = (\$1.20)\left(\dfrac{3000}{\text{month}}\right)\left(12\,\dfrac{\text{month}}{\text{yr}}\right)(1 - 0.45)$

$$
= \$23{,}760/\text{yr}
$$

For annual depreciation, assume straight-line method.

$$
D = \frac{C}{n} = \frac{\$200{,}000}{20}
$$

$$
= \$10{,}000/\text{yr}
$$

$$
\text{annual production cost} = (\$1.10)\left(\frac{3000}{\text{month}}\right)\left(12\,\frac{\text{month}}{\text{yr}}\right)
$$

$$
= \$39{,}600\ \text{yr}
$$

$$
\begin{aligned}
\text{total after-tax annual cost} &= (-\$10{,}000)(0.45) + (\$20{,}000 + \$10{,}000 + \$39{,}600) \\
&\quad \times (1 - 0.45) \\
&= \$33{,}780/\text{yr}
\end{aligned}
$$

$$
\text{cost} > \text{revenue}
$$

$$
\boxed{\text{The factory should be shut down.}}
$$

4. Assume an 8-hour workday.

$$\text{time required} = \frac{200 \text{ units}}{\left(6 \dfrac{\text{unit}}{\text{hr}}\right)\left(8 \dfrac{\text{hr}}{\text{day}}\right)} = 4.17 \text{ days or 5 days}$$

$$\text{cost per item} = \frac{\$20 + (5)(\$100) + (200)(\$1.00)}{200} = \boxed{\$3.60}$$

5. Notice in this problem that no interest rate is given, so a rigorous (time value of money) analysis cannot be performed.

$$\text{first year: } A_1 = \text{annual cost} + \text{average drop in value}$$

$$= \$1000 + \frac{\$8000 - \$3900}{1}$$

$$= \$5100$$

$$\text{second year: } A_2 = \$1000 + \frac{\$8000 - (\$3900 - \$550)}{2}$$

$$= \$3325$$

$$\text{third year: } A_3 = \$1000 + \frac{\$8000 - (\$3900 - \$550 \times 2)}{3}$$

$$= \$2733$$

$$\text{fourth year: } A_4 = \$1000 + \frac{\$8000 - (\$3900 - \$550 \times 3)}{4}$$

$$= \$2437$$

$$\text{fifth year: } A_5 = \$1000 + \frac{\$8000 - (\$3900 - \$550 \times 4)}{5}$$

$$= \$2260$$

$$\text{sixth year: } A_6 = \$1000 + \frac{\$8000 - (\$3900 - \$550 \times 5)}{6}$$

$$= \$2142$$

$$\text{seventh year: } A_7 = \$1000 + \frac{\$8000 - (\$3900 - \$550 \times 6)}{7}$$

$$= \$2057$$

$$\text{eighth year: } A_8 = \$1000 + \frac{\$8000 - (\$3900 - \$550 \times 7)}{8}$$

$$= \$1994$$

In the ninth year, the salvage value is zero.

$$A_9 = \$1000 + \frac{\$8,000}{9} = \$1899$$

$$A_{10} = \$1000 + \frac{\$8,000}{10} = \$1800$$

> Keep the car for at least 10 years.

6. $P = -\$50,000 - (\$10,000)(P/A, 4\%, 20) - (\$1,000,000)(0.04)(P/A, 4\%, 20)$
    $- (\$1,000,000)(P/F, 4\%, 20) + \$970,000$

    $= -\$50,000 - (\$10,000)(13.5903) - (\$1,000,000)(0.04)(13.5903)$
    $- (\$1,000,000)(0.4564) + \$970,000$

    $= \boxed{-\$215,900}$

7. Using the present worth method,

$$\$1000 = (\$140)(P/F, i\%, 1) + (\$150)(P/F, i\%, 2) + (\$170)(P/F, i\%, 3)$$
$$+ (\$50)(P/F, i\%, 4) + (\$730)(P/F, i\%, 5)$$

Try $i = 6\%$.

$$P = (\$140)(0.9434) + (\$150)(0.8900) + (\$170)(0.8396)$$
$$+ (\$50)(0.7921) + (\$730)(0.7473)$$

$$= \$993$$

Try $i = 5\%$.

$$P = (\$140)(0.9524) + (\$150)(0.9070) + (\$170)(0.8638)$$
$$+ (\$50)(0.8227) + (\$730)(0.7835)$$

$$= \$1029$$

The rate of return is about 6%. The exact rate of return can be computed as follows (x is the incremental rate).

$$A_1[(P/F, 5\%, 1) - (P/F, 6\%, 1)]x$$
$$+ A_2[(P/F, 5\%, 2) - (P/F, 6\%, 2)]x$$
$$+ A_3[(P/F, 5\%, 3) - (P/F, 6\%, 3)]x$$
$$+ A_4[(P/F, 5\%, 4) - (P/F, 6\%, 4)]x$$
$$+ A_5[(P/F, 5\%, 5) - (P/F, 6\%, 5)]x$$
$$= P(5\%) - \text{investment}$$

$$(\$140)[(0.9524 - 0.9434)]x$$
$$+ (\$150)[(0.9070 - 0.8900)]x$$
$$+ (\$170)[(0.8638 - 0.8396)]x$$
$$+ (\$50)[(0.8227 - 0.7921)]x$$
$$+ (\$730)[(0.7835 - 0.7473)]x$$
$$= (\$35.88)x = \$1029 - \$1000$$
$$= \$29$$
$$x = \frac{\$29}{\$35.88} = 0.81$$
$$i = 0.05 + \frac{0.81}{100}$$
$$= 0.0581 \text{ or } \boxed{5.81\%}$$

8. Unit of time is a year.

$$a = 1000 \text{ item/yr}$$
$$\text{fixed cost: } K = \$40$$
$$\text{inventory cost: } h = \frac{\$0.80}{\text{item-yr}} + \frac{(0.12)(\$5.20)}{\text{item-yr}}$$
$$= \$1.424/\text{item-yr}$$
$$Q^* = \sqrt{2\frac{aK}{h}}$$
$$= \sqrt{(2)\left[\frac{\left(1000 \dfrac{\text{item}}{\text{yr}}\right)(\$40)}{\dfrac{\$1.424}{\text{item-yr}}}\right]}$$
$$= \boxed{237 \text{ items}}$$

9. Year-end compounding:

P_j is the present worth of year j return.

$$P_1 = (\$10{,}000)(F/P, 15\%, 0)(P/F, 5\%, 1)$$
$$= (\$10{,}000)(1)(0.9524)$$
$$= \$9524$$
$$P_2 = (\$10{,}000)(F/P, 15\%, 1)(P/F, 5\%, 2)$$
$$= (\$10{,}000)(1.1500)(0.9070)$$
$$= \$10{,}430$$
$$P_3 = (\$10{,}000)(F/P, 15\%, 2)(P/F, 5\%, 3)$$
$$= (\$10{,}000)(1.3225)(0.8638)$$
$$= \$11{,}424$$
$$P_4 = (\$10{,}000)(F/P, 15\%, 3)(P/F, 5\%, 4)$$
$$= (\$10{,}000)(1.5209)(0.8227)$$
$$= \$12{,}512$$
$$P_5 = (\$10{,}000)(F/P, 15\%, 4)(P/F, 5\%, 5)$$
$$= (\$10{,}000)(1.7490)(0.7835)$$
$$= \$13{,}703$$

$$P = \boxed{\$57{,}593}$$

Month-end compounding:

$$\phi_1 = \frac{r_1}{k} = \frac{15\%}{12}$$
$$= 1.25\%$$
$$\phi_2 = \frac{r_2}{k} = \frac{5\%}{12}$$
$$= 0.4167\%$$

Effective annual rate:

$$i_1 = (1 + \phi_1)^k - 1 = (1 + 0.0125)^{12} - 1$$
$$= 0.1608$$
$$i_2 = (1 + \phi_2)^k - 1 = (1 + 0.004167)^{12} - 1$$
$$= 0.0512$$

As in year-end compounding,

$$(F/P, 16.08\%, 0)(P/F, 5.12\%, 1) = \frac{(1 + 0.1608)^0}{(1 + 0.0512)^1}$$

$$= 0.9513$$

$$P_1 = (\$10{,}000)(0.9513) = \$9513$$

$$(F/P, 16.08\%, 1)(P/F, 5.12\%, 2) = \frac{(1 + 0.1608)^1}{(1 + 0.0512)^2}$$

$$= 1.0505$$

$$P_2 = (\$10{,}000)(1.0505) = \$10{,}505$$

$$(F/P, 16.08\%, 2)(P/F, 5.12\%, 3) = \frac{(1.1608)^2}{(1 + 0.0512)^3}$$

$$= 1.1600$$

$$P_3 = (\$10{,}000)(1.1600) = \$11{,}600$$

$$(F/P, 16.08\%, 3)(P/F, 5.12\%, 4) = \frac{(1.1608)^3}{(1.0512)^4}$$

$$= 1.2809$$

$$P_4 = (\$10{,}000)(1.2809) = \$12{,}809$$

$$(F/P, 16.08\%, 4)(P/F, 5.12\%, 5) = \frac{(1.1608)^4}{(1.0512)^5}$$

$$= 1.4145$$

$$P_5 = (\$10{,}000)(1.4145) = \$14{,}145$$

$$P = P_1 + P_2 + P_3 + P_4 + P_5$$

$$= \$9513 + \$10{,}505 + \$11{,}600$$
$$+ \$12{,}809 + \$14{,}145$$

$$= \boxed{\$58{,}572}$$

10. The corrected inflation rate is i'.

$$i' = i + e + ie$$
$$= 0.10 + 0.08 + (0.10)(0.08)$$
$$= 0.188$$

$$P = A(P/A, 18.8\%, 10) + G(P/G, 18.8\%, 10)$$

$$= A\left[\frac{(1+i)^n - 1}{i(1+i)^n}\right] + G\left[\frac{(1+i)^n - 1}{i^2(1+i)^n} - \frac{n}{i(1+i)^n}\right]$$

$$= (\$100,000)\left[\frac{(1+0.188)^{10} - 1}{(0.188)(1+0.188)^{10}}\right]$$

$$+ (\$10,000)\left[\frac{(1+0.188)^{10} - 1}{(0.188)^2(1+0.188)^{10}} - \frac{10}{(0.188)(1+0.188)^{10}}\right]$$

$$= \boxed{\$574,300}$$

11. DDB: depreciation in any year $j = \left(\dfrac{2C}{n}\right)\left(1 - \dfrac{2}{n}\right)^{j-1}$

$$\frac{2C}{n} = \frac{(2)(\$40,000)}{12} = \$6667$$

$$\left(1 - \frac{2}{n}\right) = \left(1 - \frac{2}{12}\right) = 0.8333$$

$$D_1 = (\$6667)(0.8333)^{1-1} = \$6667$$
$$BV_1 = \$40,000 - \$6667 = \$33,333$$
$$D_2 = (\$6667)(0.8333)^{2-1} = \$5556$$
$$BV_2 = \$33,333 - \$5556 = \$27,777$$
$$D_3 = (\$6667)(0.8333)^{3-1} = \$4629$$
$$BV_3 = \$23,148$$
$$D_4 = (\$6667)(0.8333)^{4-1} = \$3858$$
$$BV_4 = \$19,290$$
$$D_5 = (\$6667)(0.8333)^{5-1} = \$3215$$
$$BV_5 = \$16,075$$
$$D_6 = (\$6667)(0.8333)^{6-1} = \$2679$$
$$BV_6 = \$13,396$$
$$D_7 = (\$6667)(0.8333)^{7-1} = \$2232$$
$$BV_7 = \$11,164$$
$$D_8 = (\$6667)(0.8333)^{8-1} = \$1860$$
$$BV_8 = \$9304$$
$$D_9 = (\$6667)(0.8333)^{9-1} = \$1550$$
$$BV_9 = \$7754$$
$$D_{10} = (\$6667)(0.8333)^{10-1} = \$1292$$

$$BV_{10} = \$6462$$
$$D_{11} = \$463$$
$$BV_{11} = \$5999$$

present worth of the depreciation recovery of the year $j = D_j(P/F, 6\%, j)$

$$P_1(DR) = D_1(P/F, 6\%, 1) = (\$6667)(0.9434) = \$6290$$
$$P_2(DR) = D_2(P/F, 6\%, 2) = (\$5556)(0.8900) = \$4945$$
$$P_3(DR) = D_3(P/F, 6\%, 3) = (\$4629)(0.8396) = \$3886$$
$$P_4(DR) = D_4(P/F, 6\%, 4) = (\$3858)(0.7921) = \$3056$$
$$P_5(DR) = D_5(P/F, 6\%, 5) = (\$3215)(0.7473) = \$2402$$
$$P_6(DR) = D_6(P/F, 6\%, 6) = (\$2679)(0.7050) = \$1889$$
$$P_7(DR) = D_7(P/F, 6\%, 7) = (\$2232)(0.6651) = \$1484$$
$$P_8(DR) = D_8(P/F, 6\%, 8) = (\$1860)(0.6274) = \$1167$$
$$P_9(DR) = D_9(P/F, 6\%, 9) = (\$1550)(0.5919) = \$917$$
$$P_{10}(DR) = D_{10}(P/F, 6\%, 10) = (\$1292)(0.5584) = \$721$$
$$P_{11}(DR) = D_{11}(P/F, 6\%, 11) = (\$463)(0.5268) = \$244$$

$$\sum P_j(DR) = \$27,001$$

$$P = -\$40,000 + (\$30,000 - \$23,000)(1 - 0.52)(P/A, 6\%, 12)$$
$$+ (\$27,001)(0.52) + (\$6000)(P/F, 6\%, 12)$$
$$= -\$40,000 + (\$7000)(0.48)(8.3838) + \$14,040 + (\$6000)(0.4970)$$

$$= \boxed{\$5192}$$

12. The available fund is \$7120. Unused funds will be invested at 8%.

$$EUAC(\$4500) = (\$4500)(A/P, 8\%, 10) + \$2700 = (\$4500)(0.1490) + \$2700$$
$$= \$3371$$

$$EUAC(\$5700) = (\$5700)(A/P, 8\%, 10) + \$2500 = (\$5700)(0.1490) + \$2500$$
$$= \$3350$$

$$EUAC(\$6750) = (\$6750)(0.1490) + \$2200$$
$$= \$3206$$

$$EUAC(\$7120) = (\$7120)(0.1490) + \$2030$$
$$= \$3090$$

> The \$7120 precipitator should be recommended.